New Eyes on the Universe

Twelve Cosmic Mysteries and the Tools We Need to Solve Them

Stephen Webb

New Eyes on the Universe

Twelve Cosmic Mysteries and the Tools We Need to Solve Them

 Springer

Published in association with
Praxis Publishing
Chichester, UK

Dr. Stephen Webb
University of Portsmouth
Portsmouth
U.K.

SPRINGER–PRAXIS BOOKS IN POPULAR ASTRONOMY
SUBJECT *ADVISORY EDITOR*: John Mason, M.B.E., B.Sc., M.Sc., Ph.D.

ISBN 978-1-4614-2193-1   ISBN 978-1-4614-2194-8 (eBook)
DOI 10.1007/978-1-4614-2194-8
Springer New York Heidelberg Dordrecht London

Library of Congress Control Number: 2012934535

Cover design: Jim Wilkie

Printed on acid-free paper

Springer is part of Springer Science+Business Media (www.springer.com)

To my brother, Peter

Contents

Preface

These are amazing structures: thousands of optical sensors deployed on kilometer-long strings, distributed throughout a volume that dwarfs the largest office block and interred deep in Antarctic ice; a gigantic tank of liquid argon, surrounded by ancient Roman lead and state-of-the-art photodetectors, sitting at the bottom of one of the world's deepest mines; a host of giant antennae placed on an almost inaccessible mountain top; a group of cables, each about the length of the Empire State Building and containing hundreds of photomultipliers encased in glass spheres, anchored to the Mediterranean seabed by underwater robots; and satellites – lots of them – orbiting Earth while they stare, unblinking, out into space. These constructions are cathedrals of science, all of them examples of a new type of astronomical telescope.

For a couple of years as a schoolkid I was obsessively interested in two things: telescopes and cricket. I joined the local amateur astronomy society, which gave me the chance to observe with a half-decent instrument. On those occasions when the telescope's availability coincided with a cloud-free sky (unfortunately, being in England, these were almost non-intersecting sets of events) I marvelled at the sights such an instrument afforded. And each time was the thought: "If I can see all this using a mirror that's the width of a cricket wicket, what could I see with a mirror that's the width of a cricket pitch?" (For US readers, a cricket pitch is 10 feet, or 3.05 m, wide.)

Later, when I began to study physics at university, I learned that professional astronomers already had access to a telescope bigger than the width of a cricket pitch: the 5 m Hale telescope at Palomar had been completed decades earlier. The view through Hale was indeed impressive: astronomers had already used it to study distant galaxies, to get glimpses of enormously energetic objects, and to firm up the notion of an expanding Universe. But

I learned that the view through even the world's largest telescope was never really sharp enough. In studying physics I was learning about a subject in which experiments could test theory to eight decimal places. Cosmology, on the other hand, seemed hopelessly imprecise. Basic parameters were quite uncertain, with cosmologists bickering over whether the Universe was ten billion years old or twenty. The telescopes of my boyhood imagination, it turned out, simply weren't powerful enough to do the job required.

And then it changed. About two decades ago astronomy and cosmology entered a Golden Age. Space-based telescopes such as Hubble and COBE transformed the field. Cosmology became a precision science. This Golden Age continues, and it's going to get even more glittering over the next few years: a plethora of giant telescopes – some of them in space; some of them hidden deep beneath ice, sea or rock; most of them bearing no resemblance at all to the traditional optical telescope – will soon start to observe. These marvels of technology will give humankind new eyes through which to study the Universe. And there's much to study. One of the lessons from the Golden Age is that the Universe is much stranger than previously thought. Mysteries abound: what's causing the Universe to blow itself apart? Why can't we see most of the matter in the Universe? Where *is* everybody?

This book is an introduction to a dozen of the most interesting mysteries over which astronomers are puzzling – and it's a guide, too, to the powerful new telescopes that will help solve those mysteries. Since the subject matter is so fast-moving, it's inevitable that there'll be interesting developments even as the book is being printed. For the latest on these questions please visit my website, stephenwebb.info, where I'll post regular updates.

* * *

I'd like to thank several people who have helped me in the writing of this book. Clive Horwood and everyone at Praxis Publishing have, as always, been extremely supportive. I'd particularly like to thank Dr John Mason, who made innumerable insightful comments and suggestions on an early draft; needless to say, any errors that remain are entirely of my own making. Several people helped me to source photographs and images; in particular I would like to acknowledge Hanny van Arkel, Steve Criswell, Emma Grocutt, Bill Harlan, Luigi Ottaviani & Roberta Antolini, and Martin White.

Finally, I would like to thank Heike and Jessica for showing so much patience with me over recent months. I would never have been able to finish this book without them.

1

Introduction

What is causing the Universe to blow itself apart? How come most of the matter in the Universe is invisible? Why haven't we heard from extraterrestrial intelligences? Astronomers are trying to answer these difficult questions – and solve many other cosmic mysteries – by building bigger and better telescopes. Not just telescopes that capture visible light, but telescopes that capture light from all parts of the spectrum – from radio waves all the way up to gamma-rays. By analyzing that radiation, astronomers can deduce vast amounts of information. In recent years astronomers have even begun to study the Universe in ways that don't depend on light at all. The result of all this activity is that, for the first time, we have a good understanding of the history and eventual fate of the Universe. And one by one those cosmic mysteries are starting to be solved...

Eyes on the Universe – New windows, new views – Let there be light – Barcoding the Universe – Telescopes for more than light – The Universe: a potted history

S. Webb, *New Eyes on the Universe: Twelve Cosmic Mysteries and the Tools We Need to Solve Them*,
Springer Praxis Books, DOI 10.1007/978-1-4614-2194-8_1, © Springer Science+Business Media, LLC 2012

Throughout most of history and prehistory our view of the heavens has been limited by the capacity of our eyes. The Sun dominates the daytime sky, of course, but when darkness falls people have surely always marvelled at the beauty of the Moon, puzzled over how five planets seem to wander across the celestial sphere, and wondered about the nature of the constant stars that stud the night sky. The rare appearance of a bright comet or the even rarer appearance of a nova heralded a brief period when the sky contained something out of the ordinary. For the most part, however, our Neanderthal and Denisovan cousins would have enjoyed much the same view that the Greeks or the Romans or the Mongol hordes enjoyed much later. For millennia, humankind's view of the sky was unchanging. And then came the telescope.

Eyes on the Universe

No one knows for certain when or where the telescope was invented, nor who invented it. The magnifying glass has been around for a long time and by the end of the Crusades people were making spectacles, so it's likely that many individuals at various times, when playing around with lenses, found that by holding a convex lens (one that causes light rays to converge) in front of a concave lens (one that causes rays to diverge) they had a simple device for seeing at a distance: faraway objects appeared closer and bigger. Place the convex **objective** lens and the concave eyepiece in a tube, about an arm's length apart, and you have a simple refractor – a telescope that works by bending, or refracting, the paths of light rays. Various refractors were developed in the late sixteenth century and the list of supposed inventors of the telescope is a long one. What's known for certain is that in 1608 Johannes Lippershey, a spectacle maker in the Dutch coastal town of Middelburg, became the first to both demonstrate a working model of a refracting telescope and apply for a patent. It makes sense then to call Lippershey the inventor of the telescope and 1608 the year of invention.

News of the invention spread quickly. Soon after Lippershey's patent application, Galileo Galilei built an instrument that magnified distant objects by a factor of three; a few weeks later and he had a telescope that magnified objects by a factor of eight; a few months later and he had a refractor that magnified by a factor of 30. And with that instrument he saw things no one else had ever seen. He saw the details of lunar craters and the phases of Venus. He saw the four largest satellites of Jupiter and objects that we now know were the rings of Saturn. He saw spots on the surface of the Sun. Galileo transformed humankind's understanding of the Universe.

It was Isaac Newton, surely the greatest of all scientists, who made the next major advances in optics. Newton passed sunlight through a prism and showed that a spectrum forms: white light splits into a rainbow of different colors – red through to violet. This was a crucial finding for several reasons, some of which we'll discuss later in this chapter. Purely in terms of telescope design Newton's discovery was important because it prompted a solution to what seemed a fundamental problem with refracting telescopes: the objective lens bends different colors by different amounts, thus limiting the sharpness of the image formed. It's a problem called chromatic aberration. Well, Newton showed experimentally that the reflection of light by a mirror was not subject to chromatic aberration and, with his characteristic single-mindedness, he set about constructing a reflecting telescope. Newton's design was unique: a curved main objective mirror brought light to a focus and then a diagonal secondary mirror, placed near the focus, reflected the image through a right angle into an eyepiece mounted on the side of the telescope. With the very first telescope of this design he repeated Galileo's observations. His second telescope, which he presented to the Royal Society in 1672, exceeded the power of Galileo's instruments and magnified objects by a factor of 38.

One of the functions of a telescope is of course to magnify *small* objects, which is why I mentioned the magnifying power of those early refractors and reflectors, but far more important for astronomers is a telescope's capacity for gathering light from *faint* objects and thus its ability to image objects that are otherwise too dim to see. Now, the light-gathering capacity of a telescope is related to the diameter of its objective lens or mirror: just as a larger bucket can catch more falling raindrops than a smaller bucket, so a larger objective can catch more light than a smaller objective. A larger objective also has the advantage of being able to resolve more fine detail than a smaller objective. So, all things being equal, bigger is better when it comes to telescopes. (As we'll see in later chapters, a smaller telescope can sometimes outperform a larger instrument. For example, the location of a telescope plays an important role in its effectiveness. Nevertheless, as a rule of thumb, a larger aperture is better than a smaller one.) Thus it was that a sort of astronomical 'arms race' took place in the eighteenth, nineteenth and twentieth centuries, with observatories around the world building ever-larger refracting and reflecting telescopes.

The limit for refracting telescopes was reached in 1897, with the construction of an instrument at the **Yerkes Observatory**; the Yerkes refractor had an objective lens that was 1.016 m in diameter. It's probably impossible to make a refractor with a lens much bigger than this because the lens must be held in place around its edge: gravity causes the center of the lens to sag and its images are inevitably distorted. The situation is better with a reflector since the mirror can be supported all over its surface. In 1917, the **Mount Wilson Observatory** opened the Hooker telescope, which had a mirror with a diameter of 2.5 m (the old measurement of 100 inches somehow sounds more evocative). Three decades later and the **Mount Palomar Observatory** opened the huge Hale telescope, which had a mirror with a diameter of 5.08 m. Modern giant telescopes are twice as large again. See figure 1.1.

Astronomers devoted decades of their working lives to these construction projects, not to get their names into the record books but so they could peer further and deeper into the Universe. They used these telescopes to discover more planets in our Solar System; to understand the place of the Solar System in the Galaxy; and to learn that there are hundreds of billions of galaxies in the Universe. In just four hundred years of telescopic observation, astronomers replaced a cosmological model in which Sun, Moon and planets revolved around Earth to one in which the Solar System is just an insignificant part of a much larger Universe. It was quite a change.

The optical telescope may have transformed humankind's view of Earth's importance, but other types of telescope have become even more important than refractors and reflectors in the ongoing quest to understand the Universe and our place in it. The key to the development of these new types of telescope lay in Newton's discovery that he could split sunlight into a spectrum of colors – it opened up completely new windows on the Universe.

Figure 1.1 *Telescopes have increased in sophistication and become much bigger over time. (a) Newton's 1672 reflector, of which this is a replica, had a diameter of 33 mm. (b) The Hale telescope, completed in 1948, is just over 5 m in diameter. (c) The ESO Very Large Telescope, based in Paranal, Chile, has four 8.2 m telescopes that can be linked to form an interferometer. The particular telescope shown here is called Antu. (d) An artist's impression of the TMT as it will look on Mauna Kea. This telescope, when it is completed in about 2018, will have a diameter of 30 m. ((a) Andrew Dunn; (b) Commons; (c) ESO; (d) TMT Observatory Corporation)*

New windows, new views

William Herschel constructed some of the best telescopes of his day, including the first 'giant' reflector – an instrument built in 1789 that had a 1.24 m mirror. He used those telescopes quite brilliantly. For example, even with one of his smaller telescopes he was able to discover Uranus, which was the first new planet found since ancient times. Herschel also found something that led eventually to a completely new kind of telescope.

In 1800, Herschel was engaged in a program of observing the Sun through different colored filters and, in making these observations, he came to suspect that different colors passed different levels of heat. To investigate this he passed sunlight through a glass prism – just as Newton had done more than a century earlier – and created a spectrum. He then used a thermometer to measure the temperature of each color of the spectrum, and compared the result with two identical thermometers that he placed well away from the spectrum. He measured the temperature of each color of the rainbow – violet, indigo, blue, green, yellow, orange, red – and in each case the recorded temperature was higher than the temperatures on the control thermometers. Furthermore, he found that different colored filters did indeed pass different amounts of heat: temperatures increased steadily from violet through to red. He decided, perhaps in a moment of idle curiosity, to measure the temperature just below the red part of the spectrum, a place where there was no visible sunlight. Rather than measuring the same temperature as the control thermometers, he found that the region below the red end of the spectrum had a temperature higher than any of the colors.

Herschel, being the thorough scientist that he was, experimented further on this invisible radiation beyond the red part of the spectrum. Apart from the fact that he couldn't see it, this radiation behaved in exactly the same way as visible light: it was transmitted, absorbed, reflected and refracted just like light. Herschel had discovered what eventually became known as **infrared** radiation (the prefix 'infra' meaning 'below', since the radiation lies below the red part of the spectrum).

The field of infrared astronomy was not long in arriving. In 1856, the astronomer Charles Piazzi Smyth combined his honeymoon with a scientific voyage to the mountain peaks of Tenerife. I'm not sure what his new wife thought of his activities, but Smyth began the modern practice of putting telescopes at high altitudes in order to observe under optimum conditions. Whilst on the peak of Guajara he used a **thermocouple** – a device that converts heat into electric current – to measure infrared radiation from the full Moon. Smyth's instrument didn't make an image in the way that optical telescopes usually do; the development of an infrared telescope had to wait. Nevertheless, his instrument enabled him to gain useful information about a celestial object – information that a 'normal' telescope could not provide. Smyth thus showed that it was possible to study the cosmos using something other than our aided or unaided eyes.

Soon after Herschel's discovery of infrared radiation, scientists demonstrated the existence of radiation beyond the blue part of the spectrum. What prompted the discovery in this case was the observation that certain substances, such as paper soaked in silver chloride, would darken when exposed to sunlight. Well, in 1801 a pharmacist by the name of Johann Wilhelm Ritter found that the violet end of the spectrum was more effective than the red end at darkening silver chloride paper – and that invisible radiation beyond the violet end of the spectrum was more effective still. Ritter had discovered what eventually became known as **ultraviolet** radiation (the prefix 'ultra' meaning 'beyond', since the radiation lies beyond the violet part of the spectrum). Newton's spectrum was thus richer than he could have known: radiation extended past both edges of the visible spectrum, into the infrared at one end and the ultraviolet at the other.

The full explanation of these various radiations came two hundred years after Newton's work, when in the 1860s James Clerk Maxwell showed that visible light, infrared and ultraviolet are all manifestations of the same thing. Maxwell argued that a changing electric field, generated for example by an oscillating electric charge, creates a changing magnetic field in a direction perpendicular to the original electric field; and a changing magnetic field creates a changing electric field in a direction perpendicular to the original magnetic field. See figure 1.2. This is therefore a self-propagating phenomenon: changing electric and magnetic fields generate each other – and move outward as a wave in a direction perpendicular to both of the fields. Furthermore, the wave moves through a vacuum with a particular speed – a speed that Maxwell calculated to be the speed of light. Visible light, infrared and ultraviolet are all examples of an **electromagnetic wave**.

Electromagnetic waves can be described and classified in the same way as any other type of wave. For example, any wave has a **wavelength**. Just as the wavelength of an ocean wave can be defined as the distance between successive crests or successive troughs, so the wavelength of an electromagnetic wave can be defined as the distance between successive crests or successive

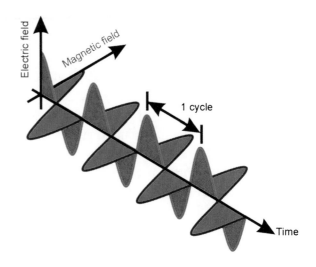

Figure 1.2 An electromagnetic wave consists of oscillating electric fields (red) and magnetic fields (blue). The electric and magnetic fields are perpendicular to each other; the direction of propagation of the wave is perpendicular to both. The wavelength is the distance between successive crests or successive troughs. The wave moves with the speed of light.

troughs of the electric (or magnetic) field. The difference in wavelength distinguishes the different colors of light and distinguishes visible light from infrared or ultraviolet light. The electromagnetic radiation that our eyes can detect has a wavelength between about 400 nm (violet light) to about 700 nm (red light). (One nanometer, or 1 nm, is one billionth of a meter. To give some indication of size, 1 nm is roughly the radius of the DNA helix.) Infrared radiation possesses a wavelength that's longer than 700 nm and ultraviolet radiation has a wavelength that's shorter than 400 nm, but neither type of radiation is different to visible light in a fundamental way. The reason we think of them as being different phenomena is because of a quirk of biology: evolution happened to provide us with optical sense organs that are sensitive to radiation in the 400–700 nm region. The fact that we don't directly see other electromagnetic wavelengths doesn't make those wavelengths somehow less important; it does mean, however, that we need instruments if we wish to observe them. (Some creatures are sensitive to other wavelengths. Bees, for example, are attracted by the ultraviolet properties of plants. The colors of flowers, which seem so beautiful to human eyes, may be entirely incidental: the main requirement of a flower is how it looks to bees in the ultraviolet rather than to humans in the visible.)

Another way of describing a wave is to use **frequency**. The frequency of a wave, whether ocean wave or electromagnetic wave, is simply the number of crests or troughs that pass by a given point every second. If one wave crest passes a given point every second then its frequency is one hertz (1 Hz).

Electromagnetic waves in the optical region have rather large frequencies when written in terms of hertz: violet light has a frequency of about 750 trillion hertz (7.5×10^{14} Hz) while that of red light is about 4×10^{14} Hz. A simple relationship exists between wavelength and frequency: multiply the two numbers together and you get the wave speed. For any electromagnetic wave, if you multiply its frequency and wavelength you get the speed of light. The speed of light is a universal constant so it follows that frequency and wavelength are inversely proportional: a longer wavelength means a lower frequency while a shorter wavelength means a higher frequency. Thus red light, with its relatively long wavelength, has a lower frequency than the shorter wavelength violet light.

Maxwell's work prompted an obvious thought: just as infrared and ultraviolet radiation extend beyond the ends of the visible spectrum, perhaps radiation exists with a wavelength longer than infrared or shorter than ultraviolet. We now know that the electromagnetic spectrum does indeed span a vast range: wavelengths can be longer than the distance from Earth to Moon or shorter than the diameter of an atomic nucleus. The part of the electromagnetic spectrum to which our eyes are sensitive is a minuscule fraction of the whole. See figure 1.3.

Although there is no *fundamental* difference between long-wavelength/low-frequency electromagnetic radiation and short-wavelength/high-frequency radiation – it's all just oscillating electric and magnetic fields – it turns out that radiation in different parts of the spectrum interacts with matter in quite different ways. This fact means it does make sense to distinguish between different types of radiation. Thus infrared radiation has a wavelength in the range between about 750 nm up to 1 mm. (Even within the infrared part of the spectrum there are subregions – far, mid and near – just as the visible part of the spectrum is subdivided into different colors). Radiation possessing a longer wavelength – between say 1 mm up to a few tens of centimeters – lies in the **microwave** region. Radiation with an even longer wavelength is in the **radio** part of the spectrum. The ultraviolet part of the spectrum possesses a range of wavelengths between about 400 nm down to 10 nm (and, as with the infrared and visible parts of the spectrum, the ultraviolet is divided into subregions: near, middle, far and extreme). If the wavelength is shorter than 10 nm then it's in the **X-ray** part of the spectrum; wavelengths shorter than about 0.01 nm belong to the **gamma-ray** region. All these wavelengths, from radio all the way up to gamma-rays, contain a trove of information for astronomers – if they can detect them.

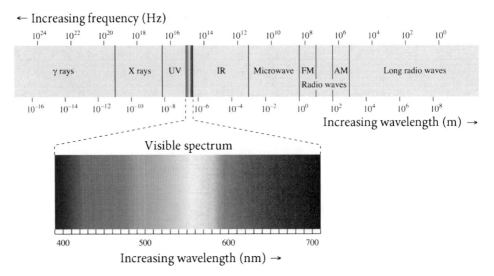

Figure 1.3 *The electromagnetic spectrum spans a vast range of frequencies. Compare the situation with a piano: each octave on the keyboard represents a doubling of frequency, which means that the frequency increases greatly in just a few octaves. A modern piano has a range of just over seven octaves. The human ear, for comparison, has a range of nine octaves. The electromagnetic spectrum has a range of 50 octaves! Take note of the small width of the optical frequencies compared to the overall spectrum; visible light is a tiny part of the spectrum.*

It didn't take physicists long to demonstrate experimentally the existence of these other regions of the electromagnetic spectrum. On the long-wavelength side of the infrared, Heinrich Hertz in 1886 showed how to generate and detect radio waves. On the short-wavelength side of the ultraviolet, Wilhelm Röntgen in 1895 systematically studied X-rays and in 1900 Paul Villard discovered gamma radiation. As the twentieth century dawned, physicists were investigating all parts of the electromagnetic spectrum, from radio waves (long wavelength/ low frequency) all the way up to gamma-rays (short wavelength/high frequency). Using the full spectrum for astronomy, though, turned out to be a difficult task.

Although Smyth was quick to make astronomical observations in the infrared, astronomers soon found that observations in the rest of the spectrum almost always encountered a problem: the atmosphere. Earth's atmosphere is effectively transparent to electromagnetic radiation in the visible part of the spectrum – which is, of course, why optical telescopes are useful for astronomy. The atmosphere is also transparent to many radio wavelengths, a few ultraviolet wavelengths (as you yourself can testify if you've ever been sunburned) and a some infrared wavelengths (although in general you have to go high in a balloon or, as Smyth did, climb a mountain to detect these wavelengths). But the atmosphere blocks most wavelengths quite effectively. It presents a barrier to astronomy throughout most of the electromagnetic spectrum.

Since the atmosphere is transparent at many long wavelengths, radio and some microwave astronomy can take place with ground-based instruments. Thus radio astronomy has a relatively long history. It can be said to date back to 1931, when Karl Jansky discovered that the Milky Way was a source of radio emission; six years later, Grote Reber had built the first dedicated radio telescope. However, astronomy at the shorter wavelengths – ultraviolet, X-ray and gamma-ray astronomy – could not start in earnest until the development of reliable satellite technology. It was only in the 1960s, then, that astronomers could put their instruments above the blanketing effect of Earth's atmosphere and observe at all wavelengths.

With the latest generation of telescopes astronomers are studying the Universe through the radio, microwave, infrared, optical, ultraviolet, X-ray and gamma-ray windows. It's a multi-wavelength view and, as we shall learn, the vista is stunning. The change in worldview brought about by being able to observe throughout the spectrum is almost as large as that wrought by Galileo's first telescopes.

Let there be light

Astronomy at the shorter wavelengths tells us we live in a Universe that's home to violent explosions and collisions, to processes involving incredible temperatures and vast releases of energy. Astronomy at the longer wavelengths enables us to understand the way in which cosmic structures – and even the Universe itself – came into being. But how can astronomers deduce all this information just from studying electromagnetic waves?

In order to squeeze as much information as possible from light, astronomers need to understand how electromagnetic waves are created. With that understanding comes the ability to use telescopes to discover not just objects too faint to see with the naked eye, but also to learn how hot those objects are, what they are made of, their surface gravity, how fast they are moving… Galileo could never have dreamed of how much information a telescope can give.

The rough-and-ready explanation of how electromagnetic waves come into being is that matter generates them. In particular, the waves are created by the atoms of which matter is composed.

An atom consists of one or more negatively charged **electrons** swarming around a central massive nucleus, which in turn consists of one or more positively charged **protons** and zero or more electrically neutral **neutrons**. In any atom, the number of protons is the same as the number of electrons so that the atom overall is electrically neutral. Since the electrons are orbiting the nucleus, those negative electrical charges are constantly being accelerated. One's first thought is likely to be that these electrons are the source of electromagnetic waves: Maxwell's theory says that accelerating electrons emit electromagnetic radiation. However, if the atomic electrons radiated they would lose energy, and thus they would spiral almost immediately into the nucleus. Clearly this doesn't happen: atoms are stable enough to form matter. In fact, it turns out that atomic electrons occupy only certain orbits – those orbits, or energy levels, allowed by quantum mechanics – and a basic principle of quantum mechanics is that electrons occupying allowed orbits do not radiate energy. But if atoms don't radiate because of orbiting electrons, how do they generate electromagnetic waves?

Matter can generate electromagnetic waves in several different ways. For example, **synchrotron radiation** is generated whenever the paths of fast-moving charged particles are bent by a magnetic field; such a situation occurs when the intense gravitational field close to a black hole whips matter around its event horizon, or when cosmic rays encounter magnetic fields. Quite different mechanisms for producing electromagnetic radiation arise in certain events that take place at the atomic and nuclear level; we shall discuss these in more detail in the next section. However, perhaps the most important way in which matter generates electromagnetic waves occurs simply because matter is hot.

If we heat a body it will radiate. We all know that. We talk, for example, about things being 'red hot' or 'white hot' – objects that are radiating visible light simply because they are at a high temperature. The mechanism behind the radiation, roughly speaking, is that heat causes the atoms and molecules in a solid to vibrate. Some electrons become detached from individual atoms and are then able to move freely throughout the solid. The distributions of electrical charges are then complicated to describe, but in essence it's the vibration of these charge distributions, rather than the oscillations of orbiting electrons, that generates electromagnetic radiation.

Physicists long sought to describe this radiation in a quantitative fashion and they realized that in order to predict how a body will *emit* radiation it pays to understand how radiation is *absorbed*. Well, absorption depends on the body in question. If you're talking about glass then most of the light seems to go right through without being absorbed. If you're talking about a shiny, metallic surface then much of the light falling on it gets reflected without being absorbed. But if you're talking about a material such as the black vinyl seats you might find inside a car, then you find that such material readily absorbs light and becomes warm. (I can personally attest to this, having once jumped into my car wearing only shorts. The car had been parked in direct sunlight. I jumped out quicker than I jumped in.) What accounts for these differences?

In the case of glass – again, roughly speaking – the structure of the material is such that the charge distributions are only able to oscillate at certain frequencies. Since none of those frequencies happen to correspond to the frequencies of visible light, a light wave passing through the glass loses no energy because it can't get those charges to oscillate. Glass is thus transparent to visible light (which is why we use it for windows, as if that needed spelling out). Note that the situation is different at different frequencies: glass can be opaque in the ultraviolet and in the infrared, where the frequencies involved are such that the charges *can* oscillate. In the case of a shiny metal surface, it turns out that any light waves falling on it force the free electrons into large oscillations and those oscillating electrons emit electromagnetic radiation – the light is reflected, in other words. In the case of light waves falling on a dark surface it turns out that the electric field can drive the electrons into motion (unlike in glass) but the process is not particularly effective (unlike in metal). Any unattached electrons that are driven into motion quickly collide with atoms, and they transfer their kinetic energy into heat. Thus a dark surface absorbs the light wave with very little reflection, and energy is transformed from the light wave into heating up the object.

Now, a good absorber of radiation is a good emitter of radiation. Again, the reason for this is to do with the distribution of free charges in the material. The electrons inside a black object such as coal bounce around and each time they collide with something they are accelerated and emit radiation. Coal is a good emitter. The electrons in a metal, however, move around more freely before colliding and undergoing acceleration; metals are therefore less efficient than dark objects at emitting radiation. The charges in glass are tied down tightly, so glass is a poor emitter. In fact, at any given temperature and frequency a body emits radiation exactly as well as it absorbs radiation. That's why an understanding of how radiation is absorbed allows one to understand how radiation is emitted.

It follows from the above discussion that if we want to know precisely how a body will radiate then we need to understand its detailed structure. Well, that's a difficult problem so physicists did what they always do in such situations: they invented a simplified, idealized situation that they *could* analyze, and then compared that model with the real world in the hope of learning something. Thus it was that the physicist Gustav Kirchhoff introduced the idea of a **blackbody** – a perfect absorber (and hence 'black' at room temperatures) and a perfect radiator – in order to learn more about radiation.

Kirchhoff and his colleagues soon took up the challenge of measuring the amount of energy emitted from a blackbody at different wavelengths as its temperature changes. They learned several interesting things. First, they found that the energy spectrum of blackbody radiation has a quite specific shape. A blackbody emits radiation of all wavelengths, so the blackbody or thermal spectrum is continuous, but there's always a wavelength at which the radiated energy is a maximum. (Synchrotron radiation also has a continuous spectrum, but it has a very different shape to a blackbody spectrum.) Second, the characteristic shape and the location of the peak wavelength depend *solely* upon the temperature of the blackbody. It doesn't matter what the body consists of, or how big it is, or any other complicating factor: every blackbody behaves in precisely the same way. Third, as the temperature rises the peak wavelength moves to shorter wavelengths/higher frequencies. (We know this from everyday life – a red-hot poker becomes orange and then yellow as it gets hotter.) Fourth, as the temperature increases, the height of the peak becomes very much higher. (Again, we know this from everyday life: an object becomes *very* much brighter as it gets hotter.) See figure 1.4.

There's no such thing as a perfect blackbody but it turns out that many celestial objects produce excellent blackbody spectra. Moons, planets and stars emit electromagnetic radiation simply because they are hot, and this is the way that much of the electromagnetic radiation in the Universe, and much of the radiation picked up by our telescopes, is generated. If an object is really cold then it will be dim and will radiate mainly in the radio or microwave regions. A cool object – such as you, gentle reader – radiates mainly in the infrared region. A hot object, such as a star, with a temperature measured in thousands of degrees, will emit significant amounts of infrared and ultraviolet radiation but its radiation will peak in the visible region –

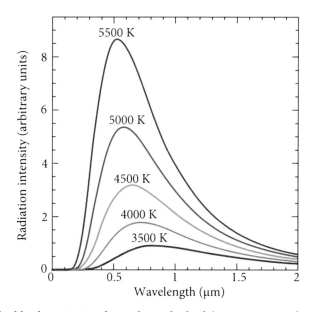

Figure 1.4 *Blackbody emission depends on the body's temperature, but not on factors such as the body's chemical composition or shape. At higher temperatures, the radiation peak moves to shorter wavelengths. At 3500 K the peak is at 0.83 μm; at 5500 K the peak shifts to 0.53 μm.*

and the hotter the star, the brighter and bluer it is. The radiation peak of an extremely hot object such as an intergalactic gas cloud, with a temperature measured in the hundreds of thousands or even millions of degrees, is in the ultraviolet or even the X-ray region. (As we'll see, the even shorter-wavelength gamma-rays are created in other, highly energetic, processes.)

With appropriate telescopes, then, electromagnetic radiation of all wavelengths can be detected – and by locating the peak of the radiation astronomers immediately know the temperature of the object that generated the radiation. That's important information.

But much more than a knowledge of an object's temperature can be gleaned from a study of electromagnetic radiation.

Barcoding the Universe

The radiation coming from a blackbody contains every wavelength, even if most of the wavelengths occur in minuscule amounts. There are no gaps in a thermal spectrum. Look closely at the solar spectrum, however, and you

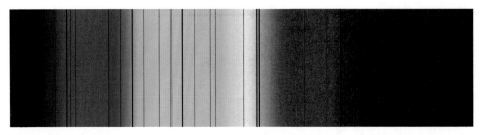

Figure 1.5 *This absorption spectrum shows some of the absorption lines seen in the Sun's spectrum.*

Figure 1.6 *This emission spectrum shows some of the emission lines of iron. By identifying the characteristic lines in an object's spectrum, astronomers can determine the presence of chemical elements in that object.*

see lots of dark lines superimposed on the continuous spectrum – wavelengths where the energy has been absorbed by relatively cool gas in the Sun's outer layers. Such lines are called, naturally enough, **absorption lines**. See figure 1.5. A hot, low-density gas cloud (which is a collection of widely separated atoms, and is thus nothing like a blackbody) will have a discrete spectrum, a series of bright lines called **emission lines**. See figure 1.6. What's going on in these cases?

Well, as already mentioned, the electrons in any given atom occupy only certain well-defined orbits. Each orbit corresponds to a particular energy level. An electron occupying the orbit closest to the atomic nucleus has the lowest energy; the farther out from the nucleus, the more energy the electron possesses. When an electron moves from one level to another it must do so by either absorbing or emitting a particular quantum of energy, corresponding to the difference in energy between the two orbits. For any particular atom the differences between different energy levels are fixed and unchanging; for example, an electron jumping from the third allowed energy level in a hydrogen atom to the second allowed level will always emit *precisely* the

Figure 1.7 *The electrons in an atom exist only at certain energy levels. Different atoms possess different levels. The farther from the nucleus an electron is, the greater the energy it has. An electron in the third energy level (in other words n = 3) thus has more energy than an electron in the second energy level (n = 2). When an electron drops from a higher energy level to a lower energy level the excess energy is carried away by the emission of a photon – in other words, the atom emits radiation. Since particular energy levels are involved, the photon carries a quite specific energy: energy is quantized.*

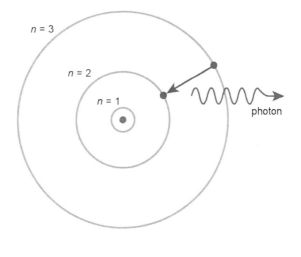

same amount of energy. An emission line therefore occurs when an electron falls from a higher level to some lower level and radiation carries away the difference in energy. See figure 1.7. (An absorption line occurs when radiation possesses just the right energy to kick an electron from one level to another.) Thus emission in atomic processes is another way in which electromagnetic radiation is generated.

A couple of crucial points follow from this quantization of energy levels. The first is that the electromagnetic radiation generated in an atomic process is created at a particular position in space and a particular instant in time; similarly, absorption takes place at a localized point. This is behavior more typical of a particle than a wave. It's an example, of course, of the famous wave–particle duality of quantum mechanics. Electromagnetic radiation is a wave: as it propagates through space it moves as a wave, and can undergo phenomena typical of a wave such as diffraction, reflection and refraction in the same way as any other wave. But electromagnetic radiation is also a particle: when radiation is emitted or absorbed by matter it does so as a particle. The particle of electromagnetic radiation is called the **photon**.

The second crucial point is that photons carry energy in an amount that's directly related to the frequency of the radiation: the higher the frequency the greater the energy of the photon. So just as electromagnetic waves can be classified in terms of their frequency and wavelength, they can also be classified in terms of the energy they carry. Physicists usually measure the masses and energies of particles in terms of a unit called the **electron Volt**, which has the symbol eV. The energy of a radio photon is extremely small, so small that it makes little sense to dwell on its particle aspects; it almost always makes more sense to discuss radio waves in terms of wavelength and frequency. For wavelengths shorter than the microwave, however, it's as convenient to talk about photon energy as about wavelength or frequency. Photons of visible light carry energies between about 1.8 eV (red light) to about 3.4 eV (violet light). Ultraviolet photons carry energies between a few eV up to about 100 eV. X-ray photons carry energies in the range 100 eV up to 100 000 eV (or 100 keV). Gamma-rays are photons with energies greater than 100 keV. (Gamma-rays are usually generated in processes involving the atomic nucleus rather than atomic electrons; such processes generally take place at much higher energies.)

Now, since each atom can absorb or emit radiation only at certain wavelengths, a particular pattern of lines acts like the identifying barcode that manufacturers put on their products. Find a set of lines in an object's spectrum – either dark absorption lines in a continuous spectrum of a star, say, or bright emission lines in the spectrum of a gas cloud – and you can infer the presence of a particular chemical element. It doesn't matter how far away the object lies: if you can identify the lines then you can learn something about its chemical composition. To me, that's amazing!

The unique pattern of lines belonging to a particular element won't always show up in the same place in the spectrum. If the object is moving then the spectral lines will shift away from their usual position; this is the **Doppler effect**, which occurs with any wave phenomenon when motion is involved. (We've all heard the Doppler effect in the changing pitch of a siren as an emergency vehicle first moves towards you and then recedes.) Sometimes the spectral lines will be shifted towards the shorter (bluer) wavelengths; in other words, they'll be blueshifted. A blueshift denotes that the object is approaching. Sometimes the lines will be shifted towards the longer (redder) wavelengths; in other words, they'll be redshifted. A redshift denotes that the object is receding. Furthermore, the size of the observed shift tells you how fast the relative motion is: the faster the motion, the larger the shift.

It's clear, then, why the ability to capture electromagnetic radiation is so important to astronomers. By building bigger and better telescopes, and increasingly sensitive instruments with which to study spectra, astronomers can see objects that are too faint to see with the unaided eye; they can determine an object's temperature; they can investigate its chemical composition; they can tell whether it's moving away from us or towards us; they can ascertain the sizes of stars and the strengths of their magnetic fields; they can find out about the orbiting of planets around stars and accretion disks around black holes; there's a host of other information they can glean too. And they can do all this even if the object lies halfway across the Universe. By capturing electromagnetic radiation at the various wavelengths, astronomers learned how the Universe was born and how it is put together. The new generation of telescopes are going to provide even more detail.

In recent years, however, it has become increasingly clear that these instruments for studying electromagnetic radiation aren't the *only* tools in the astronomers' toolkit.

Telescopes for more than light

Almost all of humankind's current knowledge of the cosmos comes from information carried by electromagnetic waves. These waves are such important information carriers for several reasons: the Universe is awash with them; they are detectable over a vast range of wavelengths and thus provide information on many different types of source; they are stable and electrically neutral so they can travel for billions of light years until something – possibly a telescope – blocks their path; and they can contain detailed information on many aspects of the objects with which they have come into contact along the way. However, it's not only electromagnetic radiation that rains down on Earth. Other stuff reaches us from the depths of space. If it can be stopped and studied, if we can construct a 'telescope' with which to 'observe' it, then astronomers have yet another tool with which to learn something about the Universe. And those tools will complement the information provided by electromagnetic radiation: different messengers will deliver different sorts of information.

For example, if you have an instrument that can detect radioactivity then you'll soon notice that your instrument registers a certain level of radiation even in the absence of obvious radioactive sources: there's always some background radiation. This phenomenon puzzled the early investigators of radioactivity. In 1912, in an attempt to learn more about the source of this background radiation, Victor Hess willingly risked life and limb (and those of two probably less-willing assistants) by making radiation measurements while airborne in a balloon. The general expectation at the time was that the background radiation came from Earth itself, and so the higher Hess flew the less radiation he expected to find. The level of radiation did indeed decrease up to about 1 km above the ground, but at higher altitudes the radiation level *increased*. At an altitude of 5 km Hess detected radiation at a level about twice that on the ground. Later ascents in free balloons equipped with recording instruments showed that at a height of 10 km the radiation was more than 40 times as intense as on the ground. Hess explained this finding by postulating the existence of a source of radiation entering the atmosphere from above, a conclusion confirmed in 1925 by Robert Millikan. Hess was awarded the Nobel prize in 1936 for his discovery.

Millikan coined the term **cosmic ray** for what Hess discovered, but the word 'ray' turned out to be a poor choice. During the 1930s physicists showed that cosmic rays are affected by Earth's magnetic field, which implied that the 'rays' are electrically charged. They must be *particles*, arriving discretely rather than in the form of a beam or a ray.

What could these particles be? Well, by the 1930s physicists had a good understanding of the atom so there were at least three candidates: cosmic rays might be electrons, protons or ions – atoms in which one or more of the electrons have been stripped away, leaving an excess positive charge. All three types of particle are indeed present in cosmic rays.

By capturing particles direct from the cosmos, rather than the electromagnetic radiation produced by particles, astronomers would surely be able to learn much about the Universe. That was the hope, anyway, but it's not so simple. Space is filled with magnetic fields, which bend the path of slowly moving charged particles; and Earth itself is surrounded by magnetic fields, which funnel those particles into certain paths. Thus even if we detect a cosmic ray we can't necessarily tell where it came from. Nevertheless, it might be possible to deduce various things by discovering the precise form of the cosmic ray rain. How many of the rays are protons, how many are electrons, how many are heavy ions? How much energy does each of these types of particle carry? Such information would provide clues about the cosmic objects that created the rays. And it turns out that cosmic rays of *very* high energy would be relatively unaffected by magnetic fields: the really energetic objects should point straight back to their source and thus cosmic ray astronomy becomes possible. Cosmic ray 'telescopes' hold the promise of providing a completely different view of the Universe. That's why some of the largest telescopes being built – instruments that cover an area the size of a small country – are there to study cosmic rays rather than electromagnetic waves.

Astronomers have built another type of 'telescope': an instrument for detecting the **neutrino**. The neutrino is a subatomic particle that's almost as insubstantial as a ghost: it carries no electric charge, it has very little mass and it shuns most interaction with matter. Such particles are fiendishly difficult to spot. On the other hand, immense numbers of neutrinos are produced every second in the nuclear reactions taking place within stars, and they are produced in even greater numbers when a star explodes. Vast numbers of them bombard Earth every second. Neutrino telescopes – instruments that require many tons of pure material to catch these bits of almost-nothingness, and kilometers of rock to shield that material against unwanted radiation – are able to catch only a handful of those countless quadrillions of particles. But when they do catch neutrinos they provide astronomers with a glimpse of cosmic environments that are otherwise hidden from view.

There may be yet other windows on the Universe. Astronomers are building dark matter observatories and gravitational wave telescopes, and in doing so are attempting to overcome technical challenges even greater than those facing neutrino hunters. If those challenges can be met then something very special might come into view: these new types of telescope can provide information about what happened in the first fraction of a second after the Universe was born.

In the future scientists will even go out and directly sample the stuff of the Universe. This has already happened on a small scale: in 2010, the Japanese **Hayabusa** probe became the first spacecraft to land on an asteroid. It stayed on the surface of the near-Earth asteroid 25143 Itokawa for about 30 minutes, took a sample of dust, and returned to Earth with its cargo for analysis. And in 2016, a NASA mission by the unwieldy name of **Origins Spectral Interpretations Resource Identification Security Regolith Explorer** (OSIRIS-REX) will begin its journey to the asteroid 1999 RQ36. The intention is for OSIRIS-REX to map the asteroid's surface for minerals and then collect up to 2 kg of material and return with it to Earth in 2023.

For the main part, however, scientists will continue to study the Universe by using instruments to detect those waves and particles from the cosmos that have happened to reach our planet. It's those instruments – the various types of telescope – that are the focus of this book.

The Universe: a potted history

There is much interesting science to look forward to when the new generation of powerful telescopes come online. Even with existing instruments, however, astronomers have in recent years transformed our view of the Universe. One of the most significant outcomes is vastly improved knowledge of how the Universe was born and has subsequently evolved.

Thirteen and three-quarter billion years ago, give or take 110 million years, a quantum fluctuation gave birth to our Universe. In its earliest stages, in an epoch we don't fully understand, the fledgling Universe underwent a brief period of exponential expansion, a momentary spurt of inflation that increased its volume by a factor of at least 10^{78}. When inflation ended, the mysterious field that was causing this exponential expansion decayed into 'ordinary' subatomic particles. The Universe continued to expand, but it did so in a far more gentle fashion: the traditional expansion of the **Big Bang**.

At the end of inflation a soup of incredibly energetic, fast-moving elementary particles filled the Universe. Much later in the history of the Universe – perhaps a millionth of a second after the Big Bang – things had cooled sufficiently for particles called quarks and gluons to combine to form more familiar objects such as protons and neutrons. From that point on, the history of the Universe is much better understood. The Universe was still so hot and dense that nothing, not even the notoriously aloof neutrinos, could travel far before smashing into some other particle. One second after the Big Bang, however, things had cooled enough for neutrinos to start travelling freely. It's an amazing thought: neutrinos that set off a second after the Big Bang are to this day travelling unhindered through space.

By about ten seconds after the Big Bang, most particles had mutually annihilated and the temperature of the Universe had fallen sufficiently to ensure that no new particle–antiparticle pairs could be created. For every particle of matter there was about a billion photons: radiation dominated. Particles of matter – electrons, protons, neutrons and so on – could not travel far before slamming into a highly energetic photon. Then something interesting happened: between 3–20 minutes after the Big Bang, the temperature of the Universe became a mere billion kelvin. At this temperature protons and neutrons could combine to form atomic nuclei. Roughly 75% of protons stayed solitary, as hydrogen neuclei, but deuterium nuclei (one proton and one neutron) and then helium nuclei (two protons and two neutrons) were also formed, along with traces of heavier atomic nuclei such as lithium.

At 20 minutes after the Big Bang, then, the Universe contained a soup of energetic photons, electrons, and light atomic nuclei. Photons could not travel far before being scattered by an electrically charged particle. Things stayed this way for the next 380 000 years (give or take a few thousand years). But throughout this time the Universe was expanding and, as it did so, it continued to cool. When the Universe was 380 000 years old its temperature had dropped to about 3000 K. Suddenly it was cold enough for the atomic nuclei to grab electrons and form stable, electrically neutral hydrogen and helium atoms. Photons were no longer fated to be absorbed or scattered by charged particles. As happened with neutrinos one second after the Big Bang, now photons could travel freely through space.

How do we know all this? How can we possibly speak with confidence about events that happened more than 13 billion years ago? Well, when photons decoupled from matter they had a wavelength corresponding to a blackbody temperature of about 3000 K. The subsequent expansion of the Universe has stretched those wavelengths so that they now correspond to a temperature of 2.75 K. *Microwave telescopes can detect those photons!* In recent years, orbiting microwave observatories have studied in exquisite detail these messengers from a time soon after the Big Bang. In turns out that traces from earlier times, perhaps even from inflation itself, are imprinted in the microwave background. Thus microwave telescopes, which are the subject of chapter 2, may be able tell us what happened soon after the Universe itself came into existence.

Tiny density fluctuations at that key point in history, 380 000 years after the Big Bang, were the seeds from which galaxy clusters later formed. But where did those fluctuations come from? The story sketched above omits a key component: dark matter. If matter consisted purely of 'ordinary' particles – protons, neutrons, electrons, and so on – then small density fluctuations in the early Universe could not have produced the lumpy Universe we see today. In order to explain the evolution of large structures astronomers are led to postulate the existence of dark matter. This mysterious substance is far more abundant than 'normal' matter. In fact, most of the matter in the Universe must be dark: this stuff dominated the evolution of the early Universe. Normal matter, which makes up you and me and everything that's important to us, is hardly worth mentioning on a cosmic scale. A worrying thing for science (or a really exciting thing – it depends on your viewpoint) is that no one knows what dark matter is. Chapter 3 discusses dark matter 'observatories', and the latest hunt for these elusive particles.

As already mentioned, after inflation the Universe continued to expand – but the rate of expansion was ever-slowing. The self-gravitation of all the matter in the Universe, both dark and ordinary matter, acted as a brake on the expansion. As the Universe cooled, gravitational collapse on small scales caused the formation of extremely bright, massive stars and active galaxies called quasars. As time passed, more stars and galaxies formed and the galaxies eventually clustered to form a cosmic web whose structure was determined by dark matter. Between about 3–6 billion years after the Big Bang (the precise time is uncertain) our own Milky Way galaxy formed. And then, about 8 billion years after the Big Bang, something exceedingly peculiar began to happen: the expansion of the Universe, which had been slowing, began to accelerate. Some phenomenon – cosmologists call it dark energy – began to blow the Universe apart. Instead of expanding at an ever-slower rate, as gravitational attraction pulled matter in on itself, the Universe began to expand at an ever-increasing rate.

It seems there is some property of space itself that causes this behavior. Dark energy was always there, even at the beginning, but it played a negligible role early on because there was so little space just after the Big Bang. As the Universe expanded, though, more space was created; that meant there was more dark energy, which tended to increase the expansion; which meant there was more space, and hence more dark energy, which tended to increase the expansion... until, about 8 billion years after the Big Bang, dark energy began to play a dominant role in the Universe. Dark energy now constitutes almost three-quarters of the mass-energy inventory of the Universe.

No one knows what dark energy is; this mystery is one of the strangest in all of science. An understanding of dark energy probably requires a deeper understanding of gravity, which is why it's appropriate that one of the properties of gravity – its ability to curve the paths of light rays, and thus generate what are in effect gravitational lenses – is being used to survey dark energy. Chapter 4 discusses gravitational lenses, and how they may shed light on dark energy.

Thus chapters 2–4 discuss three peculiar ideas: inflation, dark matter, and dark energy. With these elements, though, cosmologists can concoct a standard model of the Universe that agrees with all the observations they've ever made. The model goes by the name of ΛCDM (with the symbol Λ referring to the presence of dark energy and the acronym CDM referring to the presence of 'cold' dark matter). The ΛCDM model is surely one of the most magnificent examples of what can be achieved with input from great telescopes.

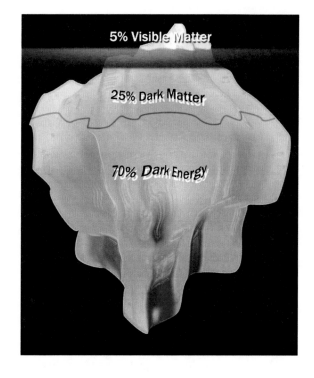

Figure 1.8 *In the same way that the tip of an iceberg is only a small part of the total structure, so visible matter is only a small part of the total mass–energy inventory of the Universe. The latest figures from surveys of the cosmic microwave background suggest that dark energy constitutes 72% of the energy density of the Universe; dark matter accounts for 23%; and only 4.6% is in the form of 'normal' matter, or atoms. The rest comes from a background of cosmic neutrinos. (SLAC/Nicolle Rager)*

The interplay of gravity with radiation, matter, and dark energy accounts for the large-scale evolution of the Universe. As the Universe matured, however, gravity's effect on matter was also responsible for more local phenomena. For example, gravity causes matter to contract and form dense objects. Massive black holes were common in the early Universe, and probably influenced how galaxies evolved; nowadays, supermassive black holes are found at the centers of most galaxies. Dense objects of much smaller mass formed when large stars reached the end of their lives: these stars became supernovae and blasted material into interstellar space, leaving behind neutron stars or stellar-mass black holes. These dense objects may be responsible for the most powerful explosions in the Universe: gamma-ray bursts.

A black hole possesses a gravitational field so intense that not even light, once it's past the event horizon, can escape its grip. How, then, can we learn more about black holes and the host of interesting phenomena that involve black holes? Well, chapter 5 discusses X-ray telescopes; observations at X-ray wavelengths can tell us what happens as material spirals into a black hole. Chapter 6 discusses the new generation of cosmic ray telescopes while chapter 7 discusses neutrino telescopes; together, particle detectors such as these are helping to provide a more complete picture of the nature of black holes and the phenomena they power. Chapter 8 is devoted to gamma-ray telescopes, which are helping elucidate the nature of gamma-ray bursts. Chapter 9 describes how a completely new type of astronomy is about to open up: gravitational wave telescopes will not only 'see' the events that give rise to gamma-ray bursts, they might even see waves that came from the inflationary epoch of the Universe.

As the Universe evolved, the processes of star birth and star death continued. Stars have always formed by condensing out of contracting clouds of dust and gas; they shine by converting light elements into heavier elements by nuclear fustion; and if they die in a supernova explosion they disperse those heavy elements back into interstellar space. Thus younger stars, which can incorporate the debris from earlier stellar generations, tend to possess a relatively high proportion of heavier elements. This recycling of material is a complex process, but even without knowing the details it's clear there's a problem with our understanding: much of the 'normal' matter that was around in the early Universe doesn't seem to be around in the present-day Universe. It's one thing not to understand the 95% of the Universe that's in the dark sector; it's quite another thing to be ignorant of the 5% of the Universe that we think we understand! Chapter 10 discusses ultraviolet telescopes, and how they can help us understand the whereabouts of the missing 'normal' matter.

About 9 billion years after the Big Bang our Solar System formed. A natural question is: are there are other planetary systems out there? In recent years it has become clear that, as a general rule, planets accompany stars. Chapter 11 describes how observations with infrared telescopes are clarifying how other planetary systems form. An even more natural question is: are there are other Earth-like planets out there? Chapter 12, which covers the latest advances in optical telescopes, reviews the status of the search for an Earth twin.

About 10 billion years after the Big Bang, life started on Earth. And about 13.75 billion years after the Big Bang – a nanosecond, in cosmic terms – an advanced technological civilization arose that could probe the history of the Universe. Have such civilizations arisen elsewhere in the Universe? That's the question tackled in chapter 13, which discusses how radio telescopes are being used in the search for extraterrestrial intelligence.

Twelve more chapters, each devoted to a cosmic mystery – and how the new generation of telescopes will provide the tools we need to solve them.

2

The oldest light in the Universe

A key part of the standard cosmological model is the concept of infla-
tion, the notion that the very early Universe underwent a period of
exponential growth. But did inflation really occur? And if it did,
how did it start and why did it stop? The latest orbiting microwave
observatories are poised to answer these questions – and, what is
quite astonishing, they may even be able to do so by detecting the
imprint of events that occurred when the Universe was just a tiny
fraction of a second old.

The temperature of space – Cosmic ripples – Inflation – MAPping
the Universe – The coldest thing in space – Beyond the Universe –
Rumbles from the start of time

S. Webb, *New Eyes on the Universe: Twelve Cosmic Mysteries and the Tools We Need to Solve Them*,
Springer Praxis Books, DOI 10.1007/978-1-4614-2194-8_2, © Springer Science+Business Media, LLC 2012

One of the most important discoveries in modern cosmology came about almost by accident. In 1964, Arno Penzias and Robert Wilson were experimenting with receivers intended for use in the testing of communications satellites. Their most sensitive instrument was a horn antenna which, in order to eliminate interference from the antenna itself, they cooled to a temperature just a few degrees above absolute zero. However, even after cooling the antenna they detected a faint background hum spread evenly over the entire sky. They heard noise at a wavelength of 7.35 cm, a hum that was present day and night, summer and winter.

Penzias and Wilson systematically went about eliminating possible sources and explanations for the noise. They found that it did not come from New York City (the antenna was based in Holmdel, New Jersey, which was close enough to New York for them to suspect the city as the source of the interference). It did not come from the droppings of pigeons that had nested in the antenna (a substance tastefully described by Penzias and Wilson as being a 'white dielectric material'). It did not come from possible after-effects of atmospheric nuclear testing; and they ruled out any other terrestrial source. Eventually, they concluded that the electromagnetic noise in their antenna must emanate from outside our Galaxy.

Radio astronomers typically characterize their signals by fitting them to a blackbody spectrum – even if they are unsure whether the source is actually emitting blackbody radiation. The excess radiation seen by Penzias and Wilson was thus described as having a temperature of 3.5 K. It turned out that Penzias and Wilson had discovered a simple and basic fact about the Universe: wherever and whenever you look into space you see excess microwave radiation corresponding to a particular temperature. That's why their discovery was so important: if a particular cosmological model could not account for this radiation then that model had big problems.

While Penzias and Wilson were still investigating their findings, a team of astrophysicists led by Robert Dicke at a university department less than an hour's drive away was independently planning a search for low-temperature radiation from the sky. Dicke's team believed such radiation must fill the Universe if the Big Bang theory of cosmology were correct. A mutual colleague put the two teams in touch, and it was then that Penzias and Wilson understood the nature of the noise in their antenna: they had discovered the **cosmic microwave background** (CMB) – relic radiation from the Big Bang itself. In 1978, Penzias and Wilson were awarded the Nobel Prize for their momentous discovery.

Figure 2.1 *Penzias and Wilson were using this horn antenna in 1964 when they discovered radiation that turned out to be remnants of the Big Bang. (Commons)*

The temperature of space

Why should an expanding Universe, which came into being in a Big Bang that happened billions of years ago, cause the present-day cosmos to be filled with microwave radiation?

For the first 380 000 years or so after the Big Bang the Universe would have been too hot to allow neutral atoms to exist. Instead, matter would have been in the form of a soup of freely moving electrons, protons and neutrons. Any photons moving around back then would have interacted vigorously with the electrons in this soup, the result being that the early Universe was opaque: light could not travel far before being scattered. As the Universe expanded, it cooled. Roughly 380 000 years after the Big Bang the temperature of the Universe would have dropped to about 3000 K – cool enough for neutral atoms to form. This time is called the **recombination** epoch. (It's a strange choice of term: nothing was 're' combining – this was the first time that electrons, protons and neutrons were combining to form atoms.) After this event the Universe became transparent: no longer were there free electrons to get in the way, so photons suddenly began to move unimpeded – which they have been doing ever since (until, roughly 13 billion years later, astronomers are catching them in their telescopes). The photons in the CMB – and there are about 400 of them in each thimbleful of space – are the oldest photons in the Universe: it's impossible to 'see' further back in time because, prior to recombination, photons couldn't travel freely. In a sense, then, the CMB is the most distant 'object' that telescopes can see. But why does this oldest light consist of *microwave* photons?

The Big Bang scenario predicts more than just the existence of cosmic photons: it predicts what their temperature will be. If the Big Bang was a 'smooth' event (homogenous and isotropic, as the jargon goes) then the gases at recombination would have been in perfect thermal equilibrium and there'd be no reason for the spectrum to be anything other than that of a blackbody. At the recombination temperature of $3000\,K$ the peak in the radiation spectrum would have been at around $1\,\mu m$, which is in the infrared. (There'd have been significant amounts of visible radiation too. If you'd been there to see it then your eyes would have perceived the Universe to have a uniform, dull orange color.) Those cosmic photons, though, have been stretched as the Universe expanded. The Universe has expanded by a factor of about 1000 since the recombination epoch, so the wavelength of those photons too has increased by a factor of about 1000. Thus the Big Bang model predicts that the peak in the radiation spectrum is at about $1000\,\mu m$, or 1 mm in other words, which is in the microwave region.

So it's not just the existence of the background radiation that's important, it's whether the radiation has the form of a blackbody. If the Universe is filled with blackbody radiation then the existence of the Big Bang is pretty much nailed down: competing cosmological theories just can't explain the CMB with elegance. However, what Penzias and Wilson discovered was not blackbody radiation. Astronomers *assumed* that this radiation came from a blackbody – but Penzias and Wilson had observed the radiation from only one location and at only one wavelength. (They observed at 7.35 cm simply because that was the signal wavelength Bell Labs had been using to communicate with orbiting satellites. But 7.35 cm was many times longer than the peak wavelength of the background radiation.) Astronomers could not *prove* this was blackbody radiation without measuring the background at a variety of wavelengths and from different locations. Many subsequent observations tended to confirm that the spectrum was that of a blackbody.

Unfortunately, however, microwaves with a wavelength less than about 3 cm are absorbed by water molecules in Earth's atmosphere. Astronomers would have to get above the atmosphere if they wished to study the CMB in detail and thus confirm, beyond doubt, that the Big Bang theory was correct.

Cosmic ripples

In 1989, NASA launched a satellite called the **Cosmic Background Explorer** (COBE) in order to make precise measurements of the CMB at infrared and microwave wavelengths. See figure 2.2. The satellite carried three instruments. One of them, the Far Infrared Absolute Spectrophotometer (FIRAS), was there to compare the CMB spectrum against a precise, laboratory blackbody over the wavelength range from 0.1 mm to 10 mm. By the time FIRAS had come to the end of its life ten months after launch it had surveyed the sky 1.6 times. Its measurements showed that the CMB is indeed a blackbody and it has a temperature of 2.725 ± 0.002 K. The FIRAS result astonished scientists: they were expecting to see a blackbody spectrum, but not one *this* good. In any scientific experiment you're going to see some 'noise' in the data, some points on the graph that don't align exactly with the theoretical curve. Not so here. If the FIRAS data points were plotted on a theoretical blackbody curve, similar to one of the curves in figure 1.4, then you wouldn't be able to see them: the error bars would be smaller than the thickness of the line used to draw the curve. The CMB blackbody spectrum was almost perfect. Nature had produced a blackbody more precise than anything made by human hands – and COBE had measured it.

The CMB was an *almost* perfect blackbody; but it could not be an *absolutely* perfect blackbody. The early Universe could not have been *perfectly* smooth and featureless because there's such a lot of structure around today; we see galaxies, and clusters of galaxies, and clusters of clusters of galaxies (to say nothing of really small-scale stuff such as stars and planets and people). That structure had to come from somewhere. Cosmologists thus argued that some regions of the very early Universe must have been slightly denser than average and some regions must have been slightly less dense. Those primordial density variations would subsequently evolve through gravitational attraction into the structures we see today.

It turns out that any density variations that existed back at the time of recombination would leave an imprint on the CMB. A full understanding of why this should happen requires general relativity, but a rough argument

Figure 2.2 *An artist's impression of the* COBE *satellite, which was launched by* NASA *in November 1989. The satellite orbited Earth at a height of 900 km with a period of 103 minutes. The data taken by* COBE *instruments during the four-year mission transformed cosmology.* (NASA/COBE *Science Team*)

goes as follows. Before recombination, the higher-density regions acted as gravitational traps, and any photons falling into these traps gained energy – just as a falling apple gains kinetic energy. Since those gravitational traps were dense with freely moving electrons, the photons were unable to escape before interacting with them. After recombination, however, those gravitational traps became transparent: photons could escape and in the process of escaping they expended energy – just as you have to expend energy to throw an apple upwards. Now, if you drop an apple from waist height, and later pick it up and bring it back to the same height, then over the whole cycle the

Figure 2.3 In two flights (the first in 1998, the second in 2003) the BOOMERANG mission measured the anisotropies imprinted on the CMB. The photograph here shows the balloon preparing to launch, with Mount Erebus in the background. (NASA/NSBF)

apple neither gains nor loses energy. You might expect the same argument to apply to those photon traps. But between the time a photon fell into one of the overdense regions and the time it escaped, the Universe got bigger: so a photon had to expend *more* energy escaping from the trap than it gained when it fell in. In this particular scenario the net result is a loss of energy – so regions of the sky that were denser than average would appear to be slightly colder than average. If this effect existed it would be observable as anisotropies – tiny variations or ripples – in the temperature of the CMB. That was the theory, anyway, and these ripples on top of a blackbody spectrum were something for which COBE was designed to hunt.

A second instrument on board COBE, the Differential Microwave Radiometer (DMR), was there to map the microwave background with exquisite sensitivity. The DMR did indeed find anisotropies, those ripples that theory suggested must be there. The ripples were small – at the level of 1 part in

100 000 – so if they were ripples on a pond you'd never notice them. But these ripples were important. The DMR couldn't detect variations on angular scales smaller than about 7° so it was unable to see the ripples that would later form clusters or even superclusters of galaxies. But cosmologists were able to extrapolate the data and they found that the observed ripples were large enough to produce the clustering that's seen today (so long as plenty of dark matter is added to the mix). So by taking a snapshot of the Universe as it appeared when it was 380 000 years old, COBE told astronomers a tremendous amount about our present-day Universe.

In the years BC – before COBE – cosmologists used to argue over the most important, fundamental underpinnings of their science. The COBE instruments showed that cosmology could be a precision science. Even more than its results, which firmly fixed the Big Bang theory, COBE's key legacy was how it paved the way for precision cosmology. Two of the COBE personnel – John Mather (who led the team that confirmed the blackbody form of the CMB) and George Smoot (who led the team that found the anisotropies) – were awarded the 2006 Nobel Prize in physics.

Even though COBE was an unqualified success, astronomers were frustrated that they could not study anisotropies on a scale of about 1° or less. The search soon turned to finding those important small-scale ripples.

In 1999, a mission called **Balloon Observations of Millimetric Extragalactic Radiation and Geophysics** (BOOMERANG) flew for 10.5 days above Antarctica. See figure 2.3. Polar winds carried the balloon in a circle and returned it to its launch site at McMurdo Station; the telescope's trajectory seems to have prompted its acronym and thence the name. This international collaboration led by American and Italian physicists made exquisitely detailed measurements of fluctuations in the microwave background. The BOOMERANG experiment took almost a billion measurements of temperature variations in the CMB over a large part of the sky. The subsequent analysis revealed precisely the form of ripples on small angular scales that cosmologists expected. Cosmologists now knew they had a model of the birth of the Universe that made sense. Their model, however, was not without problems.

Inflation

Astronomers highlighted several difficulties with the Big Bang idea soon after the discovery by Penzias and Wilson of the CMB. Three problems turned out to be particularly troubling.

First, there was the monopole problem: a Big Bang should have produced huge numbers of massive particles called magnetic monopoles. Physicists have searched long and hard for these particles but without success. If magnetic monopoles exist then they are scarcer by far than Big Bang cosmology predicts.

Second, there was the flatness problem: the geometry of the Universe seems to be flat. In a Big Bang cosmology, though, curvature grows with time. For a Universe to be as flat as we see it today, and yet not be exactly flat, would require conditions in the early Universe to have been fine-tuned to an unbelievable degree.

Third, and most troubling of all, there was the horizon problem: the microwave background radiation is just too smooth. The puzzle arises because patches of sky separated by a couple of degrees of arc – that's just a few times the diameter of the full Moon – were not in causal contact back at the time of recombination. In other words, those regions of sky were so far apart at recombination that no signal, not even one travelling at lightspeed, had enough time to pass between them. So 380 000 years after the Big Bang those thousands of causally disconnected regions of the Universe could not possibly 'know' what the other regions were doing. How could it happen, then, that all those regions have the same temperature?

The cure for all these problems is **inflation**.

The inflationary paradigm was proposed in the early 1980s by Alan Guth and, independently, Alexei Starobsinki. Soon after, Andreas Albrecht and Paul Steinhardt and, independently, Andrei Linde fixed a problem with the initial formulation of the paradigm.

The basic idea behind inflation is simple: the early Universe had a growth spurt. A really impressive growth spurt. Inflation started about 10^{-36} s after the Big Bang; it ended about 10^{-33} s after the Big Bang. In that unimaginably brief time the Universe increased in volume by an incredible factor of at least 10^{78}. Space itself was therefore, in a sense, inflating much, much faster than the speed of light. This phenomenon is allowed in general relativity. By contrast, matter must obey the 'local' speed limit imposed by special relativity – the lightspeed limit.

How does inflation cure the problems with Big Bang theory? Well, it cures the monopole problem because, although magnetic monopoles could have been created in abundance by the Big Bang, while the Universe is inflating the density of monopoles gets smaller and smaller. By the time inflation has stopped the density of monopoles can be so small that they become undetectable. Inflation solves the flatness problem because, even though the Universe could have had any curvature to start out with, the exponential increase in the size of the Universe stretches any curvature to effective flatness. (If you stand on a basketball you know you're standing on a curved surface. Stand on the curved surface of the Earth and it's as if you're standing on a flat surface. Inflation did the same thing to the Universe, except that the ratio between the size of the Universe before and after inflation is much, much greater than the ratio of the size of a basketball to Earth.) Finally, it solves the horizon problem because, prior to inflation, distant regions of the Universe were much closer together than the standard Big Bang expansion assumes. Before that brief phase of exponential growth everything that we currently see – all the matter that forms stars, galaxies, clusters – was packed into an incredibly tiny volume. Everything that we see *could* have been in causal connection after all and different regions *could* have attained a uniform temperature. Inflation elegantly explains why the Universe looks so uniform.

The inflationary paradigm rapidly established itself as a standard part of cosmology because not only did it address problems with traditional Big Bang theory it did so using ideas from a completely different area of physics. Guth's original formulation of inflation used concepts from grand unified theories of particle physics and quantum theory – ideas that had been developed to explain the world of the very small, with no conception that they might apply to cosmology. This notion that theories developed to explain the microworld of particles could solve problems in theories describing the Universe itself was immensely appealing.

Inflation became a key element in the standard model of cosmology (we'll consider the other key elements in chapters 3 and 4). But how can cosmologists possibly learn more about inflation, a fleeting event that happened a vanishingly short time after the birth of the Universe itself? How can they even prove that inflation took place?

It's possible that the answers to these questions, as with so many others to do with the Universe, are imprinted on the cosmic microwave background.

MAPping the Universe

Following the huge success of COBE, a group led by Charles Bennett pitched a mission to NASA called the Microwave Anisotropy Probe (MAP). The idea was to develop a spacecraft for carrying instruments that would be 45 times

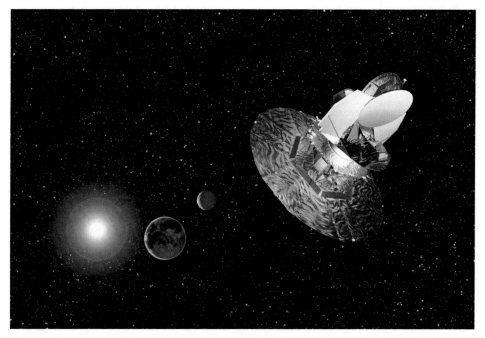

Figure 2.4 *An artist's impression of* WMAP *leaving Earth orbit for* L2, *the second* **Lagrange point** *of the Earth–Sun system. This region of space, about 1.5 million km distant from Earth, can be thought of as a gravitational 'saddle point'. By making only small maneuvers, mission controllers can keep spacecraft at a roughly constant distance from the Earth throughout the year. It's thus an ideal place for spacecraft carrying astronomical observatories.* (NASA/WMAP *Science Team*)

more sensitive than those on COBE and possess 33 times the angular resolution of its predecessor. By studying the tiny temperature *differences* in the microwave background radiation at small angular scales – about a quarter of a degree, a finer scale than even BOOMERANG could observe – the team knew they would be able to learn huge amounts about the age, contents and structure of the Universe. The idea got approved in 1997, and in 2001 the rocket carrying the MAP mission was launched. In 2003, following the death of David Wilkinson, one of the MAP team's leading scientists, the mission was renamed the **Wilkinson Microwave Anisotropy Probe** (WMAP). The satellite ceased its observations in 2010.

The instruments on WMAP looked at two different points of the sky simultaneously, and at five different microwave frequencies. By doing this often enough, WMAP could measure the differences in the CMB temperature across the whole sky. The top diagram in figure 2.5 shows the temperature of the microwave background of the whole sky as measured by seven years' worth of observations by WMAP; different temperatures are represented by different shades of color. The fact that you can see only one shade represents the fact that the Universe is so uniform. WMAP confirmed the COBE result: the microwave background radiation is an astonishingly accurate blackbody. The average temperature of the CMB is 2.725 K and the difference between the hottest part of the sky and the coldest is just 0.003 K. This result has surely confirmed, if confirmation were needed, that the early Universe was in a hot, dense state: there can surely be no doubt that the Universe began in a Big Bang.

The point of the WMAP mission was to find temperature *differences*, however, and not just to measure the average microwave background with more accuracy. So when the mission team subtracted this average 2.725 K temperature from the figure, along with noise due to the Milky Way, what did they see? Well, once they turned up the contrast they saw something similar to what the COBE team saw, as shown in the middle diagram of figure 2.5.

The purplish region to the bottom-left of this middle diagram is dark because it is represents a blueshift: this part of the sky is hotter than average. The orange region to the top-right represents a redshift: this part of the sky is colder than average. These two regions are exactly opposite each other in the sky – they form two poles. This is interesting because it shows we are in motion, but it's hardly a great discovery: we know that Earth orbits the Sun, the Sun orbits the center of the Milky Way, the Milky Way moves within the Local Group of galaxies, and the Local Group itself is being pulled towards the Virgo cluster of galaxies. So the effects of this motion – the dipole – must be subtracted (and the contrast turned up yet again) in order to see temperature differences that are intrinsic to the cosmic background.

And what did the WMAP mission team see once they did *this*? They saw the result shown in the bottom diagram of figure 2.5. This is a map of the minuscule temperature fluctuations in the cosmic microwave background, seen for the first time in high-definition detail.

The different colors in the bottom diagram of figure 2.5 represent tiny temperature variations. These variations are what later became clusters of galaxies; they are quantum fluctuations writ large across the entire sky, and they are really quite beautiful. The map is not only beautiful, it encodes a vast amount of information about the early Universe. Various different physical causes will have generated temperature fluctuations in the CMB.

Figure 2.5 *In these diagrams, different colors represent different temperatures in the cosmic microwave background. (The maps created by COBE and WMAP have the characteristic oval shape of a Mollweide projection: this sort of map provides a more accurate depiction of area than other types of map.) Top: seven years of WMAP data. The single color here demonstrates the uniformity of the CMB! Middle: the dipole pattern seen by COBE. Bottom: intrinsic temperature variations in the cosmic microwave background as determined by seven years of WMAP data. In this diagram the regions shown in red are fractionally hotter than the regions shown in blue. (NASA/WMAP Science Team)*

For example, density variations at recombination will have had an effect, as will variations in gas velocity and gravitational potential. But these different effects will have generated fluctuations at different angular scales. The *sizes* of these various physical effects are in turn determined by the value of fundamental cosmological parameters. By plotting the angular fluctuation spectrum – which is just a curve showing the size of the temperature fluctuations as a function of angular scale – scientists can deduce the value of the fundamental cosmological parameters: they simply tweak the parameters until they find a set of values that generate the curve that's actually observed.

Figure 2.6 shows the angular fluctuation spectrum that WMAP measured. If the microwave background had been perfectly uniform then the spectrum would have been a straight line: there would have been no variation of temperature at any angular scale. Instead, of course, WMAP saw 'wiggles' in the spectrum. Variations on large angular scales, represented by the left-hand side of the spectrum, are the anisotropies that COBE first saw back in 1992. What WMAP was able to detect were variations at smaller angular scales – and thus a treasure trove of information.

Consider the first peak in the angular fluctuation spectrum, which occurs at about 1°. Why does this peak appear?

Well, consider an initial overdense region. Gravity will have pulled dark matter, 'normal' matter, and photons towards the region. As we shall explain in more detail in chapter 4, the dark matter falling towards that overdense region will have felt nothing but the gravitational force; not so the photons and 'normal' matter. The photons will have exerted an outward pressure on electrons and protons. A wave of photons and 'normal' matter will have bounced away from the overdense region, moving outward together like a ripple on the surface of a pond. At recombination, however, the photons began to move freely and the waves of matter they were pushing would suddenly have frozen. That first peak in the spectrum, then, represents waves – sound waves, essentially – that were just starting their first compression when recombination occurred. There is thus a preferred distance scale in the microwave background. (Later peaks, to the right of the first peak, represent higher-frequency sound waves that happened to be in compression at the time of recombination.) Regions that were in compression at the recombination time will seem to be hotter than average; regions in rarefaction will be colder than average. What is particularly interesting is that the location of the first peak provides information on the geometry of the Universe – whether it's flat, open or closed. (It's flat.)

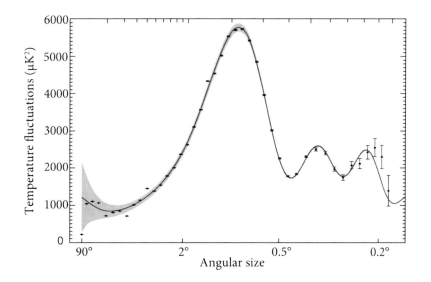

Figure 2.6 *Temperature fluctuations of the sky at different angular scales as measured by WMAP. The size and location of the peaks and troughs in this diagram contain information on key parameters of the Universe, as discussed in the text. (NASA/WMAP Science Team)*

The other peaks contain equally important information. For example, the relative heights of the first and third peaks compared to the second peak is governed by the relative density of protons to dark matter; the WMAP angular fluctuation spectrum therefore tells astronomers how much dark matter there is and how much 'normal' matter. There's even more: the microwave photons that WMAP observe are affected by the gravitational fluctuations they encountered on their way from recombination to telescope. If the Universe were flat because dark matter dominates then these gravitational fluctuations would cancel each other. If the Universe were flat because of the presence of dark energy, however, then WMAP would see extra fluctuations. In this way it turns out that WMAP can provide information on the amount of dark energy in the Universe.

The impact of an analysis of seven years of WMAP data has been staggering. Thanks to the WMAP team we now know some of the key parameters that describe the Universe with an accuracy that was unimaginable just a few years ago. For example, WMAP tells us that the age of the Universe is $13.75\,(\pm 0.11)$ billion years. The rate at which the Universe is currently expanding, the so-called Hubble constant, is $70.4\,(\pm 1.4)\,\mathrm{km\,s^{-1}\,Mpc^{-1}}$. The geometry of the

Universe is flat, to within 0.5%. At present, only 4.56 (±0.16)% of the mass–energy content of the Universe is in the form of baryons; 72.8 (±0.5)% is in the form of dark energy and 22.7 (±1.4)% is in the form of dark matter. See figure 2.7. The form that dark energy takes remains uncertain but, as we shall see in chapter 4, it seems increasingly likely that it's the presence of a cosmological constant that generates dark energy. And so on and so on.

But what about inflation? Can WMAP tell us whether inflation took place. Well, the fact that measurements indicate a flat Universe is indirect evidence that a period of inflation took place, but WMAP has found further evidence to support the idea. The locations of the peaks in the angular fluctuation spectrum are spread out in a way that is consistent with inflation but inconsistent with other mechanisms of structure formation that astronomers have proposed. Furthermore, before the idea of inflation was introduced into cosmology, theorists argued that quantum fluctuations in the early Universe would produce irregularities of the same strength at all angular scales. In other words, the strength of the temperature anisotropies would be the same on all distance scales and a measure of the relative strength on small and large scales – the so-called 'scalar spectral index' – would be precisely equal to 1. The simplest models of inflation, however, suggest that the initial quantum fluctuations would give rise to slightly weaker anisotropies at larger angular scales: the scalar spectral index would be close to, but measurably less than, 1. Well, the WMAP team measured this index and found a value of 0.96 (±0.01).

These results from WMAP suggest that inflation really did take place. But is it possible to find *direct* evidence that inflation took place? If so, what was the physics behind inflation? When did inflation begin? How did it stop? If the CMB can be examined in even more detail than WMAP has managed perhaps cosmologists could answer these questions. That was the hope behind an ESA orbiting observatory, which continued in the tradition of the two NASA missions, COBE and WMAP.

The coldest thing in space

Unless there is some other intelligent, technologically advanced species out there in the cosmos, at the time of writing the coldest object in space is the **Planck Observatory**. The observatory, which is named after the German physicist Max Planck, who won the Nobel prize in 1918, was launched in 2009. Fifty days after launch it reached its operating temperature: a frigid

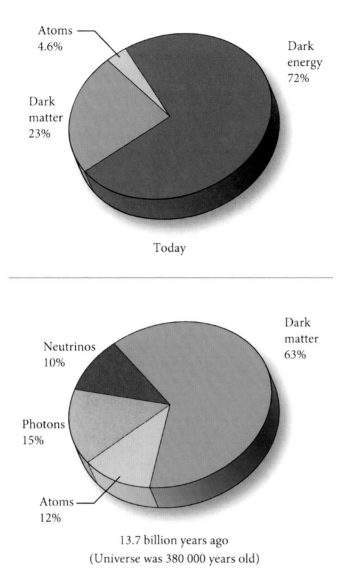

Today

13.7 billion years ago
(Universe was 380 000 years old)

Figure 2.7 *The atoms that make up our everyday world constitute only a fraction of the 'stuff' of the Universe. An analysis of the WMAP angular fluctuation spectrum has determined precisely how much of each ingredient is found in the present-day Universe (top part of the diagram) and how much was present back at the time of recombination (bottom part). Dark energy is now the dominant component of the Universe; 'normal' matter constitutes less than 5% of the total. (The small neutrino component is not shown in the diagram). Dark matter, not dark energy, was the dominant component in the Universe at the time of recombination.* (NASA/WMAP Science Team)

0.1 K. There's nothing colder in space: specialist laboratories on Earth can reach lower temperatures, but no natural object in space can reach a temperature that's just one tenth of a degree above absolute zero – the background radiation, after all, is 2.7 K. (Planck won't be that cold forever. By January 2012 the coolant for one of its detectors had already run out.)

It's not easy to build a space-based instrument that can be cooled to a fraction of a degree above absolute zero and then maintained at that temperature while observations are made. Unsurprisingly, then, a number of different teams were involved in developing techniques for refrigerating the satellite. A passive cooling system initially brought the temperature at the heart of Planck down to about 43 K simply by radiating heat into space. Three active cooling systems then took over: an American-designed cooler brought the temperature down to 20 K; a British cooler took over and got the temperature down to 4 K; the final cooler, a French system, reached the temperature of 0.1 K. The controllers in charge of Planck help maintain the necessary frigid conditions by ensuring the satellite is always pointing away from the Sun. This bothersome procedure is worthwhile because it enables Planck to study the microwave background radiation with unparalleled sensitivity.

Planck carries a telescope that focuses radiation onto two detectors. One of them, the high frequency instrument (HFI), whose coolant supply has now been exhausted, studied the background radiation at six frequencies between 100 –857 GHz. The 52 HFI detectors – or **bolometers** – detect the tiny amount of heat released when a photon from the microwave background hits electrons in the device. The bolometers must be cold because the intention is to search for temperature variations in the CMB at the level of about a millionth of a degree. The Planck mission team compare this sensitivity to measuring here on Earth the heat produced by a rabbit sitting on the Moon. (Presumably a rabbit that's given an air supply.) Only by cooling the bolometers to temperatures approaching absolute zero can such tiny variations be detected. The other detector, the low frequency instrument (LFI), which at the time of writing is still operational, consists of 22 receivers that operate at a slightly higher temperature. These receivers are studying the background radiation at 30 GHz, 44 GHz and 70 GHz.

Planck spins about its Sun-pointing axis once every minute in order to stabilize its attitude, and as it spins its bolometers scan small strips of the sky and measure minuscule temperature differences in the CMB. By one year after its launch it had observed every part of the sky and began its second all-sky survey. (Some Planck results from the first survey were published

Figure 2.8 *An artist's representation of the Planck spacecraft. As with WMAP, the Planck spacecraft orbits around L2. (ESA – D. Ducros)*

in 2010. These results are 'trash' – but useful trash! The Planck measurements are contaminated with microwave radiation from our Galaxy and extragalactic objects, and this 'trash' must be removed for the purposes of cosmology. To facilitate this, Planck takes measurements at several different frequencies. Although for cosmology all this nuisance foreground radiation can simply be dumped, this data contains a treasure trove of information for astronomers interested in other things.) As already mentioned, the cooling systems on Planck are beginning to fail and some time in 2012 the mission will be over. Final results from Planck are expected towards the end of 2012.

Planck's angular resolution – in other words, its sharpness of vision – is better than WMAP. Furthermore, its bolometers can detect signals that are ten times fainter than WMAP could pick up and it studies more wavelengths than WMAP could examine. Taken together, all this means that Planck will squeeze even more information from the microwave background radiation than WMAP could. If it does nothing else, Planck will pin down values for the important parameters of the Universe – its age, its mass-energy inventory, its rate of expansion, and so on – with even greater precision than WMAP. But can Planck possibly see signs of inflation? There's a chance that it can. Here's why.

The cosmic background radiation consists of electromagnetic waves. These waves now peak at a wavelength that means we define them to be microwaves, but they possess exactly the same qualities as all other electromagnetic waves – including visible light. And one of the familiar properties of visible electromagnetic radiation is that we can often see **polarized light**. Polarization can occur because light is a transverse wave with both electric and magnetic components: although in most cases these components vibrate in many different planes, so that the light is unpolarized, various effects can cause the vibrations to occur in a single plane – and the light is then said to be polarized. For example, light emitted by the Sun, a lamp or a flame is unpolarized; passing this light through a Polaroid filter, however, will polarize it. More importantly in this context, the scattering of light off free electrons is another phenomenon that can cause polarization. So although the light in the very early Universe would have been unpolarized, it could have became polarized at the moment just before radiation decoupled from matter – when it struck the freely moving electrons that existed just before they combined with protons to form stable atoms. The standard model of cosmology thus predicts that the microwave background became polarized by matter in density fluctuations.

If astronomers could create maps of this microwave polarization then they would not only be providing yet another confirmation of the standard model, they could deduce the location of density variations at the time of decoupling and also, critically, *how matter was moving at that time*. The CMB thus has even more information imprinted upon it than originally thought. Density fluctuations would have imparted only a small degree of polarization to the CMB – so measuring it is even more of a challenge to technology than is the measurement of temperature fluctuations. Nevertheless, in 2002 a team led by John Carlstrom used the **Degree Angular Scale Interferometer** (DASI) – an instrument based at the South Pole – and found that the microwave background is indeed polarized. This key result by DASI showed that important features of the CMB could be studied by ground-based observatories, and other observatories have confirmed the finding. Subsequently, of course, WMAP has been able to generate much more detailed maps of the polarization pattern. Figure 2.9 shows the polarization pattern that WMAP has seen around hot and cold spots in the CMB and compares them with what the standard model predicts.

What has this to do with inflation? Well, the polarization found by DASI, WMAP and other experiments was generated ultimately by density fluctuations and these produce polarization patterns of a particular type: the pattern is called **E mode polarization**. The two diagrams on the left in figure 2.10 illustrate the typical symmetric patterns of the E mode; by comparing with figure 2.9 you can see that these are precisely what WMAP found around hot and cold spots. However, there's something else that could have polarized light: gravitational waves. We'll learn much more about gravitational waves in chapter 9, but the key point here is that a gravitational wave is a disturbance of space itself: at the same time that a gravitational wave *stretches* space in one direction it *squeezes* it in the perpendicular direction. These waves are generated whenever large masses move suddenly (in the same way as electromagnetic waves are generated whenever electric charges accelerate) so gravitational waves would have been generated in the very early Universe, and inflation would then have stretched these disturbances in spacetime.

Light waves moving through a distorted space – stretched in one direction, squeezed in another – would be forced to vibrate in particular directions: they would be polarized. In fact, two different polarization patterns would be produced: an E mode pattern, similar to that produced by density variations, and the so-called **B mode polarization** pattern. The B mode polarization pattern – which has a swirling sort of aspect (see the two diagrams on the right of figure 2.10) – can *only* be produced by gravitational waves. Furthermore, unlike electromagnetic waves, gravitational waves would have been free to move through the fog of electrons, protons and neutrons that filled the Universe prior to the recombination epoch. So at the time when the Universe became transparent to electromagnetic waves, which subsequently streamed freely into space, gravitational waves were able to leave an imprint on the CMB. This imprint is inevitably going to be tiny; the search for it is like trying to find the fingerprint left behind by a ghost. Nevertheless, the search is worth the effort since the B mode pattern left behind by gravitational waves can tell cosmologists much about inflation.

The precise strength of the B mode pattern depends on how big the gravitational waves were. According to the inflationary model, the longer the wavelength the stronger the gravitational wave; the strength of the gravitational waves depends on how rapidly the Universe expanded during inflation and this, in turn, depends on the precise moment that inflation began.

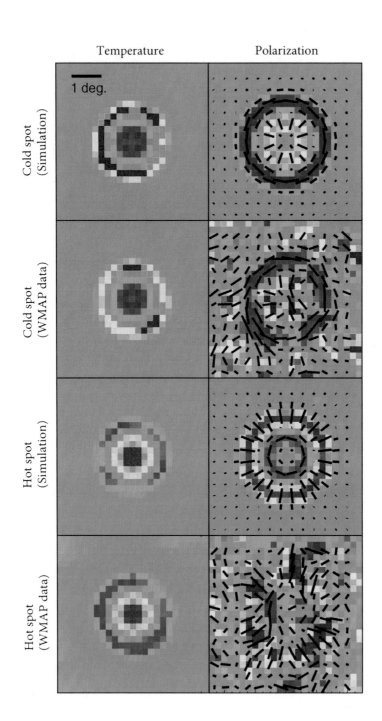

Figure 2.9 *The standard model of cosmology makes predictions for the temperature patterns and the polarization patterns that should be found around hot and cold spots in the cosmic microwave background. (In this figure, these predictions are marked with the word 'simulation'.) The messy, real-world WMAP observations can be seen to follow the theoretically expected pattern.* (NASA/WMAP Science Team)

Figure 2.10 *The two patterns on the left are typical of E mode polarization: they possess symmetry and are the sorts of pattern found by WMAP. The two patterns on the right are typical of B mode polarization: they have a particular 'handedness'.*

If inflation began sooner rather than later in the history of the Universe (remembering that both 'sooner' and 'later' here refer to astonishingly early periods of time) then there is a chance that the waves will have been large enough for Planck to see them imprinted on the polarization of the background radiation. There's no certainty that Planck can observe the B mode – but there's a tantalizing prospect that it will. And other experiments have joined in the hunt. The **Q/U Imaging Experiment** (QUIET), for example, was an international collaboration that employed four telescopes and a thousand polarization detectors based high in the Atacama desert in Chile. The QUIET collaboration tried to detect the B mode polarization by observing the microwave background for long periods of time. (The term 'Q/U' in its name refers to parameters that are often used to describe the polarization of electromagnetic radiation.) A similar experiment, also at Atacama, is the **Polarization of the Background Radiation** (POLARBEAR). Starting in 2012, POLARBEAR will spend three years trying to detect the B mode using a dedicated telescope and a large imaging camera; after that first run, the experiment will be upgraded in 2014 with more detectors and improvements to the telescope. Yet another microwave background experiment based high in this Chilean desert is the **Atacama B-mode Search** (ABS), which in 2011 started to measure the patterns of polarization anisotropy at large angular scales. Clearly Atacama is the place to be for these experiments!

If and when the B mode is detected in the cosmic microwave background – whether by Planck, POLARBEAR, ABS or some other observatory yet to be built – physicists would have their first probe of grand unification physics, well beyond the reach of any accelerators that can be constructed with our present level of technology. And the rest of us would be able to look at the effects of something that happened when the Universe was only a trillionth of a trillionth of a trillionth of a second old. It's an incredible prospect. With luck, we won't have too long to wait.

Beyond the Universe

One of the commonest questions asked of astronomers is: how big is the Universe? The short answer is that no one knows. It could be finite or infinite in extent. The size of the *observable* Universe, however, is calculable and it depends on the age of the Universe. Since we know thanks to WMAP that the Universe came into being 13.75 billion years ago, give or take 110 million years, then by definition the farthest we can possibly observe is from signals travelling at the speed of light that have taken 13.75 billion years to reach us. (In fact, if we are talking about electromagnetic radiation, then we can see no further than the CMB.) But what does that convert to in terms of distance?

It turns out that distance is a tricky concept in an expanding Universe. Consider, for example, a galaxy whose light was emitted 8 billion years ago and that we are only now detecting. How distant is that galaxy? The simplest, and most common, way of giving its distance is to use the light travel time; in this case the galaxy would be said to be 8 billion light years away. However, when people talk about 'distance' they usually mean separation in space at a common time; and because the Universe is expanding, when the light from that galaxy began its journey it was much closer to us than 8 billion light years. Similarly, it is now (assuming that it still exists) much further away from us than 8 billion light years. Astronomers therefore have several different measures of distance, and the appropriate measure to use depends on the context. It turns out, though, that in the standard cosmological model with the parameters measured by WMAP you can easily calculate the **proper distance** to an object – the distance it is from us, *right now* – if you know how long its light has taken to reach us. Light from the CMB, for example, took about 13.7 billion years to reach us; that converts to a proper distance of 46 billion light years. So the distance to the edge of the visible Universe – in a sense, the answer to the question 'how big is the Universe' – is 46 billion light years.

However, since the cosmos is expanding, the radius of the observable Universe is ever increasing. Depending upon the details of the expansion, it is entirely possible that there are objects presently 'beyond the horizon' of the observable Universe that will eventually come into view. Indeed, the inflationary idea suggests that the observable Universe is just a small part of a much wider cosmos. When inflation caused spacetime to smooth out by growing exponentially, the result was presumably a cosmos that was much bigger than the patch we can actually observe. (It's even possible that different patches of spacetime underwent their own period inflation. That would give rise to Universes beyond our own with quite different properties and laws of physics, and with whom we could never communicate.)

Such ideas are speculative, perhaps, but is it possible to see hints of something that lies beyond our Universe? Once more, the answer to this may lie in a study of the CMB.

In 2008, a team of astronomers led by the NASA scientist Alexander Kashlinsky published an analysis of the motion of 700 very distant galaxy clusters. The gas clouds in these clusters are so hot that they emit X-rays, which can be detected by X-ray telescopes. After locating the clusters, the team then looked at the same spots in the CMB using early results from WMAP. When microwaves from the background radiation passed through the clusters they would change temperature slightly, depending on how the clusters moved with respect to the microwave background. The motion of these clusters should have been randomly distributed in all directions. Instead, the team found that the clusters tended to have a common motion, travelling at about $600 \, \text{km s}^{-1}$ in a 20° sector of sky between the constellations of Centaurus and Vela. Kashlinsky called this motion the **dark flow** (thus joining dark energy and dark matter as enigmatic features of our Universe).

The existence of the dark flow was disputed by some cosmologists, and it any case it could simply have been a statistical fluke. In 2010, however, Kashlinsky and his team published the results of a similar analysis but this time based on five years' worth of WMAP data and 1400 clusters. They found the same result, and argue that this finding cannot now be considered to be a result of chance.

But what could cause such a dark flow? If the flow really exists, then some have suggested that the explanation might lie in something outside our observable Universe: a huge distribution of mass – or even a whole different Universe, one of those distant patches of spacetime that underwent its own bout of inflation – tugging on us.

Rumbles from the start of time

Are microwave telescopes, searching for patterns imprinted upon the CMB, the only way to look for signs of inflation? Maybe not. It's possible that astronomers may one day find other means to test inflation.

There was a time in the early Universe when stars, galaxies and quasars had not yet switched on. Rather than being bright, and filled with light, the

Universe instead contained clouds of neutral hydrogen. Not surprisingly, this era is known as the **dark ages**. (It seems that everything in astronomy is 'dark'.) Now, hydrogen clouds emit radiation at a particular wavelength, namely, 21 cm; the emission occurs when the spin of the electron in the hydrogen atom flips direction. By the time such radiation reaches us from the dark ages, it will have been redshifted deep into the radio region. For example, such radiation from 550 million years after the Big Bang, which corresponds to a redshift of 9, will now have a wavelength of 2.1 m. One of the main goals of the next generation of radio telescopes (these instruments are the focus of chapter 13) is to detect the whispers of this radiation. And the relevance to inflation? Well, gravity waves from inflation will have left an imprint on the dark ages – and it's possible that, some day, radio telescopes might detect it.

Even further in the future, there is a possibility that astronomers might detect directly the rumbles from inflation. There's only one known form of information that can reach us in undistorted form from the era of inflation: gravitational waves. As we shall see in chapter 9, astronomers have proposed a space-based gravitational-wave observatory with the aim of capturing these signals from the very start of time. Perhaps, one day, it will happen.

3

Through a glass, darkly

The study of the cosmic microwave background radiation tells us that 23% of the total mass–energy inventory of today's Universe is in the form of dark matter. But what exactly is dark matter? No one knows. Soon, though, astronomers hope to catch particles of dark matter – using 'observatories' buried deep underground.

Bright guy, dark matters – The Bullet Cluster – Of WIMPs and will o' the WISPs – To catch a WIMP – Closing in on dark matter – Seeing the invisible

S. Webb, *New Eyes on the Universe: Twelve Cosmic Mysteries and the Tools We Need to Solve Them*, Springer Praxis Books, DOI 10.1007/978-1-4614-2194-8_3, © Springer Science+Business Media, LLC 2012

The results from WMAP tell us that most of the matter in the Universe is in some as-yet unknown form – **dark matter**. It's a sobering thought: the part of the Universe we think we understand – the totality of electrons, protons and neutrons that make up living matter and planets and stars; the material that physics has so far been concerned about – constitutes only a fraction of the matter in the Universe. The rest is unidentified.

Dark matter is strange stuff. It possesses mass – so it both acts as a source of gravitational attraction and it responds to the gravitational interaction – but it doesn't interact electromagnetically. That means we can't see, touch, taste, or smell it. A cloud of the stuff could be wafting through you right now and you'd never know. Given these elusive properties the term 'dark matter observatory' seems an oxymoron. Nevertheless, several dark matter observatories are trying to find this stuff right now and more observatories are being planned. And as befits their search for such inaccessible particles, these observatories themselves are in some of the most inaccessible locations – generally scientists like to place them down mineshafts, the deeper the better. Such observatories are built for the *direct* detection of dark matter particles. However, before we begin to discuss the direct detection of dark matter it's worth looking at the *indirect* evidence that points towards its existence. After all, in order to trust in the dark matter hypothesis we need more than just hints from the cosmic microwave background. And if scientists weren't sure that dark matter exists, then the construction of such peculiarly situated observatories would be rather quixotic.

Bright guy, dark matters

Fritz Zwicky was a brilliant astrophysicist. He also had a reputation for rubbing his colleagues up the wrong way. Perhaps it's appropriate, then, that the oldest unsolved problem in astronomy – a problem that irritates astronomers to this day – was first highlighted by Zwicky.

In 1933, Zwicky was studying the Coma Cluster – a large grouping of galaxies that lies more than 300 million light years away. Zwicky estimated the total mass of the cluster in two ways. First, he estimated its mass based on the number of galaxies he could see and the total brightness of the cluster. Second, he made an estimate based on the radial, or line-of-sight, velocities of the galaxies. The average energy of motion in a closed system such as a cluster is related to the average potential energy due to the cluster's gravitational mass. So measuring the velocities enables one to estimate the mass. Neither method gave the precise mass of the cluster, of course. The first method was problematical because there might be galaxies that were too dim to see. The second method was problematical because it depended on assumptions that might not hold. Nevertheless, the two estimates should have been in the same ballpark. They weren't. Put simply, the galaxies in the cluster were moving too fast. The second method implied that the Coma Cluster contained almost 500 times more mass than Zwicky could see.

Zwicky concluded that the Coma Cluster must contain matter that does not shine brightly, which would explain why he couldn't see it, but that possesses sufficient mass to provide the gravitational attraction to hold the fast-moving galaxies in the cluster together. In other words, there had to be some form of dark matter.

For the next four decades, astronomers made little or no progress in taking this finding further. Then, in the early 1970s, Vera Rubin began to study how interstellar matter moves in spiral galaxies. (Spirals are those that have a flat disk of stars surrounding a central bulge. Figure 3.1 shows a spiral galaxy with its disk 'edge on' to our line of sight.) The interstellar regions of spiral galaxies contain vast clouds of ionized hydrogen gas, and gravity causes these clouds to orbit the massive center of the galaxy in which they reside. Such clouds of ionized hydrogen emit radiation at a precise frequency – the so-called H-alpha line – and this radiation is easy to detect. So by measuring the Doppler shift of the H-alpha line in any particular cloud, Rubin was able to determine the orbital velocity of the cloud. She studied the variation of such orbital velocities with distance from the center of the galaxy.

Rubin knew what to expect from her studies because there's an example of a similar situation much closer to home. Just as dust, gas clouds and stars orbit the massive center of a spiral galaxy, so asteroids, comets and planets orbit the massive center of the Solar System, the Sun. It's all due to gravity. The situation within a spiral galaxy is of course more complicated than in the Solar System, but the overall features of orbital motion should be similar. In particular, in the Solar System the orbital velocities of planets vary inversely as the square root of their distance from the Sun. For example, by definition Earth is one astronomical unit from the Sun and it orbits in one year; Saturn has a mean distance from the Sun of just over 9.4 astronomical units and thus has an orbital velocity that's just one third that of Earth's. See figure 3.2. Such variation of orbital velocity with distance from a massive central body is known as Keplerian motion, and it's precisely what Newton's theory of gravity predicts should happen.

Figure 3.1 *An edge-on spiral galaxy such as this one, NGC 7814, can have its rotation curve determined quite easily. According to Newtonian ideas, the farther away a star or gas cloud is from the bright nucleus the slower it should orbit. (NASA)*

Rubin expected that any **galaxy rotation curve** she measured would exhibit the same overall Keplerian decline. The curve would undoubtedly be messier than the beautifully smooth curve of the Solar System, but the overall behavior would be the same.

It wasn't.

Rubin plotted rotation curves for lots of spiral galaxies and found that the orbital velocities of the gas clouds held more or less constant with increasing distance from the galactic center. In other words, a gas cloud at ten thousand light years from the center of a galaxy would generally have the same orbital velocity as a gas cloud at twenty thousand light years from the center. In some cases the velocities even *increased* with increasing distance from the galactic center.

The bright matter in spiral galaxies is thus moving more quickly than can be explained by the Newtonian gravitational attraction of the central region. However, such 'flat' rotation curves are precisely what the theory would predict if galaxies were spheres rather than disks. Since we see no signs of large amounts of material surrounding the bright disks of spiral galaxies, the matter would have to be non-luminous – dark, in other words.

Rubin's observations – and the hundreds of subsequent observations that confirmed her findings – thus provided empirical evidence for Zwicky's dark matter. Since then, several other lines of evidence have pointed to the existence of dark matter. For example, computer models of how large-scale structures formed in the early Universe work just fine if dark matter is added to the mix; without this extra mass it's difficult to understand how stars and galaxies could have formed. And the amount of dark matter in the Universe is staggering: there's five times more dark matter than 'normal' matter. As we

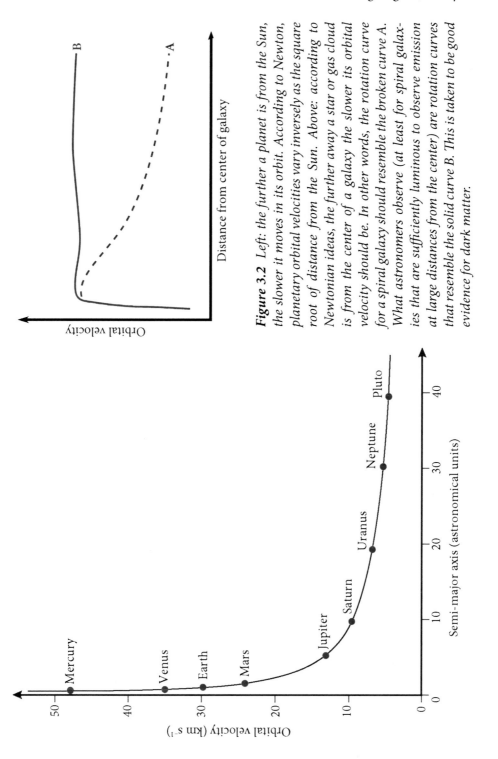

Figure 3.2 *Left: the further a planet is from the Sun, the slower it moves in its orbit. According to Newton, planetary orbital velocities vary inversely as the square root of distance from the Sun. Above: according to Newtonian ideas, the further away a star or gas cloud is from the center of a galaxy the slower its orbital velocity should be. In other words, the rotation curve for a spiral galaxy should resemble the broken curve A. What astronomers observe (at least for spiral galaxies that are sufficiently luminous to observe emission at large distances from the center) are rotation curves that resemble the solid curve B. This is taken to be good evidence for dark matter.*

saw in figure 1.8 the bright matter in the Universe is just the visible tip of an iceberg, most of which – dark matter and dark energy – is hidden from view.

So dark matter is there. But why should all this material – this vast amount of matter that purportedly surrounds galaxies and fills clusters – be *dark*? Here's one suggestion: suppose this material consists of 'normal' matter – protons, neutrons, and electrons; the stuff that constitutes the atoms that in turn make up the world around us – and that this matter has condensed into small, compact objects in the halo surrounding a galaxy. In other words, suppose the material was able to form a **massive astrophysical compact halo object** (macho), which might be found in large numbers around galaxies. Possible machos would be black holes, neutron stars or dwarf stars; in each case those objects would not shine brightly and would therefore be difficult to detect. So an obvious proposal is that dark matter consists of machos, and it's not an unreasonable proposal because we know that objects such as black holes, neutron stars and dwarf stars do indeed exist.

However, over the past four decades astronomers have used pretty much all the weapons at their disposal to study machos and it's now clear there simply aren't enough of them to form the bulk of dark matter. This conclusion has been strengthened in recent years by astrophysicists studying the early history of the Universe: their models indicate that the Big Bang simply couldn't have produced enough protons and neutrons to solve the dark matter conundrum. The most common substance in the Universe can't be 'normal' matter.

Dark matter, it seems, really is *dark*: since it doesn't interact with the electromagnetic force it neither emits nor reflects photons that we can subsequently collect with telescopes. Dark matter is thus likely to be something completely new to physics. The elusive nature of dark matter has led some to compare it to 'fairy dust' – the accusation being that astronomers are invoking some magical, undetectable stuff to explain their observations. The comparison is specious. Physicists already know of particles that are 'invisible', in the sense that those particles do not interact easily with normal matter: neutrinos are so aloof that every second trillions of them pass through Earth as if it wasn't there. Billions of neutrinos are continuously passing through your body without bothering you in the slightest. Yet despite the disinclination of neutrinos to interact with anything, physicists have managed to investigate them in detail; they've even managed to build neutrino telescopes, as we shall see in chapter 7. The hope is that the same thing will eventually be possible for dark matter.

Despite all the indirect evidence for dark matter, it's surely worth us asking the question: could there be some other explanation for the observations? Could there be some cause that does not require us to postulate the existence of a new type of matter?

Well, Rubin's observations only constitute a problem under the assumption that Newton's ideas about gravity and dynamics apply on very large distance scales. Newtonian ideas undoubtedly work well on small distance scales – they describe motions in the Solar System with exquisite accuracy – but we don't know for *sure* that the same ideas apply on much larger distance scales. It's possible to construct a theory that explains those troublesome galaxy rotation curves by assuming that masses respond differently when the gravitational force is tiny. The theory, first proposed by Mordehai Milgrom in 1983, is called **modified Newtonian dynamics**. More recently John Moffat, in his search for a unified theory of gravity and the other forces, has developed ideas that modify Einstein's general relativity; such a **modified gravity theory** also has the potential to describe flat galaxy rotation curves without the need to postulate the existence of dark matter.

However, the vast majority of the astronomical community are happy to maintain their faith in the theories of Newton and Einstein. Most astronomers accept the existence of dark matter since the idea explains not only Rubin's observations but also all those other phenomena mentioned above. If you're still be in doubt, however, it's worth briefly mentioning some images that – *almost* – let you see dark matter.

The Bullet Cluster

The galaxy cluster 1E 0657-56, better known as the Bullet Cluster, is actually *two* clusters of colliding galaxies. As viewed from Earth, the smaller cluster punched its way through the larger cluster about 150 million years ago in a high-speed collision: the two clusters were moving at several million kilometers per hour, relative to one another. Galaxy clusters contain galaxies of stars – obviously – but they also contain huge clouds of extremely hot intergalactic gas (the gas in the smaller cluster was at a temperature of about 70 million degrees and the gas in the larger cluster was at a temperature of about 100 million degrees). These clouds are *vast!* Their total mass exceeds by far the total mass of the stars in the galaxies. The clusters also contain gigantic amounts of dark matter (if you believe in the existence of dark matter, that is). See figure 3.3.

Figure 3.3 *A false color (and composite) image of the Bullet Cluster. Two galaxy clusters are colliding. The region on the right is a shock front; it's this distinctive bullet-shape of the front that gives rise to the name of the cluster. Individual galaxies can be seen from optical images; hot gas clouds, shown here in red, can be seen from X-ray images; gravitational lensing observations, shown here in blue, can determine where most of the mass of the system resides. (X-ray: NASA/CXC/CfA/M. Markevitch; optical and lensing map: NASA/STSCI, Magellan/U. Arizona/D. Clare; lensing map: ESA WFI)*

The three components – galaxies consisting of stars, clouds consisting of hot gas, and dark matter consisting of who knows what – will have interacted quite differently during the collision. The stars within each galaxy will essentially be unaffected by the collision. After all, apart from the central region of a galaxy, stars are widely separated in space. The gentle tug of gravity will cause their courses to alter but the stars themselves won't otherwise be particularly affected by the collision. As far as the stars are concerned, the clusters just pass through each other.

The hot intergalactic gas clouds behave quite differently. The gases interact not only gravitationally but also electromagnetically – and the electromagnetic force is much stronger than gravity. This interaction causes the clouds to decelerate and form a large central mass, which becomes so hot it emits X-rays.

Now, suppose that dark matter doesn't exist. In that case the distribution of mass in the Bullet Cluster would be as follows: most mass is in the central region, where the hot gas clouds are, while to the left and right of the central region, where the galaxies are, there's relatively little mass. And if dark matter exists? Well, dark matter in the two clusters will have affected each other even less than did the normal matter in the galaxies; dark matter clouds will have sailed through each other and followed the paths of the visible galaxies. So if dark matter exists the mass distribution in the Bullet Cluster would be as follows: to the left and right of the central region, where the dark matter and the galaxies are, lies most of the mass while in the central region, where the hot gas clouds are, there's relatively little mass. The two mass distributions are completely different.

Astronomers can tell where most of the mass exists by using a gravitational telescope, with the Bullet Cluster as the gravitational lens. (The idea of a gravitational lens is described more fully in chapter 4. Here, all we need to know is that large masses can bend the paths of light rays from more distant objects – and that astronomers can measure this bending, or lensing.) The amount of lensing depends upon the intervening mass: the greater the mass, the greater the amount of lensing that will be observed. If most of the mass of the cluster pair were in the form of dark matter then we would expect to see stronger gravitational lensing in the separated regions near the visible galaxies. If a theory such as Milgrom's modified Newtonian dynamics were correct then we would expect to see stronger gravitational lensing in the central region where the hot gas is. In 2006, astronomers published a gravitational lensing map of the Bullet Cluster. The result? The lensing is strongest near the visible galaxies. It's good evidence that dark matter exists.

One example might be written off as a statistical aberration. Evidence that the Bullet Cluster is not some fluke came in 2008, when similar observations were published of the cluster MACS J0025.4-1222. Again, gravitational lensing shows that a collision between two galaxy clusters has separated the dark matter from the hot gas. And in 2011 astronomers studied the so-called Pandora Cluster, which is a collision of four small galaxy clusters. Yet again, similar techniques show that dark matter contributes most of the mass.

Still not convinced about the existence of dark matter? In 2010 astronomers studied the images of 42 distant galaxies that had been distorted by the intervening mass of the giant galaxy cluster Abell 1689. Gravitational

Figure 3.4 *The distribution of dark matter (shown as a light blue tint) in the galaxy cluster Abell 1689. (NASA/ESA/E. Jullo/P. Natarajan/J.-P. Knieb/D. Coe/N. Benítez/ T. Broadhurst/H. Ford)*

lensing created 135 lensed images of these background galaxies. In order to explain the lensing, astronomers *had* to invoke dark matter. They were even able to generate a map of the dark matter concentrations within the cluster. See figure 3.4. The densest concentration of dark matter, as might be expected, was in the cluster's core.

It's hard to avoid the conclusion that dark matter is as real as 'normal' matter. But what *is* it? What's it made of?

Of WIMPs and will o' the WISPs

The direct detection of dark matter would constitute a huge advance (and provide a guaranteed Nobel prize for the discoverers). But how can scientists design experiments to search for something if they don't know precisely what it is they're searching for?

If dark matter doesn't emit or absorb photons then there's no chance of 'seeing' it using any type of conventional telescope – dark matter is essentially invisible. However, there's every chance that dark matter can make its present felt via another of nature's fundamental forces, the **weak interaction**. The neutrino is an example of a subatomic particle that has precisely these characteristics – it interacts with matter weakly but not electromagnetically. Indeed, could it be that neutrinos *are* dark matter?

The answer, perhaps unfortunately, is that neutrinos are unlikely to be the dark matter we're looking for. Neutrinos are a form of hot dark matter – 'hot' in this sense meaning that they're travelling at speeds close to that of light. As already mentioned, cosmologists have developed successful models of how structures such as galaxy clusters evolved after the Big Bang. Dark matter is a necessary component of those models, but it has to be slow-moving: it has to be **cold dark matter** (CDM). The existence of a large amounts of hot dark matter is ruled out by these models: if large numbers of such fast-moving particles *had* existed they would have smeared out the structures that subsequently evolved into the forms we see today. The dark matter we seek cannot be in the form of neutrinos.

So dark matter particles may be able to feel the weak interaction and since they are slow moving, or cold, they must be relatively massive. In other words, perhaps dark matter is some form of **weakly interacting massive particle** – or WIMP. That physicists have been able to directly detect neutrinos is cause for optimism here: neutrinos are also weakly interacting particles but they have extremely little mass. They are about as close to nothingness as it's possible to come and still have something. If neutrinos can be detected then surely so can WIMPs. The difference is that physicists know where nearby sources of neutrinos can be found – in the interior of the Sun and in nuclear reactors – so they know where to look; they also know beforehand what properties the neutrino possesses. The same conditions don't hold true for dark matter particles.

It would aid the search if physicists knew precisely what particle the WIMP was, what properties it possessed. That knowledge isn't available, unfortunately, but there are at least several trails that physicists can follow. For example, one of the crowning achievements of modern science is the **standard model of particle physics**. It successfully describes essentially all of the particles and the particle interactions that have ever been observed. Nevertheless, there are good reasons for thinking that the standard model is incomplete and various extensions to the model have been proposed. So although the standard model itself contains no plausible cold dark matter candidate, the model extensions contain particles that *are* plausible candidates.

Consider, for example, **supersymmetry**. This is a hypothetical symmetry that relates the 'matter' particles of the standard model (six different types of **quark**, plus the electron, muon and tau and their associated neutrinos) with the 'force' particles of the standard model (the photon, the gluons and the W and Z bosons). The difference between the two types of particle could not be starker. Matter particles can never get too close to one another, which means they take up space; force particles can pile on top of one another, which means they give rise to fields, such as the electromagnetic field, that propagate through space. Supersymmetry links these two quite different worlds: it postulates that for every 'matter' particle there's a corresponding 'force' partner, and vice versa. For example, the electron has a superpartner called the selectron while the photon has a superpartner called the photino.

Supersymmetry is, so far, a speculative notion. No supersymmetric particle has ever been found. Furthermore, the idea has the drawback of doubling the number of fundamental particles. On the other hand, it's such a beautiful symmetry that it would be a shame if Nature hadn't used it. On a more practical level, it solves several problems that physicists grapple with whenever they try to unify the standard model with gravity. Plus, it's an absolute requirement of superstring theory. For these reasons many theoretical physicists have, for the past three decades, thought it worthwhile to investigate the properties of supersymmetry.

One of the first conclusions they reached is that supersymmetry is not a perfect symmetry of our everyday world. If it were, then the selectron would have precisely the same mass as the electron and we would have seen it a long time ago. The very fact that we haven't seen a supersymmetric particle tells us that the symmetry is imperfect. Instead, the breaking of the symmetry must push the masses of the superpartners up to such a large value that physicists haven't yet been able to detect them. Indeed, the **lightest supersymmetric particle** must be at least a thousand times more massive than the proton.

Another interesting conclusion follows if 'superness' is conserved in the same way that electric charge, say, is conserved. In other words, suppose there's some quantity that measures whether a particle is a 'standard model' particle or a 'supersymmetric partner' particle and that this quantity is conserved – so in any interaction the overall amount of the quantity can't change. Then, just as conservation of electric charge forbids an electrically charged particle from decaying into a neutral particle (electric charge can't just disappear) so conservation of 'supersymmetric charge' would forbid a superpartner particle from decaying into one of the familiar standard model particles. It would then follow that the lightest supersymmetric particle is stable since there would be no particle into which it could decay.

Different models make different predictions for the lightest supersymmetric particle. One favored candidate is a particle called the **neutralino**. If this is indeed the lightest superparticle then theory says three important things about it: it's stable; it's massive – perhaps many thousands of times more massive than the proton; and it feels the weak interaction. The neutralino is thus an excellent WIMP candidate. It possesses all the properties expected of a dark matter particle.

It's worth reiterating that the idea of supersymmetry was not invoked to answer the observations of Zwicky, Rubin and others. Rather, the idea came about as a possible solution to some deep problems in particle physics; the same idea just happens to resolve the dark matter conundrum, too – so supersymmetry is an elegant solution.

The neutralino may be an excellent dark matter candidate – perhaps the best one so far – but it's not the only one. The fertile minds of particle theorists have produced another dark matter candidate: the **axion**.

The axion is a hypothetical particle introduced to solve a problem in the theory of the strong interaction, the force that binds quarks together to form protons and neutrons. (That the axion addresses the dark matter mystery is, as with the neutralino, a bonus.) The problem relates to the neutron. Even though the neutron carries no overall electric charge, the theory of the strong interaction suggests that the charged quarks inside the neutron should give rise to an electric dipole moment: in essence, the positive and negative charge distributions within the neutron won't coincide perfectly and thus won't quite cancel. To date, however, no one has ever found a neutron electric dipole moment and the experimental limits on this are astounding. Think of the neutron electric dipole moment as being a pear-shaped deviation from an otherwise spherical form; then, if a neutron were the same size as the Earth, the distortion would be less than the length of a red blood cell. How does one explain the difference between theory and experiment?

In 1977, Roberto Peccei and Helen Quinn suggested that there might be a new symmetry in the standard model of particle physics. An important feature of this symmetry is that, when broken, it would automatically send the neutron electron dipole moment towards zero. The following year, Frank Wilczek and Steven Weinberg, both of whom later won the Nobel prize for unrelated work, pointed out that the breaking of this symmetry would generate a new particle. Wilczek called this particle the axion after Colgate's *Axion* laundry detergent – the idea being that the particle cleaned up the problems with the theory.

If the axion exists then theory demands it has zero electric charge, zero spin and an interaction with ordinary matter that is weak. Furthermore, it will possess an *extremely* small mass, perhaps even less than that of the neutrino. Indeed, the mass of the axion must be less than 1 eV, which makes it a sub-eV particle. This constellation of properties makes it a **weakly interacting sub-eV particle** – or WISP – which is a wonderful name since a WISP can be so insubstantial that it's even less of a 'something' than the neutrino.

The reason axions are a plausible dark matter candidate is that they could have been created in vast quantities soon after the Big Bang; although each individual axion must have a miniscule mass – they are a type of light dark matter – the total mass of axions in the Universe could be enormous. Note that neutrinos were discounted as dark matter candidates because they have little mass and therefore move too fast to constitute the dark matter we seek. Why doesn't the same argument apply to axions? The answer is that the mechanism that gives rise to axions is completely different to the mechanism that gives rise to neutrinos or WIMPs. In the early Universe the axion field was not interacting with all the other stuff that was around; and when the Universe expanded and cooled, that field turned into a host of axion particles with essentially zero velocity.

The **Axion Dark Matter Experiment** (ADMX) has been searching for axions since 1996. It hasn't found them, but the experiment has already ruled out the existence of axions with a certain range of masses. The ADMX set-up is currently being upgraded, so that it will become even more sensitive to the presence or otherwise of axions.

So here are two (as yet hypothetical) dark matter candidates: the neutralino (a type of WIMP) and the axion (a type of WISP). Other candidates arise from speculative ideas in physics. String theorists, for example, argue that our Universe must have several more spatial dimensions than the three of which we are aware. In some models, a fourth dimension can be curled up tightly so that every point in our familiar three-dimensional space is actually a small ring. Any fundamental particle moving around such a ring would appear to us as a more massive version of a traditional standard model particle. Since the least massive of these particles can be stable, some researchers consider them to be viable dark matter candidates. Other dark matter candidates – such as Q-balls, wimpzillas, mirror matter – have in common only the fact of their exotic names. (It's always worth reminding ourselves that there's yet another candidate – 'none of the above' – that has not been ruled out.) Thus the good news is that there's a variety of plausible explanations for dark matter, with perhaps the favored outcome being that some of it is neutralinos and some is axions; that way physicists get to solve lots of problems in one go! The bad news is that an experiment designed to find a neutralino, say, is quite different to an experiment such as ADMX that is designed to find an axion. And in both cases the experiments are tricky to carry out. Finding dark matter won't be easy.

The most popular explanation for dark matter is that it consists of WIMPs. So in the rest of this chapter, let's look at how astrophysicists are trying to develop WIMP observatories.

To catch a WIMP

The best way to study WIMPs would be to produce them in a laboratory. It's possible that particle collisions at the **Large Hadron Collider** (LHC) could produce WIMPs and, since those WIMPs won't interact with 'normal' matter, their presence could be deduced through missing energy and momentum that they would carry away from collision sites. If the LHC finds WIMPs it will be a key discovery, of course, but it won't necessarily prove that WIMPs are the explanation for dark matter in the cosmos. Astronomers really need to detect WIMPs from space in order to convince themselves that these particles constitute the bulk of dark matter.

If dark matter does indeed consist of WIMPs then trillions of them will pass through Earth every second. Very, very, *very* occasionally one of those WIMPs will come close to the nucleus of an atom. If it's a true head-on collision, and not just a near miss, then the WIMP and the nucleus can interact via the weak force. Several observatories are running experiments to detect dark matter directly by observing such collisions.

Physicists are trying a variety of technologies in the hope of observing these collisions, but the effort is focused primarily on two types of detector.

First, let's look at cryogenic detectors.

Cryogenic detectors consist of a crystal absorber, such as germanium or calcium tungstate, cooled to less than a tenth of a degree above absolute zero. At this low temperature deposits of even tiny amounts of energy (the amounts that would occur if a particle from outside collided with the atoms in the crystal) lead to quite large and measurable effects. So the hope is to detect the tiny amount of heat generated when a WIMP hits one of the atoms in the absorber.

A crystal absorber such as calcium tungstate also exhibits **scintillation** – it emits a small flash of light when an 'ordinary' particle smashes into it. However, it would emit little or no light if a WIMP hit one of its atomic nuclei. Measuring both the heat and light emitted in a collision thus allows scientists to distinguish between dark matter and ordinary particles.

Since the early part of the century, several experiments employing cryogenic detectors have been running. For example, the **Cryogenic Rare Event Search with Superconducting Thermometers** (CRESST) is a European collaboration based under about 1400 m of rock at an Italian National Laboratory. Scientists took the chance to build the observatory as engineers were cutting twin tunnels through Gran Sasso, the highest mountain in the Apennines of Central Italy, as part of a road connecting Rome to Teramo. A French–German experiment called **Expérience pour Détecter les WIMPs en Site Souterrain** (EDELWEISS) is situated below 1800 m of rock in a laboratory in the Fréjus road tunnel linking France and Italy. The **Cryogenic Dark Matter Search** (CDMS) is an American collaboration based about 714 m underground in a laboratory housed in the Soudan mine in Minnesota. The bottom two photographs in figure 3.5 show some of the technology used in the EDELWEISS and CDMS experiments. Just a few meters away from CDMS is its smaller cousin: **Coherent Germanium Neutrino Technology** (COGENT). Although COGENT is less well shielded from background events than CDMS, it is more sensitive than CDMS to WIMPs that are relatively small in mass.

Figure 3.5 Instruments used in four different dark matter experiments. (a) CDMS. (b) XMASS. (c) XENON100. (d) EDELWEISS. In some experiments, such as XMASS and XENON100, photomultipliers are used to try to detect the scintillation that will occur when a WIMP scatters off a xenon nucleus. In other experiments, such as CDMS and EDELWEISS, cryostats are used to cool crystal absorbers to temperatures close to absolute zero; the hope is to detect energy when a WIMP is absorbed. These photographs show the CDMS cryostat and the EDELWEISS germanium bolometers. ((a) Fermilab; (b) Kamioka Observatory, ICRR (Institute for Cosmic Ray Research), The University of Tokyo; (c) ASPERA/XENON Collaboration; (d) EDELWEISS Collaboration)

The common theme with these detectors, as you will of course have noticed immediately, is that they are all based deep underground. Their location helps shield them against the continuous shower of high-energy cosmic ray particles that bombards Earth. Since these particles are both copious and penetrative, the atoms in a cryogenic detector are far more likely to interact with cosmic rays than they are with WIMPs. The idea is that, by shielding the setup with hundreds of meters of rock, this inevitable background noise can be reduced enough to let scientists hear the WIMP signal. For example, the flux of cosmic ray muons at the EDELWEISS detector is a million times less than it would be at the surface. That's an impressive reduction, but even more care has to be taken. Scientists working on these projects must also suppress signals emanating from the natural and induced radioactivity in the immediate surroundings, and thus various shields must surround the detectors. The EDELWEISS detector, for example, has a 30 cm thick shield of paraffin to reduce the speed of fast-moving neutrons; a 15 cm thick shield of lead helps to block gamma radiation; and close to the detector itself lies a shield of lead that comes from ingots recovered from the wreck of a Roman ship. Why is Roman lead used? Well, when lead is freshly extracted it naturally contains a radioactive isotope called lead-210. The radiation this isotope emits when it decays would swamp the sensitive EDELWEISS detectors. Fortunately, lead-210 has a half-life of about 22 years so in ancient lead all this troublesome stuff has already decayed. A shield of Roman lead does not add to the background noise with which the scientists must contend. Other experiments also make use of ancient lead, so when relevant archaeological finds are made physicists are keen to get their hands on the material. If the ingots contain surface inscriptions then these are generally saved for historical purposes. The bulk of the ingots, though, are simply melted down.

Figure 3.6 *Dark matter observatories look nothing like traditional astronomical observatories! (a) The entrance to the CRESST experiment is from a busy road tunnel running through the highest mountain in the Apennines. (b) The Davis Cavern, shown here as it appeared in 2009 during refurbishment, is now home to LUX. Previously, this was where Ray Davis conducted his pioneering solar neutrino experiment. (c) A home for the JinPing laboratory has been carved out 2.4 km below mostly marble rock. (d) The Soudan Mine is home to the CDMS and COGENT experiments, which operate 714 m beneath the surface. ((a) Luigi Ottaviani; (b) Bill Harlan/Sanford Underground Laboratory; (c) Qian Yue; (d) Fermilab)*

Even with all this shielding, some background radiation will get through and be picked up by the detectors; so whenever an interaction is spotted, various tests must be undertaken in order to reveal whether a 'normal' particle was involved or whether dark matter could have been responsible.

The teams at EDELWEISS, CRESST, CDMS, and COGENT are essentially trying to maximize their chances of finding a needle in a haystack by getting rid of as much hay as possible and then examining closely what's left. They've been doing this for the past several years. And the results? They're tantalizing.

In late 2009, the CDMS team announced they had detected two interaction events with a signature one would expect from WIMPs. In early 2011, the EDELWEISS team reported they had seen five interaction events with a WIMP signature. And in late 2011, the CRESST team announced they had seen 67 events that were consistent with WIMP collisions having taken place in their crystal absorbers; about half of them couldn't readily be explained by known background events. So, has WIMP dark matter been discovered? Unfortunately not. At least, scientists *may* have detected dark matter but they can't know for sure. The problem is that both experiments see lots of events that are due to other causes, such as a cosmic ray barrelling in, or the radioactive decay of atoms near the detector, or whatever. The teams involved can certainly estimate how many of these 'non-WIMP' background events they are likely to see, but it's often the case in science that an experiment sees an interesting effect that later turns out to have been just a 'fluke' – a chance bunching of background events. Before the scientific community accepts that some result might be a true discovery it quite reasonably demands that there's only a small chance the events are due to other causes. Neither EDEL-WEISS, CDMS nor CRESST saw enough events to claim a discovery.

The conclusion to draw from these results is perhaps the following: if scientists are going to detect WIMPs they need even larger detectors with even greater levels of background suppression. It's a conclusion that many had already reached. In 2005, scientists from the EDELWEISS and CRESST projects proposed a successor project called the **European Underground Rare Event Calorimeter Array** (EURECA). The idea is to build a cryogenic detector employing a range of detector materials with a total absorber mass of one ton and to cool it to just a few thousandths of a degree above absolute zero. The background suppression is also to be improved. That's the plan, anyway. If things go well, and if engineers and scientists overcome the formidable technical challenges in their way, then EURECA will be built in the road tunnel that currently houses EDELWEISS. The aim is to start operations as soon as possible after 2013. And if the neutralino is the particle responsible for most of the dark matter in the Universe then it won't stay hidden forever: supersymmetry models suggest that, even in the worst possible case, the event rate would be detectable – *just* – by EURECA.

So far we've considered cryogenic detectors. The second type of direct dark matter detector employs a tank of noble gas (typically argon or xenon) that's been cooled so that it's in the liquid form. With these noble liquid detectors scientists look for signs of a dark matter particle hitting an atomic nucleus. If an energetic particle hits an argon or xenon atom, the struck atom can both scintillate and ionize. This scintillation – just a brief flash of light – can be picked up by photomultiplier tubes that surround the tank. If scintillation is detected then the energy involved can be measured. The energy of ionization can also readily be gauged. By comparing the amount of scintillation to the ionization energy scientists can identify the particles involved in the collision: the signal generated when a WIMP hits a nucleus is quite different to the signal caused when other kinds of particle (photons, say, or neutrons) hit a nucleus.

Noble liquids such as argon, neon and xenon are particularly suitable for this application since they are dense and they emit plenty of light when particles hit their nuclei. Furthermore, as with the germanium in cryogenic detectors, xenon and neon are 'clean' materials – they have no naturally occurring long-lived radioactive isotopes to complicate matters. (Argon is less clean: a form of radioactive argon is produced when cosmic rays hit the atmosphere. However, underground sources of argon may have much less of this troublesome isotope.) As with the cryogenic detectors, experiments using noble liquid detectors must be shielded from cosmic rays and other radiation: thus they, too, are to be found deep underground.

Just as there are several cryogenic detectors around the world competing to be the first to detect WIMPs, so there are several noble liquid detectors taking part in the race. The XENON100 experiment, for example, is a collaboration of institutions from eight countries that, as with CRESST, is located at the Gran Sasso National Laboratory. As its name implies, the XENON100 experiment uses xenon as a WIMP target – about 62 kg of it. Similar experiments are now coming online. The **Large Underground Xenon Experiment** (LUX) employs 350 kg of xenon. LUX lies in a disused gold mine in South Dakota; in addition to more than a kilometer of rock above it, the experiment is further shielded by huge tanks of water. A Japanese experiment called the **Xenon Detector for Weakly Interacting Massive Particles** (XMASS), which is located in a mine under a mountain in western Japan, uses 800 kg of xenon. And to show this is a truly global search, the United Kingdom plays host to an international dark matter search programme called **Zoned Proportional Scintillation in Liquid Noble Gases** (ZEPLIN). The programme is housed in Europe's second deepest mine, the Boulby potash mine, 1100 m below the beautiful North Yorkshire moors. The latest detector in the programme, ZEPLIN III, uses 12 kg of liquid xenon with a thin layer of xenon gas on top. Although there's not so much target mass, the construction of the ZEPLIN III detector is such that it is extremely sensitive to possible dark matter events.

So there's a lot of xenon around the world just waiting to interact with WIMPs. The hope is that soon there'll be lot more argon doing the same job. The **Argon Dark Matter Experiment** (ArDM), a European collaboration currently based at Geneva, is one example of an experiment that will use argon rather than xenon. The technology used in the various detectors is similar, but because argon is much cheaper than other noble gases the hope is that these detectors will have much more target mass. The finished ArDM detector, for example, will be able to employ a *ton* of material. And the greater the target mass researchers can use in their detectors, the more events they can capture and study. If ArDM is successful then the technology should readily scale to a ten-ton detector. And researchers are currently working on designs to upgrade the existing **Dark Matter Experiment using Argon Pulse-shape Discrimination** (DEAP) at a Canadian–American collaboration based in a laboratory down a nickel mine in Sudbury, Ontario (one of the deepest of all underground laboratories). The plan is to build a detector there that uses 3600 kg of argon.

There's so much target mass hanging about deep underground – crystal absorbers in the case of cryogenic detectors; argon and xenon in the case of noble liquid detectors – that if WIMPs exist they should be detected in the next few years (with luck, perhaps even before this book is published). But there is a claim, beyond those made by EDELWEISS, CDMS and CRESST, that dark matter has *already* been detected.

The Italian **Dark Matter Project** (DAMA) based at Gran Sasso has been running for years. One of its first detectors used nine 9.7 kg crystals of the salt sodium iodide doped with thallium. Photomultipliers surrounding the crystals caught the scintillating light flashes from any particle collisions. The successor to that experiment, called DAMA/LIBRA, uses similar technology but with 250 kg of target mass. Unlike the other experiments mentioned above, DAMA/LIBRA doesn't try to account for all the flashes of light that its

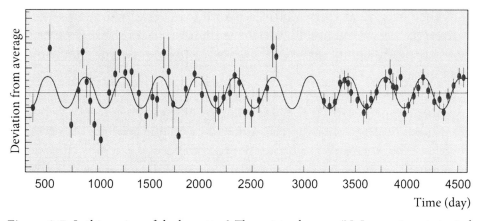

Figure 3.7 *Is this a sign of dark matter? The original DAMA/NaI experiment started taking data in 1996. Since then, the experiment has been upgraded. The graph above shows the deviation from the average signal in the days since the start of the experiment. Both the original and the upgraded experiments see a seasonal modulation in signal; the difference is that, as the experiment has been improved, the error bars have become smaller. (DAMA Collaboration, redesign C. Iezzi/Aspera)*

photomultipliers detect. Great care has been taken to shield the detectors, of course, but no attempt is made to explain every event caused by cosmic rays, stray radioactivity or impurities in the environment. Instead, the DAMA/LIBRA scientists argue as follows.

Every year, DAMA/LIBRA is going to observe tens of thousands of events – nearly all of which will *not* be due to collisions with WIMPs. However, if our Galaxy is embedded in a large cloud of WIMP dark matter – as is the prevailing understanding – then the number of events that DAMA/LIBRA detects will change depending upon the time of year. This will happen because our Solar System moves through the WIMP cloud at about 232 km s^{-1} as we rotate about the galactic center. Furthermore, in December the Earth moves at about 30 km s^{-1} *against* the motion of the Solar System, while in June it moves at the same speed *with* the motion of the Solar System. Now, just as you're going to hit more raindrops each second if you run into a squall than if you run with it, so DAMA/LIBRA expects to see more flashes of light – more WIMP collisions – in June than in December. And that is precisely what DAMA/LIBRA sees and has continued to see. See figure 3.7. Furthermore, in 2010 the COGENT team announced that their set-up was detecting hundreds of bursts of charge inside its germanium detector, with a seasonal rise and fall. The observations were consistent with relatively low-mass WIMPS. The COGENT team were not able to claim a discovery – the observations were not statistically significant – but they seemed to lend credence to the DAMA/LIBRA results.

So has dark matter been discovered? Perhaps. But most researchers think not. The problem is that the DAMA/LIBRA and COGENT results appear to be inconsistent with results from other dark matter searches. In 2011, for instance, the XENON100 collaboration announced the results of 100 days of observation between January and June 2010. The experiment found no evidence that WIMP collisions had taken place in their target mass. If the DAMA/LIBRA and COGENT experiments have been observing WIMPs then so too should the XENON100 collaboration. So the situation is currently rather murky. The annual modulation that DAMA/LIBRA observes is undoubtedly there. But the prevailing opinion is perhaps that the explanation lies in some other form than dark matter.

Closing in on dark matter

If the bulk of dark matter consists of WIMPs such as neutralinos then physicists will sooner or later detect those particles. A noble liquid detector such as ZEPLIN III probably already has the sensitivity to detect dark matter particles; if this detector doesn't find them then projects such as XMASS, ArDM and DEAP will possess so much target bulk that they surely *must* detect dark matter – if WIMP dark matter exists. And discovering the nature of dark matter is such an important problem that other experiments, often using complementary techniques, are pressing to find WIMPs: the Korean KIMS experiment, COUPP in America, PICASSO in Canada and the Italian WARP program (I leave the reader to work out what the various acronyms stand for). In 2011 in China, meanwhile, physicists got access to the world's deepest underground laboratory. The JinPing Underground Laboratory, which lies 2.4 km beneath a mountain of sandstone and marble, will house both cryogenic and noble liquid dark matter detectors. So some experiment, somewhere, will surely detect WIMPs – if such particles exist. And once one detector has detected a WIMP it will immediately become much easier for all the other detectors to do so since, for the first time, physicists will know exactly where to look: the detectors will be fine-tuned to study dark matter. Once this stage is reached, physicists will have wonderful instruments with which to study something that makes up 23% of the mass–energy inventory of the Universe. But what of astronomers? For there to be WIMP *astronomy*, astronomers would want to know not just what WIMPs are but also where they come from and how they are distributed in space.

Another dark matter collaboration has based its experiment in the Boulby potash mine: the **Directional Recoil Identification from Tracks** (DRIFT) collaboration. The second of its detectors, DRIFT-II, uses a quite different technology for catching WIMPs. It consists of a stainless steel chamber, a cubic meter in volume, which contains 167 g of the gas carbon disulphide. If a WIMP happens to bounce off one of the nuclei in the gas molecules then the recoil will produce ionization trails. These trails will drift to the edge of the chamber under the influence of an applied electric field, and information can then be gathered not only about the interacting particle but also its direction. A Japanese setup called **New Generation WIMP-search with an Advanced Gaseous Tracking Device Experiment** (NEWAGE) is another program that hopes to find not just WIMPs but also their tracks through a detector. Thus DRIFT-II and NEWAGE are *directional* dark matter detectors. Such an apparatus opens up the possibility of WIMP astronomy. For example, it should be possible to see the direction of the trails change during the year as Earth moves through the WIMP cloud.

Seeing the invisible

This chapter has focused on the attempted *direct* detection of dark matter particles – primarily WIMPs although, as we have seen, scientists are also trying to capture axions and various other plausible dark matter candidates. However, astronomers believe the dark matter problem is so important that they are throwing all sorts of resources at this. Other types of observatory are using *indirect* methods of dark matter detection.

Dark matter itself is invisible, but sometimes – very rarely – a dark matter particle will collide with its antiparticle and they will mutually annihilate. When that happens 'ordinary' particles might be created. And astronomers are exceedingly good at seeing 'ordinary' particles.

For example, orbiting gamma-ray telescopes (which are discussed in chapter 8) will search for high-energy photons coming from the annihilation of WIMPs. Radio astronomers are examining emission from filamentary structures that are close to the center of our Galaxy. The radio emission from these filaments has properties that are rather difficult to explain in a conventional framework. One hypothesis is that this emission arises when high-energy WIMPs collide. The new generation of radio telescopes (discussed in chapter 13) will be able to investigate this hypothesis in detail.

An object called the **sterile neutrino** arises naturally in some particle physics models. The sterile neutrino is hypothetical but, if it exists and if it's massive enough, it represents yet another plausible dark matter candidate. A sterile neutrino could decay into other particles and, as it did so, it would emit faint X-ray pulses. Thus orbiting X-ray telescopes (which are the focus of chapter 5) could be pressed into action as indirect dark matter detectors.

Yet other types of observation can be used to investigate dark matter. For example, astronomers can use infrared telescopes (see chapter 11) to determine how galaxies clustered together when the Universe was young. This in turn can provide information about the density of dark matter. Gravitational lensing, as we shall discuss in chapter 4, can probe the distribution of dark matter. And of course, as we saw in chapter 2, by studying minuscule temperature differences in the cosmic microwave background astronomers can infer a great deal about the amount and distribution of dark matter.

It's not just electromagnetic waves that will be used to search for dark matter. For example, it's possible that some cosmic rays might be the product of dark matter decays. A new generation of cosmic ray observatories (which are discussed in chapter 6) will search for dark matter.

Neutrinos could be another product of dark matter annihilation. Theorists have suggested that, as the Solar System moves through the dark matter background in space, the atoms that make up the Sun and Earth might hit WIMPs and cause them to lose energy. Over time this would lead to a concentration of WIMPs at the center of the Sun and Earth, and in turn this would increase the chance of WIMPs colliding, annihilating and producing high-energy neutrinos. If neutrino telescopes find high-energy neutrinos emanating from the center of the Earth or the Sun, this would be indirect evidence for the existence of WIMP dark matter. Well, a new generation of neutrino telescopes (discussed in chapter 7) is ready and waiting to find them.

And if all these direct dark matter detectors and all these indirect detection methods turn up nothing – what then? Well, astronomers and cosmologists would have to rethink almost eight decades of observation and theory: for it now seems that a Universe without dark matter would be even stranger than a Universe with it.

4

A problem of some gravity

It's the biggest puzzle in science: what's causing the Universe to blow itself apart? The key to solving this mystery may be to look at the Universe through a new type of lens – a gravitational lens.

A new type of lens – The fate of the Universe – Dark energy – The worst prediction – Through a lens, weakly – The biggest telescope of all – Sounding the Universe: a cosmic yardstick

S. Webb, *New Eyes on the Universe: Twelve Cosmic Mysteries and the Tools We Need to Solve Them*, Springer Praxis Books, DOI 10.1007/978-1-4614-2194-8_4, © Springer Science+Business Media, LLC 2012

In the latter half of the twentieth century cosmologists believed that they could describe the evolution of the Universe, on the largest scales, rather simply: the Universe began in a Big Bang and then expanded, with the rate of expansion ever-slowing due to the mutual gravitational pull of all the mass it contained. Three possibilities seemed to exist for the fate of the Universe, and they depended on how much mass there was. Too much mass and the expansion would halt; the Universe would eventually fall back in on itself. (In this case the expansion of the Universe follows a similar trajectory to that of an apple thrown upwards with insufficient speed to overcome our planet's gravitational pull; the apple falls back to Earth.) With *just* the right amount of mass the expansion would *just* stop an infinite time in the future. (Give an apple *just* the right take-off speed and it will enter orbit around Earth, never returning but never escaping either). With less than the critical mass the expansion would continue for eternity, slowing constantly but persisting nevertheless. (Throw an apple with more than the escape velocity and it goes on forever, moving more slowly all the time but nevertheless leaving the grip of Earth's gravity).

Which of these three cases applies to our Universe? Cosmologists spent decades trying to answer this question and around the turn of the millennium they finally succeeded. And the result? It turns out that none of the three cases applies. The universal expansion is *accelerating*. It's as if you throw an apple in the air and it moves away, getting faster and faster as it does so. The Universe, it seems, is blowing itself apart.

This is one of the strangest findings in modern science. All of the observational techniques discussed in this book will be brought to bear in trying to learn more about the accelerating expansion and to comprehend what it means. However, since the discovery suggests that our understanding of gravity may be incomplete, it's rather appropriate that gravity provides one of the best tools with which to investigate the problem. No longer are scientists limited to using telescopes with lenses made of glass: they can make use of lenses provided by Nature.

A new type of lens

Optical telescopes work by collecting light from distant objects and bringing the light to a focus. The focusing is done either by bending the light through a system of lenses (in the case of a refracting telescope) or reflecting it in a system of mirrors (in the case of a reflecting telescope). Whichever way the focusing is done, with a telescope we can see more than we would see with our naked eyes.

Now, one of the consequences of Einstein's theory of general relativity – his theory of gravity – is that the path taken by a light beam will bend in the presence of a massive object. The bending occurs because a light beam always follows the straightest possible path through spacetime, but mass causes spacetime to curve; indeed, gravity *is* curved spacetime. The straightest path for a light beam through curved spacetime will follow the curvature of space. In fact, it was Arthur Eddington's confirmation in 1919 that the Sun's mass can bend the path of starlight that caused the scientific community to accept Einstein's theory, and at the same time brought Einstein worldwide acclaim.

Under certain circumstances, then, it seems that it might be possible for a sufficiently large mass to act as a **gravitational lens** – after all, bending the path of light is what an optical lens does. Einstein's notebooks show that he derived the properties of gravitational lenses as long ago as 1912. He must have forgotten his work, however, along with similar work later published by others, because almost a quarter of a century later, in 1936, he was pestered into publishing a calculation of gravitational lensing. (The person doing the pestering was an amateur scientist called Rudi Mandl, who had come up with the idea that a star could act as a gravitational lens. Mandl mixed this fundamental insight with some rather unscientific speculations of the type often produced by crackpots. It seems to have been a characteristic of Einstein that he was unprejudiced and open to good ideas, even when they came from outside the formal scientific community.) Einstein re-derived his formulae for the optical properties of a gravitational lens and published the results in the journal *Nature*. He showed that the light from a distant star that grazed the limb of a nearby star would be seen as a ring – now called an **Einstein ring** – by an observer who was in perfect alignment with the two stars. However, he also argued that the probability of such a configuration occurring would be tiny and, furthermore, he showed that the radius of the ring would be so small as to be unobservable.

For Einstein these calculations were a trifling matter, but his immense reputation meant that they triggered much interest. Perhaps the most important suggestion regarding gravitational lenses came from Fritz Zwicky.

Zwicky, as hinted at in the previous chapter, was one of the more interesting personalities of twentieth century astronomy. On the one hand, reports suggest he was cantankerous in the extreme. Disputes about priority were fairly common where Zwicky was concerned, and he was well known for interrupting a talk in order to tell the speaker that the topic under discussion had already been solved – by Zwicky himself. Students and even some colleagues claimed to have been intimidated by him, and he would sometimes refer to his associates at the Mount Wilson Observatory as 'spherical bastards' – his logic being that they were bastards whichever way you looked at them. On the other hand, he came up with several ingenious ideas that were often well before their time. One of his ideas was to look for gravitational lensing effects from intervening galaxies rather than stars. The angular size of an Einstein ring increases with the mass of the lens and, since a galaxy can contain the mass of several hundred billion stars, Zwicky argued that perhaps lensing galaxies can produce observable effects.

Zwicky published his argument in 1937 but the astronomical community seems not to have taken it particularly seriously. It was not until more than four decades had passed that the first gravitational lens was discovered. In 1979, Dennis Walsh, Bob Carswell and Ray Weymann, using the 2.1 m telescope at the **Kitt Peak National Observatory**, found what at first glance appeared to be two identical objects – particularly bright galaxies of a type known as quasars – close together in the sky: the quasars had the same redshift and the same spectrum. Rather than posit absolutely identical objects, it was easier to assume that light from the distant quasar had taken two different paths on its way to us and so two images of the object were visible. A double image like this is what one would expect to see in the majority of cases of gravitational lensing. With perfect alignment of observer, foreground object and background object one sees an Einstein ring; if the alignment is not perfect, one instead expects to see multiple images of the background object. We now know that the quasar studied by Walsh, Carswell and Weymann (QSO 0957+561) is 8.7 billion light years away and that a giant elliptical galaxy (Q0957+561 G1) is 3.7 billion light years from us in the same line of sight. The intervening galaxy, along with others in the cluster to which it belongs, bends the light from the quasar and generates two images.

Since the discovery of the double quasar, astronomers have found hundreds of gravitational lenses – so many, in fact, that it pays to distinguish between three different types of lensing.

Strong lensing is the easiest type of lensing to understand. It occurs when the lens is very massive and there is alignment between us, the background object, and the lens. With a strong lens we might see an arc that is part of an Einstein ring, as in figure 4.1. The classic example of strong lensing is a

Figure 4.1 *This isn't quite an Einstein ring; it's more an Einstein horseshoe. This blueish horseshoe appears because light from a distant galaxy (which set off on its way to us about 3 billion years after the Big Bang) has been magnified and distorted by the gravitational effect of an intervening object – a very massive galaxy that just happened to be in our line of sight to the more distant galaxy. (ESA/Hubble & NASA)*

perfect, unbroken Einstein ring, but rings are rare because of the precise alignment of objects that's required. Arcs, on the other hand, are quite common. A different manifestation of strong lensing is the appearance of multiple images of the same object, as in figure 4.2.

Microlensing occurs when the lens has a small mass – such as the mass of a star, or even the mass of a planet. As Einstein had shown in his original paper, the distortion in shape of the distant object is so tiny as to be unobservable. However, when a lensing planet or star moves in front of a distant

Figure 4.2 *This stunning gravitational lens is known the Einstein Cross. It occurs because a galaxy, known colloquially as Huchra's Lens, bends the light from a quasar to form four images of the quasar. Huchra's Lens lies at a distance from us of about 400 million light years. The quasar, which happens to be directly behind the galaxy along our line of sight, is much further away. (ESA/Hubble & NASA)*

source it causes the source to brighten slightly. When such a transit event ends, the source returns to its usual brightness. Thus by studying the **light curve** of a distant object – in other words, its brightness as a function of time – it's possible to ascertain whether a microlensing event has taken place. See figure 4.3. Of course, in practice it's not so simple: the precise alignment that's required means that microlensing events are rare, which in turn means that astronomers have to survey millions of sources every few days over a period of years. And even when the light curve exhibits some change, astronomers have to rule out other explanations – after all, the source might be a variable star. Despite the formidable difficulties, the technique works. For example, in 1992 Bohdan Paczyński initiated the **Optical Gravitational Lensing Experiment** (OGLE) to search for microlensing events; the success of the method can be measured by the fact that OGLE has so far discovered 14 planets beyond the Solar System – not bad, considering that planetary discovery was not even the main thrust of the project!

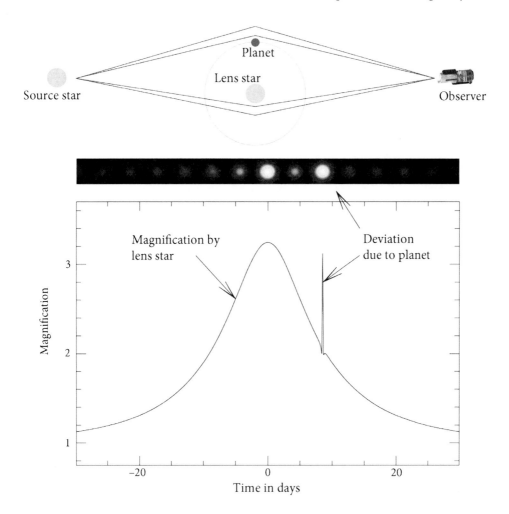

Figure 4.3 *Gravitational microlensing offers a technique for discovering exoplanets. A planet adds a small lensing effect on top of that from a star. (NASA/WFIRST)*

Weak lensing is rather different. Try the following experiment. Look at some patterned wallpaper through a magnifying glass. The amount of distortion you observe in the pattern depends on the strength of the lens – but you'll quickly discover that it also depends on how far away the lens is from both the wallpaper and your eyes. Similarly, the light from distant galaxies will be distorted by intervening mass. There's bound to be *some* matter somewhere, either dark or normal, between us and any distant galaxy – so some amount of weak lensing is inevitable. The amount of distortion

Galaxies are
randomly distributed

Galaxies exhibit
a *slight* alignment

Figure 4.4 *Astronomers have to employ statistical techniques involving many galaxies to identify cases of weak gravitational lensing. (Emma Grocutt, IfA, University of Edinburgh)*

depends upon both the amount of mass and its distance relative to us and the background objects. In contrast to strong lensing, weak lensing of distant objects generates much smaller distortions. A quasar isn't distorted into a ring, or into multiple objects, as with strong lensing. Indeed, in any individual case, it might be difficult or even impossible to detect a distortion or to attribute a distortion to weak lensing effects. However, modern technology allows astronomers to photograph thousands of galaxies at a time and, by analyzing the distortions of lots of objects, it's possible to use statistics to tease out the effects of weak gravitational lensing of distant galaxies by intervening distributions of mass. Figure 4.4 illustrates the idea.

When you look at the sky through gravitational lenses, the Universe appears to be a mildly blurry place: you see smudges and arcs and multiple images. This gravitational lensing is yet one more strand in a tapestry of evidence that shows that Einstein's theory works. Clearly, then, we can say that general relativity describes how gravity functions? Well, maybe not. That most puzzling discovery of modern science – the accelerating expansion of the Universe – has got physicists scratching their heads and wondering whether general relativity might have to be amended. Nevertheless, as we shall see, astronomers are using weak gravitational lensing – a phenomenon that depends upon Einstein's theory of general relativity – to investigate the accelerating Universe.

The fate of the Universe

Almost as soon as Einstein had developed his theory of general relativity he thought to apply it to the Universe as a whole. It made sense: gravity is the main force acting on cosmic distance scales so surely his theory should have something to say about cosmology. The point was not to try and understand how every single object in the Universe came into being and then subsequently developed; that would clearly be an impossible task. Rather, the idea was to try and understand something about the large-scale properties of the Universe – its age, its size, its eventual fate and so on.

Well, Einstein made various simplifying assumptions and concluded that there's something called a universal scale factor that determines the distance between two widely separated galaxies. To his great discomfort, he found that the universal scale factor naturally changes with time. If it increases with time then we could say that the Universe is expanding. (Even though the individual galaxies themselves might not be 'moving', the increase in the scale factor would cause the distance between them to increase; there would simply be more space between them). If the scale factor decreases with time then we could say that the Universe is contracting. This finding discomfited Einstein because he believed – as perhaps all physicists did at the time – that the Universe should be a static place: he believed the sky we see now should be much the same as would have been observed billions of years ago and as it would be observed billions of years in the future. To avoid the seemingly inevitable conclusion that the Universe is a dynamic, evolving entity he added a term to his theory – the **cosmological constant**, with symbol Λ – that acted against the usual gravitational force to produce a static Universe.

Of course, we now know that there was no solid basis for Einstein's belief in a static Universe. When Edwin Hubble showed that the Universe is, in fact, expanding – well, Einstein quickly dropped his Λ term and cosmologists rushed to couch Hubble's discovery in terms of general relativity. The picture that developed was simple.

Cosmologists assumed that on the very largest scales the Universe is homogeneous (it's the same at all places) and isotropic (it's the same in all directions). The equations governing the time evolution of the scale factor – in other words, how the size of the Universe develops over time – then became quite straightforward. It turned out that the fate of the Universe depended, as might have been expected and as we have already said, upon the amount of mass it contained. The gravitational pull of all the mass in the Universe acted continuously to slow the expansion. If the Universe contained *precisely* the amount of matter needed to halt the expansion, in other words if it possessed a critical density, then after an eternity the Universe would glide to a serene halt. That would be a *flat* Universe. If the critical density were exceeded then the spatial expansion would stop and eventually go into reverse: the Universe would shrink, galaxies would collide, and all would end in a Big Crunch. That would be a *closed* Universe. On other hand, with less than the critical density the expansion would continue forever – at an ever-decreasing rate, due to the gravitational pull of all the matter, but expanding nevertheless for all eternity until everything dies in a Big Freeze. That would be an *open* Universe. Those were the three options – similar, as mentioned above, to the three options that can occur when you try to launch an object into space from Earth.

So which Universe do we inhabit? Big Crunch, Big Freeze, or a perfectly balanced, flat Universe?

Cosmologists tried to answer this question by observation. The straight-forward approach to this was to take a typical large spherical volume of space and count the number of galaxies within that volume. The number of galaxies would provide an estimate of the mass and from there one could determine the density of the chosen volume. Whenever cosmologists did this they measured much less than the critical density. They used more sophisticated approaches, too, but always they measured less than the criti-cal density. Even when they included all the dark matter that they believe is there but can't see they found that the Universe has only about 30% of the critical density. That's close to being flat, but not close enough. It would seem on that basis that the Universe is open.

However, as mentioned in earlier chapters, lots of independent data sug-gest that the Universe *is* flat. Furthermore, there's a problem if we live in an almost-but-not-quite-flat Universe: any small departure from the criti-cal density gets larger as time passes. This is the flatness problem, which we discussed in chapter 2. If the density of the Universe is one part in three of the critical value *now*, then at the time that hydrogen and helium were first formed, about three minutes after the Big Bang, the density of the Universe must have been within one part in a thousand trillion of the critical value. The further back in time you reach, the closer the density must have been to the critical density. If the Universe started off flat then it's always flat and there's no problem; this is how the idea of inflation solves the problem. But if the Universe isn't precisely flat why should it be so close to being flat now?

There's another way of establishing the geometry of the Universe, a way that doesn't require the estimation of cosmic densities or the inferring of quantities from microwave observations. This method works instead by determining the expansion history of the Universe. Suppose astronomers can identify a cosmic **standard candle** – an astronomical object that has a known intrinsic brightness and that can be identified over very large dis-

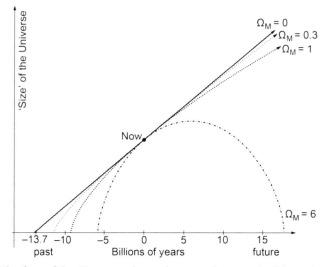

Figure 4.5 *The fate of the Universe depends on its density, Ω. (The subscript M on the diagram here indicates that we're assuming that the density depends on the mass of the Universe.) At the critical density, which is defined to be Ω = 1, the Universe will stop expanding an infinite time in the future: this is known as a 'flat' Universe. If the density were, say, six times the critical density then the Universe would eventually fall back in on itself: in this case the Universe would be 'closed'. If the actual matter density were only 0.3 of the critical value – as observations suggest is the case – then the Universe would keep on expanding, but ever slower. This is an 'open' Universe. (If the Universe contained no mass whatsoever then there would be a constant, straight-line expansion forever.)*

tances. Well, as light from the standard candle travels towards our telescopes through the expanding Universe, the cosmic expansion causes not only the distance between widely separated galaxies to increase, it also stretches the wavelength of the light. In fact, from the time when the light set off to the time it is collected by a telescope its wavelength will have been redshifted by the same factor as the Universe has been stretched. So astronomers can thus determine the universal scale factor – essentially the size of the Universe – at the time that the light set off on its journey from the standard candle. And when did that journey begin? Well, the time taken for the light to reach us is simply the candle's distance from Earth divided by the speed of light. The distance can be calculated because we know that the candle is a *standard candle* – a comparison of its apparent brightness with its known intrinsic brightness immediately provides the distance. So simply by measuring the apparent brightness of a standard candle and its redshift astronomers can determine the size of the Universe at a particular time: find enough of these candles and it's possible to plot the past expansion history of the Universe. From that, it should be possible to determine the fate of the Universe – in other words, whether it's open, flat, or closed.

Sounds easy. But what standard candle should astronomers be using? Finding a suitable standard candle has always been the difficult part. The candle must be bright enough to be seen across vast, cosmic reaches of space (which rules out 'ordinary' stars, for example). It must be *standard* so that astronomers know how bright it *should* appear (which rules out galaxies, for example, which possess too large a range in intrinsic brightness to be useful). And astronomers must be able to find reasonable numbers of them in the sky. (For a long time this ruled out those rare, exploding stars called **supernovae**, which occur too randomly to allow astronomers to know beforehand where to look. That was a shame, since a Type Ia supernova can outshine its host galaxy and thus is certainly bright enough to be seen over cosmic distances. At the same time, a Type Ia supernova has a sufficiently standard peak luminosity to be useful as a distance indicator.)

The difficulties in finding a suitable standard candle meant that astronomers were for a long time unable to pursue this idea of plotting the expansion history of the Universe. Things changed just before the turn of the millennium. The **Supernova Cosmology Project,** led by Saul Perlmutter, and the **High-z Supernova Search**, formed by Brian Schmidt and the source of an influential study led by Adam Riess, came up with a way to find supernovae 'on demand'. The idea was to take images of thousands of galaxies when the sky was dark, just after New Moon, then take images of the same galaxies three weeks later. If a supernova occurred in any of those galaxies then it would show up as a bright dot in the second set of images that was absent from the first. Although a supernova in any particular galaxy is a rare event – a galaxy such as our own hosts a supernova on average perhaps only once every five hundred years – by looking at so many galaxies at once the images will inevitably contain a few supernovae. Better yet, in the three weeks between the two sets of images a supernova will typically still be brightening towards its peak. That meant the two groups could alert the world's major observatories and ask them to take detailed, accurate measurements. At last, astronomers had a decent standard candle that worked over cosmic distances. Finally, astronomers could determine the expansion history of the Universe. The result – the strangest in science for decades –

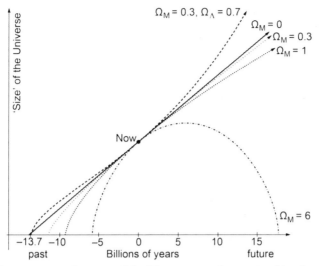

Figure 4.6 *The expansion history of the Universe, as determined by data from Type Ia supernovae, demonstrates that the traditional picture of the expansion is wrong. The Universe is flat – even though it does not contain enough matter to halt the expansion: there must be some other, currently unknown component that adds to the density. Furthermore, this new component means that the rate of expansion of the Universe is increasing rather than decreasing.*

took the community by surprise. Thirteen years later, Perlmutter, Schmidt and Riess shared the 2011 Nobel prize in physics for their findings.

The graphs that the two groups plotted, similar to the diagram shown in figure 4.6, showed that the expansion slowed as expected during the first 8 billion years after the Big Bang. And then, about 5 billion years ago, the expansion began to *accelerate*.

Dark energy

So what does it imply, this discovery that the expansion of the Universe is speeding up rather than slowing down? Does it mean that general relativity, which gives us our best and most basic understanding of gravity, is wrong? Well, possibly. In a sense, we already know that general relativity isn't the full story because it's incompatible with quantum physics. Since there's over-whelming evidence that quantum physics accurately describes the workings of Nature on the smallest scales it seems clear that sometime, somehow, general relativity will be superseded by a quantum theory of gravity. Some physicists hope that **string theory**, for example, might provide an explanation for the cosmic acceleration; others believe that a modified theory of gravity could hold the clue. But it turns out that general relativity *can* accommodate this unexpected discovery of accelerating expansion; Einstein's theory need not be dispensed with just yet.

Perhaps the simplest way to understand the accelerating expansion is to assume that the Universe is filled with **dark energy**. The word 'dark' is apt here because, as with dark matter, we don't see it and at present scientists are in the dark about what it is. However, providing that dark energy exists, and that it possesses a quality known as **negative pressure**, then it would generate the accelerating expansion observed by the Perlmutter and Schmidt teams. Note that there's nothing particularly strange about the idea of negative pressure. We know that positive pressure pushes outwards; negative pressure just pulls inwards, that's all. Most objects in their natural state possess positive pressure; but a taut rubber band, for example, possesses negative pressure. This focus on pressure is important because in general relativity it's not just matter density and energy density that can generate gravitational effects; other quantities – such as pressure – can contribute too. Now, positive pressure leads to an attractive gravitational force. But negative pressure, as I'm sure you've guessed, leads to the opposite – something that is, in effect, a repulsion.

The term Einstein added to his equations to 'fix' his theory – the cosmological constant, Λ – has negative pressure and it influences the Universe only through its gravitational effects. So could dark energy be the cosmological constant? At present, that idea seems to be the best bet: all the data since the original discovery of an accelerating Universe are consistent with there being a small Λ term. The cosmological constant can be thought of as being the 'cost of having space' – in other words, there is an intrinsic, fundamental energy associated with empty space. When the Universe was small, in the first few billions of years after the Big Bang, the volume of space was not great and so there was a relatively small amount of dark energy. The expansion was dominated by matter, and the Universe decelerated as expected. As the Universe expanded there was more and more space, and thus more and more dark energy, which caused space to expand more quickly, which meant there was more dark energy, which caused … well, about 8 billion years after the Big Bang the dark energy began to cause the expansion to accelerate. As we saw in chapter 2, astronomers believe that at the present time about 74% of the mass–energy of the Universe is in the form of dark energy.

The discovery of dark energy in some ways ties everything up quite nicely. It reconciles those observations suggesting we live in a flat Universe with those suggesting there's insufficient matter for the Universe to be flat: dark energy fills that gap. But if dark energy is indeed explained by a cosmological constant then physics has a big problem on its hands.

The worst prediction

The cosmological constant has such a profound effect on the evolution of the Universe that it seems reasonable physicists should understand why it takes the particular value it does. Well, physicists can use their best theories of nature – quantum field theories – to calculate the value that Λ should take: it turns out that Λ should have a value of roughly $10^{91}\,\mathrm{g\,cm^{-3}}$. And what value do astronomers observe (assuming, of course, that the accelerating expansion is due to a cosmological constant)? They observe a value of about $10^{-29}\,\mathrm{g\,cm^{-3}}$. The mismatch between theory and observation is thus a staggering factor of 10^{120}. This is far and away the most embarrassing discrepancy between theory and observation in all of science. In fact, it would be difficult to get any worse: such big numbers just don't appear in science. As an example of a big number in science, consider the number of particles in the entire Universe: roughly 10^{80}. Compared to 10^{120}, the number of particles in the Universe is unimaginably small.

Theorists have come up with mechanisms by which Λ might be zero. In essence, they argue that every process that contributes positively to Λ has a partner process that contributes negatively by precisely the same amount. In this case Λ should be exactly 0. But this doesn't help, because the accelerating expansion implies that the cosmological constant is *not* zero. It's tiny, but it's not zero.

So how can theory and observation be reconciled? No one knows.

Perhaps the cosmological constant really is zero, and there's some other explanation for dark energy? Well, perhaps. Some physicists have suggested, for example, that dark energy might instead be due to a field called **quintessence**. Whereas the cosmological constant is a *constant* energy density that fills the Universe, quintessence is *dynamic* – it can change over time and over space. The notion that dark energy might be dynamic leads to some interesting possibilities. For example, suppose dark energy increases in value over time; not just the total amount of dark energy – this increases simply because the volume of space is increasing – but the amount of dark energy in each thimbleful of space. Dark energy that behaves in this way is called **phantom energy**. If the observed acceleration is indeed due to phantom energy then the Universe is in for an interesting death: the phantom energy will cause space to expand ever-more quickly until everything – galaxies, planets, even atoms – is ripped apart by the huge energy density of empty space. The Universe would die in a Big Rip.

Cosmologists have introduced a number to characterize how the density of dark energy evolves as the Universe expands. They give this number the symbol w, and it's just the ratio of the average pressure of dark energy to its density. It turns out that if dark energy has a pressure that's exactly equal but opposite to its energy density then it does not change in space and time – it's constant over the entire cosmos. In other words, if w is equal to –1 then dark energy is likely to be the cosmological constant. On the other hand, if w is more than –1 (perhaps –0.9, say, or –0.8) then the dark energy density will be slowly decreasing. In other words, if w is more than –1 then traditional quintessence may well be causing the accelerating expansion. And if w is less than –1 (perhaps –1.5, say, or –2) then phantom energy may be ripping the Universe apart.

So what is the value of w?

At present, the best that can be claimed is that w lies somewhere between –0.7 and –1.3. So it's a reasonable bet that dark energy is the cosmological constant. But w could be –1.1 or –0.9 and astronomers wouldn't have noticed yet. So how can they pin down the value of w more precisely? Answering that question is the biggest project in observational cosmology.

To understand more about w, cosmologists need to measure the amount of dark energy that was present at different times in the past. One way of doing this is to use weak gravitational lensing. It's an interesting thought: perhaps cosmologists can understand more about dark energy if they view gravity through a gravitational lens.

Through a lens, weakly

As we saw earlier, weak lensing depends not only upon the mass of a lens; it also depends on the distance of the lens relative to us and the lensed object. Thus weak lensing can provide information about distances. For example, it can sometimes be used to determine the distances to galaxy clusters. Combining this information with redshift measurements of the clusters gives cosmologists another way of plotting the expansion history of the Universe.

They can calculate how much dark energy was present when light from the clusters set out on its journey to us. These observations also provide information about the masses of galaxy clusters, which in turn provides indirect clues as to the nature of dark energy – since the more rapid the cosmic expansion, the more difficult it is for gravity to attract masses together to form clusters. However, for cosmologists to tease out all this information from weak lensing they must survey the sky widely and in depth.

The largest weak lensing survey carried out to date is the **Canada–France–Hawaii Telescope Legacy Survey**. The survey ended in 2009 but astronomers will be mining the data for a while to come. This 'wide' survey covered 172 square degrees of sky, but to really unleash the power of weak gravitational lensing an even wider survey is necessary. After all, while a figure of 172 square degrees might seem a lot we need to remember that there are about 41 253 square degrees in a sphere – so this survey covered only about 0.4% of the entire sky.

A collaboration of American, Brazilian, British, German and Spanish scientists are working on the **Dark Energy Survey** (DES). The DES will, starting in September 2012, employ a 570 megapixel camera to survey 5000 square degrees of sky over a period of five years. (A megapixel is one million of those tiny, tile-like elements that constitute any digital image. Currently, the digital cameras you can buy in a shop are typically 10–12 megapixel devices, although 36 megapixel cameras are now commercially available.) The survey will provide data that can be used to study dark energy in a variety of ways including, of course, the use of weak gravitational lensing. But even 5000 square degrees is only 12% of the sky. What about a survey that covers half the sky?

Astronomers hope that the **Large Synoptic Survey Telescope** (LSST), which is being built high on the Cerro Pachón mountain in Chile, will enable the widest survey of the sky ever undertaken. The LSST is an extraordinary telescope. Its primary mirror will have a diameter of 8.4 m, making it one of the largest in the world, but it is unusual among large telescopes in that it will have a further two mirrors that will combine to give the instrument an extremely wide field of view – about 10 square degrees, which is equivalent to about 50 times the area of sky covered by the Sun. (The field of view of most other large telescopes is so needle-sharp that they can only see a minuscule patch of sky at any instant. For example, the famous Ultra Deep Field image taken by the **Hubble Space Telescope** covered only one thirteen-millionth of the full sky.) The LSST will take photographs at six different colors, from

the longest red wavelengths to the shortest blue, and it will take them much more quickly than most other telescopes, capturing pairs of images with an exposure time of just 15 seconds. Finally, the telescope will make use of the world's largest digital camera – a 3200 megapixel monster – that will image extremely faint objects from the depths of the Universe. This combination of width of view, speed and depth makes the LSST a unique instrument – and useful for much more than just weak lensing studies.

In 2010, the important and influential decadal survey of the US National Research Council selected the LSST as the highest-priority ground-based instrument for the next ten years. If all goes to plan, the LSST will see first light in about 2016 and start surveying the sky in 2018. Then each night, for the following ten years, LSST will take 800 images; it will cover the sky twice each week. After just one month of surveying it will have seen more of the cosmos than every other telescope in history combined – from the pioneering instrument used by Thomas Harriot to sketch the Moon all the way through to those magnificent images taken by Hubble. And that mountain of observational data will keep piling up: during every year of its planned ten-year existence the LSST camera will generate a **petabyte** of data every couple of months.

Greater than the engineering challenge of building a 3200 megapixel camera, greater even than the technological challenge of constructing the three large high-precision mirrors that will form the telescope, is the computational challenge of processing and making sense of such vast amounts of data. It's not surprising then that engineers from Google, who perhaps have more experience than anyone else of organizing information and turning data into knowledge, are key members of the LSST group.

Figure 4.7 *Top: an artist's rendering of the unique design of the LSST as it will appear within its dome. Bottom: an artist's rendering of the LSST building, combined with a photograph of the road leading up to the 2682 m high Cerro Pachón mountain peak. The vehicle descending from the peak gives some indication of the size of the building. Scientists chose this site because of the large number of clear nights the mountain sees each year and the clarity of images that are obtained through the atmosphere above the peak. Just as important was access to existing networking infrastructure capable of accommodating the huge amount of data the telescope will produce each night. (Top: Todd Mason, Mason Productions Inc./LSST Corporation; Bottom: LSST Corporation)*

The outcome of all this LSST activity will be a real-time movie of the Universe. Since the telescope repeatedly photographs the sky it will pick up any objects that change or move rapidly – supernovae, for example, or nearby asteroids that have the potential to devastate life on Earth, or new objects at the edge of the Solar System. These movies will be available to the public, so the potential for follow-up research – by amateur astronomers as well as professionals – will be enormous. In addition to those movies, though, the LSST will generate a detailed catalog of 10 billion galaxies and 10 billion stars. By sifting through the data in this catalog astronomers will be able to isolate the effects of weak gravitational lensing – and thus will have a probe of how much dark energy was present at different stages in the history of the Universe.

Survey missions such as DES and LSST, studying the cosmos through a weak gravitational lens, will thus certainly shed light on dark energy. However, it's likely that cosmologists will need to employ more techniques than just weak gravitational lensing before they can fully understand dark energy.

Sounding the Universe: a cosmic yardstick

As the saying goes, there's more than one way to skin a cat. Given the importance to science of understanding dark energy it's not surprising that cosmologists are not relying purely on weak lensing. They are trying to skin this particular cosmological cat in all manner of ways. One method depends on those tiny temperature fluctuations in the microwave background, which we discussed in chapter 2.

To understand how this method works, picture once again the hot plasma that constituted the very early Universe and imagine that a random quantum fluctuation caused an initially overdense region to come into existence. There would be two forces acting in such a region: gravity would attract everything – dark matter, normal matter, photons – *inwards*, towards the region, while the heat generated by collisions between matter and photons would generate *outward* pressure. Now, whenever a system has competing forces acting like this, oscillations occur. For example, oscillations occur when there are pressure differences in air: sound waves are created. The key point is that oscillations – sound waves, essentially – would have been generated by overdense regions in the early Universe.

The physics of sound waves is thoroughly understood, so cosmologists can model what would happen in this situation. Any dark matter in the overdense region would stay put since it wouldn't feel the gas pressure – dark matter, as we know, feels only the attractive force of gravity. The photons and the particles of normal matter, however, would move outward in a spherical wave at the speed of sound – and since the speed of sound in the early Universe was 57% of the speed of light they moved quickly! Note that most of the mass of the normal matter would be in the form of protons and neutrons, because these are much more massive than electrons. Particles such as protons and neutrons are known as baryons, so in this discussion of density fluctuations it's quicker to talk about 'baryons' than 'particles of normal matter'. Until recombination the photons and baryons would move outwards together. After recombination the photons would leave the baryons behind. With the pressure thus relieved, the baryon wavefront 'stalls'; there would thus be a shell of baryons at a fixed radius from the initial overdense region – and since we area talking here about sound waves the size of this fixed radius, which is known as the **sound horizon**, can easily be calculated.

Therefore, by starting at an instant after the Big Bang with an overdense region we end up, at recombination, with *two* overdense regions: one at the original site (consisting mainly of dark matter) and another at the sound horizon (consisting mainly of baryons). See figure 4.8. These overdensities would continue to attract matter, and it is at these sites that galaxies would eventually form.

If these ideas are correct we should be able to look out into the Universe now and expect to see more galaxies separated by the sound horizon than by smaller distance scales. This bunching-up of galaxies on a particular distance scale is known as **baryon acoustic oscillations** (BAO). The term 'baryon acoustic oscillation' sounds intimidating, but it's just a fancy way of saying that normal matter (baryons) moved in sound waves (acoustic oscillations) in the early Universe.

There's a problem with this simple picture. A single overdense region gives rise to a single sound wave, but the early Universe would have had a number of anisotropies each giving rise to a different sound wave. It's like dropping pebbles into a still pond: drop a single pebble and the ripple is easy to discern but drop a bunch of pebbles and the subsequent wave patterns are hugely complicated. So when cosmologists look at the separation of galaxies they won't *directly* see the preferred distance scale. However, they might be able to tease out the existence of a preferred distance scale by looking at the statistics of large numbers of galaxy separations. This is what a team led by Daniel Eisenstein did in 2005: they analyzed data from about 50 000 bright

Photons   time ⟶

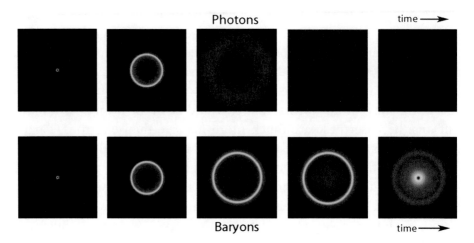

Baryons   time⟶

Figure 4.8 *Some snapshots of the process that leads to baryon acoustic oscillations. We start with a dense 'blob' of energetic photons, baryons, and dark matter. High pressure drives the photons and baryons outward together at 57% of light speed. At recombination the pressure is released: the photons continue at lightspeed while the baryons 'stall'. Later, the initial dark matter 'blob' pulls baryons back towards it. (Martin White)*

Figure 4.9 *The 2.5 meter telescope at the Apache Point Observatory is used for the Sloan Digital Sky Survey. The third phase of SDSS includes the BOSS project. (Fermilab)*

galaxies in the **Sloan Digital Sky Survey** (SDSS) and found clear evidence for BAO signal. A much larger analysis of SDSS data, carried out a couple of years later, identified the size of the BAO scale: at the present time, it's 450 million light years. In other words, galaxies are not evenly spread throughout the Universe: they tend to bunch up every 450 million light years.

The importance of this is that it gives astronomers a **standard length** or standard ruler, in the same way that supernovae provide a standard candle. And just as standard candles can be used to determine distances, so can standard lengths: if you know how big something *really* is then a comparison with how big it *appears* lets you figure out its distance. Combine these distance determinations with redshifts and you get information on the size of the Universe at different times in its history. So observations of the regular bunching of galaxy clusters, the baryon acoustic oscillations, provides a means of determining the expansion history of the Universe and thus investigating dark energy. In 2009, the **Baryon Oscillation Spectroscopic Survey** (BOSS) began a five-year mission to measure the redshifts of 1.5 million bright galaxies and the spectra of 160 000 quasars. BOSS is one of several programs in the third phase of the SDSS collaboration, SDSS-III, and uses the 2.5 m telescope at **Apache Point Observatory** in New Mexico. See figure 4.9. A proposed project called **BigBOSS** plans to retrofit a spectrograph to the 4 m Mayall telescope on Kitt Peak and use this to measure BAOs. Combining this survey information with the calibrated ruler nature has kindly provided in the form of acoustic oscillations, cosmologists will be able to place tight constraints on the nature of dark energy – and have a useful check on the results coming from other dark energy probes.

In addition to weak lensing and baryon acoustic oscillations another dark energy probe is, of course, Type Ia supernovae. After all, this was the technique that discovered dark energy in the first place. Thus new optical telescopes (which are the focus of chapter 12) will play a key role in investigating the nature of dark energy. Several supernova programs are either planned or already in operation. The DES, in addition to its weak lensing studies, will find distant supernovae out to about 7 billion light years. The **Panoramic Survey Telescope and Rapid Response System** (PANSTARRS) will also find distant supernovae while the **Nearby Supernova Factory** is studying in detail nearby (within a billion light years) supernovae. Even more ambitious supernova observatories are planned. The new generation of optical telescopes will discover and identify Type Ia supernovae at all distances and thus will allow a better determination of the expansion history of space.

The same decadal survey that assigned the LSST the highest priority to a ground-based mission selected an infrared telescope as the space-based mission with the highest priority for the forthcoming decade (infrared telescopes are the focus of chapter 11). If this **Wide-Field Infrared Survey Telescope** (WFIRST) goes ahead (and this isn't certain; the story behind the mission is rather convoluted and it's possible that there'll be other twists to come) then it might launch in 2020. The plan is for WFIRST to use the three techniques – BAOs, weak lensing and supernovae – to study the properties of dark energy. European astronomers, meanwhile, have a mission called **Euclid** to look forward to: in late 2011, ESA as part of its Cosmic Vision program gave the go-ahead for the 2019 launch of this orbiting observatory to study dark energy. Euclid will consist of a 1.2 m mirror, with three scientific instruments alongside, for making observations in the optical and the infrared. A successful launch would see the telescope cover half of the whole sky and map structures in the Universe from ten billion years ago up to the present day – the entire period over which dark energy has played a role in accelerating the cosmic expansion. As with WFIRST, Euclid will study BAOs, weak lensing and supernovae in order to unpick the nature of dark energy.

It's not just the optical and infrared parts of the electromagnetic spectrum that will be pressed into service. We have already seen how microwave telescopes can provide information about the amount of dark energy by mapping the cosmic microwave background radiation. And in 2011, an ultraviolet telescope (see chapter 10) confirmed the existence of dark energy by surveying, in conjunction with an optical telescope, more than 200 000 galaxies. It saw that the standard ruler of baryon acoustic oscillations was being stretched by a smooth, uniform force – dark energy, in other words. Radio telescopes (which are the focus of chapter 13) will map the cosmic distribution of hydrogen in order to understand how large-scale structures have changed over time, while X-ray telescopes (the focus of chapter 5) will trace how the hot gas at the centers of galaxy clusters have varied through time; both observations will help cosmologists distinguish between different theories of dark energy.

None of these observations will be treated in isolation. Astronomers will combine information from different methods to create a powerful tool for studying dark energy. For example, the DES will use *four* methods to investigate dark energy. Results from weak gravitational lensing and a method based simply on counting galaxy clusters will depend on the particular theory of gravity that is used to model the data. Results from supernovae and baryon acoustic oscillations *won't* depend on how gravity helped shape the evolution of the Universe. So if DES finds that all four methods deliver the same results then cosmologists can be pretty sure that the current ideas about dark energy are on the right lines – it will be caused by a cosmological constant or something like quintessence. If the results from the various methods differ, however, then that's a strong clue that either our present understanding of gravity is flawed or else that some hitherto unthought of explanation is required.

It's not just telescopes employing electromagnetic radiation that can address the dark energy problem. If astronomers ever get the go-ahead to construct space-based gravitational wave observatories (see chapter 9), then the observation of distant inspiral events would enable astronomers to determine cosmological parameters with extremely high accuracy. Again, such data would help uncover the nature of dark energy.

The plethora of facilities that will be used to investigate dark energy shows just how important this topic is for science.

The biggest telescope of all

Projects such as the LSST, WFIRST and Euclid are undeniably ambitious. But astronomers have occasionally dreamed of an observatory whose scope is quite staggering. There's a huge lens up there in the sky – a strong gravitational lens – that would (if we could use it) clear up pretty much all the cosmic mysteries discussed in this book. That lens is the Sun.

We know that the Sun's mass can bend light rays: after all, as previously mentioned, it was Eddington's observation of this effect that first confirmed Einstein's theory of general relativity. Well, in 1979, the Stanford University engineer Von Eshleman pointed out that the Sun's mass brings electromagnetic waves to a focus. He showed that the minimum focal length of the Sun's lens is about 550 astronomical units. In other words, the focal point is 550 times further from the Sun than is the Earth; that's 14 times further from the Sun than Pluto or, expressing this distance in another way, it would take light 3.17 days to reach the focal point after leaving the Sun. That's a long way away. Note that this distance is the *minimum* distance from the Sun; a point beyond that on the focal line will still see the Sun's gravitational lensing effect. And that effect is enormous.

Consider microwaves with a frequency of 300 GHz. Well, at that frequency the amplification factor of the Sun's lens would be ten million. For radio waves it would be like having a telescope with a collecting area about 30 000 times greater than the largest radio telescope currently in existence. An antenna placed on the Sun's focal line could detect a BBC TV transmitter at a distance of 100 light-years. Indeed, the resolution of the lens would be so great that an antenna could select for individual transmitters on the surface of nearby planets. There'd probably still be nothing worth watching on TV, but think of the possibilities for science if such an observatory existed.

So why are astronomers merely dreaming of such an observatory rather than planning it? Well, for a start it would be horrendously difficult to get to the focal point. **Voyager 1** holds the distinction of being the most distant physical object sent into the cosmos by humankind. This spacecraft was launched in 1977 and, at the time of writing, is about 110 astronomical units from the Sun: in a third of a century it has travelled one fifth of the minimum distance to the Sun's focal point. There are missions under consideration that would visit the outer reaches of the Solar System; NASA's **Innovative Interstellar Explorer** (IIE), if it is ever launched, would carry a scientific payload out to 200 astronomical units and would reach that distance in the year 2044. But that's still nowhere near far enough to make use of the solar lens. Even if it's practical to get an observatory in place, once there the navigation problems would be horrendous: the observatory would have to keep within a few meters of the focal line to benefit from the full amplification effects. The technology required for that is daunting.

So it's a nice idea – but it won't happen anytime soon. We can only hope that the puzzle of dark energy will be solved before then.

5

Where God divides by zero

How do you study something that is, by definition, impossible to see? That's the quandary faced by astrophysicists interested in black holes. The singularity itself – the point inside a black hole where the laws of physics break down – remains the most enigmatic monster in the astronomical bestiary. But as for the other properties of black holes – well, astrophysicists have learned lots about them. The key to unlocking the secrets of black holes is to look at them through X-ray telescopes.

Some like it hot – X-ray vision – Hearts of darkness – Enigmas large and small – Now available: black holes in size м – X-ray observatories: the next generation – Darkness visible

S. Webb, *New Eyes on the Universe: Twelve Cosmic Mysteries and the Tools We Need to Solve Them*, Springer Praxis Books, DOI 10.1007/978-1-4614-2194-8_5, © Springer Science+Business Media, LLC 2012

Gravity is an unrelenting force. It tugs at matter and never stops tugging. One result of this incessant pulling is that stars can form – gravity brings material closer together, the matter gets hotter, and eventually nuclear fusion reactions begin. A star switches on. Gravity may be persistent but it's also feeble; although gravity never stops pulling on the star's mass, trying to squeeze it to a point, the outward radiation pressure of a new star's hot gas is sufficient to counterbalance the inward pull. We see a star, shining. A star cannot stay forever in this equilibrium stage, however. Every star, eventually, will consume its supply of nuclear fuel and lose the outward pressure that's holding it up. What happens then?

Since gravity never stops tugging, the star will start to collapse under its own weight. The outcome of this process depends on the star's mass. If the star is less than about eight times the mass of the Sun then the temperature at its center increases so that helium can fuse to form carbon and oxygen. The star's outer layers are puffed off into space, creating a beautiful planetary nebula and leaving behind a core of mainly carbon nuclei. The electrons in the core, meanwhile, are pressed together. This halts the collapse since a basic principle of quantum mechanics prevents the electrons from getting *too* close. One way of looking at it is as follows: as the electrons become increasingly localized at the star's center, the uncertainty principle demands that their momenta become increasingly uncertain. Some of them will have a large momentum and thus a large kinetic energy. This kinetic energy causes pressure – **electron degeneracy pressure** – and once more the star possesses a source of outward pressure sufficient to counterbalance the inward pull of gravity. The collapse stops when the star becomes a **white dwarf**, an object that packs the mass of the progenitor star into a volume about that of Earth. The material of a white dwarf is so dense that a thimbleful will weigh a ton. And gravity on a white dwarf is no longer feeble: the surface gravity of a typical dwarf is about one hundred thousand times greater than Earth's and the escape velocity is about five hundred times greater. It is the fate of about 97% of the stars in the Galaxy, including our Sun, to collapse into a white dwarf.

The situation is more complex for very massive stars. In these stars the temperature at the core is high enough for carbon to fuse into other elements, including neon. When all the carbon has burned the core is squeezed further, which raises the temperature enough so that neon can fuse into oxygen and magnesium. When the neon is used up the core is squeezed further, which raises the temperature enough so that oxygen can fuse into other elements, including silicon. Finally, temperatures become so high that silicon can fuse into nickel, which then decays into iron. So towards the end of its life a massive star is a layered object, rather like an onion. On the outside there's a layer of hydrogen. Below the hydrogen there are layers of increasingly heavy elements: first helium, then carbon, neon, oxygen, silicon and then iron. But when the core consists of iron the game is up. Nuclear processes cannot squeeze energy out of iron, and the source of the radiation pressure that's been holding gravity at bay is suddenly no longer sufficient to support the star. A catastrophic collapse of the core now takes place.

During the collapse of a high-mass star, the outer parts of the core fall towards the star's center at speeds of up to a quarter of the speed of light, and the density and the temperature of the core increase even further. High-energy gamma-rays and neutrinos are created in this process and flee the scene; infalling material from the outer layers, meanwhile, rebound off the core. The energy in the shock wave that's produced is enough to blast the star's material out into space. The explosion, which thus releases high-energy photons, heavy elements, and a vast number of neutrinos, is called a core-collapse (or Type II) supernova.

And what of the core that's left behind? Again, this depends on the mass of the progenitor star. Suppose the star is less than about 20 times the mass of the Sun. In this case electron degeneracy pressure is insufficient to hold out against gravity and the core collapses further: the crushing force of gravity drives atomic nuclei into electrons. The result is an object that contains large numbers of neutrons. The same quantum mechanical principles that prevent a white dwarf from collapsing under its own weight are also at work here. The neutrons in the core are not allowed to come too close to each other, and this leads to a **neutron degeneracy pressure** that can prevent further collapse. This object is a **neutron star** and it's unimaginably dense: the entire mass of the star is contained within a sphere that's smaller than London. A thimbleful of neutron star weighs ten billion tons. The surface gravity is about one hundred billion times greater than you're used to. On Earth, if a coin falls out of your pocket it hits your foot at a speed of about 16 kilometers per hour; if it drops the same distance on a neutron star it hits the surface at about 8 million kilometers per hour. The escape velocity of a neutron star is about a third the speed of light. Gravity here is *strong*.

If the progenitor star has a mass of between about 20–50 solar masses then the core that's left behind in the collapse is so massive that not even neutron degeneracy pressure can hold out against that inexorable force of gravity. The collapse continues. At some point, as what's left of the star becomes ever smaller and the density becomes ever greater, the escape velocity of the object exceeds the speed of light. A **black hole** has formed. Inside a black hole lies a **singularity**, a point of infinite density that physicists don't yet understand; in the words of Steven Wright, it's where God divides by zero. However, surrounding the singularity is the **event horizon** – the region of space at which the black hole's escape velocity equals the speed of light. What takes place inside the event horizon may be hidden from our view – but it *is* possible to learn what happens just outside the horizon, a place where the most extreme conditions in the Universe exist.

Some like it hot

High-mass stars may be uncommon but they are certainly not unknown. The best models of stellar evolution thus predict the existence of a plethora of black holes. Since the objects left behind after core collapse possess a mass that's a few times greater than that of our Sun they are called **stellar-mass black holes**. But how can we know whether those stellar evolution models have got it right? How can we know for sure that black holes exist? How can we detect a black hole if light itself cannot escape its clasp?

Perhaps we could look for the silhouette of a black hole as it passes against a field of background stars? Well, no. A stellar-mass black hole is rather small: its event horizon is only a few kilometers across, far too small to be noticeable. However, if the black hole is close to other stars then we should be able to see the effect it has on its companions. Of all the stars in the Galaxy perhaps fewer than two thirds live alone as single stars. The remainder are in binary systems, or sometimes in multiple star systems. It follows that black holes must occasionally be found in binary systems: the high-mass star in the system will have lived fast, died young, and formed a black hole, while its smaller, more prosaic companion star continues to shine by burning its nuclear fuel. Astronomers should be able to detect the influence of the black hole on its companion star.

The obvious signal to look for is a 'wobble' in the motion of a star. The most likely cause of a to-and-fro stellar motion is that the star and an unseen companion are orbiting around a common center of mass. By measuring the details of the 'wobble', and after estimating the mass of the visible star, astronomers have enough information to estimate the mass of the unseen object. If the mass is too big for the object to be a planet, white dwarf or neutron star then the unseen object is a black hole. What else could it be?

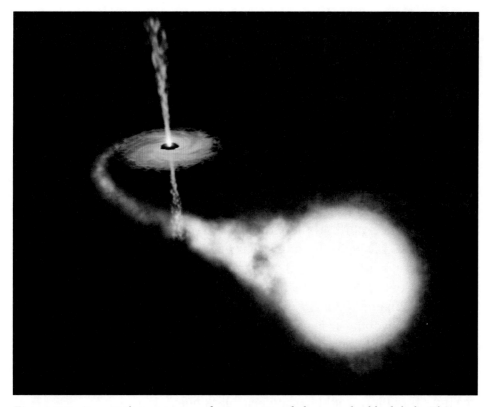

Figure 5.1 *An artist's impression of an accretion disk around a black hole. This particular depiction is inspired by the object* GRO J1655-40, *a microquasar in our Galaxy (microquasars are discussed later in this chapter). In this case material from a stellar-mass black hole is sucking material from its ageing yellow star companion. The accretion disk is shown here in blue; such accretion disks are hot enough to emit X-ray radiation. The two blue jets, at both poles of the black hole, are shooting out from the black hole at about 0.9 times the speed of light.* (ESA, NASA, & Felix Mirabel)

The problem with the 'wobble' technique is that there are many binary systems in the Galaxy. It would be quite impractical for astronomers to check every one of them in the hope of finding a black hole. What's needed is some sign that tips off astronomers to the likely candidates so they can focus their resources appropriately. Fortunately there is such a sign: the presence of X-rays.

If the objects in the binary system are sufficiently close, the collapsed star will suck some of its companion's gas towards itself. As this gas spirals onto the collapsed object a particular type of structure forms: an **accretion disk**.

Figure 5.1 shows an accretion disk around a black hole. (Such a disk can appear around neutron stars, white dwarfs, or even 'ordinary' stars. It's a structure that's typically going to form whenever a fluid falls onto a small object – think of water flowing down a bath plughole.) Gas particles in the disk scrape against each other as they orbit and, just as your fingers heat up from friction when you rub them together, so the gas in the disk gets hotter as the particles collide. The outside edge of the accretion disk won't be particularly hot, but the part of the disk closest to the collapsed object can be hot enough to emit large amounts of X-rays. And *that's* a signal that astronomers can look out for. X-ray emission is a sure sign that something interesting is happening because in order to emit copious amounts of X-rays an object has to be *really* hot – several hundred million degrees.

So here is a recipe for finding stellar mass black holes. First, find a bright X-ray source, which immediately signifies that something unusual is happening. Second, look for a companion star near the X-ray source. Third, if one's luck is in and there is a visible companion, estimate its mass and then measure the 'wobble' it makes as it moves through space. Finally, use this information to estimate the mass of the object that's causing the X-ray emitting accretion disk: if the mass is large enough then the presence of a white dwarf or neutron star is ruled out. In the absence of any other explanation, the invisible object would have to be a black hole.

Detailed observations of X-ray emission from an accretion disk can provide further reassurance that a black hole is involved. Since gas particles are unlikely to orbit the unseen object in an orderly fashion, the X-ray emissions are likely to vary with time. Sometimes there'll be particularly bright flares while at other times there'll be a relative lull in emission. However, unless one is ready to accept incredible coincidence, such fluctuations cannot take place across the whole of the accretion disk simultaneously. Since no interaction can move faster than the speed of light it follows that, by measuring how quickly the X-rays fluctuate, astronomers immediately have a value for the maximum possible size of the unseen object. The more rapid the fluctuations, the smaller the object must be. If the X-rays vary extremely rapidly then the unseen object must be very small – and if a lot of mass is contained in that small volume then what else could the object be if it isn't a black hole?

There's yet another way in which X-ray observations of accretion disks could confirm the existence of black holes. An accretion disk can form around a neutron star just as it can around a black hole; in both cases the spiralling gas emits X-rays as it gets closer to the compact object. But there's a difference between the two cases. When material from the disk slams into the surface of the neutron star the collision generates X-rays. With a black hole there's no surface to slam into. No X-rays are emitted when material falls into a black hole. Instead, the material just passes quietly through the event horizon. Thus a neutron star will be a brighter X-ray source than a black hole. Here, then, is another way of demonstrating the existence of black holes (if you need persuading that such objects exist). First, find as many compact objects as possible. Second, classify them into two groups: those with less than three solar masses and those with more. Finally, for objects in the two groups having the same orbital periods compare their X-ray brightness. Since the accretion rate should only depend on the orbital period, similar objects in the two groups should have a similar brightness. If it turns out that the more massive objects are dimmer than their less massive counterparts then black holes are involved: the relative faintness would be a clear sign that matter and energy were falling over event horizons – and perhaps disappearing from our Universe altogether.

So a study of X-ray sources would seem to be an ideal way of learning more about black holes. Unfortunately, there's a problem with trying to observe X-rays from the cosmos: Earth's atmosphere absorbs them. (This, it has to be admitted, is a Good Thing for everyone – astronomers included. Life does not prosper when bathed in X-rays.) It follows, then, that to observe cosmic X-rays we must get above the atmosphere: instruments must be placed on balloons or on rockets or on satellites.

X-ray vision

The processes that generate cosmic X-rays take place at ultra-high temperatures and they thus involve extreme phenomena – intense magnetic fields, huge gravitational potentials or vast explosive forces. So it's not just the nature of black holes that X-ray astronomy can help elucidate, it's what happens to matter in many different extreme circumstances. Small wonder that the potential of making X-ray observations has excited astronomers for decades.

It was Riccardo Giacconi who led the first team to discover an X-ray source from deep space. In 1962, astronomers placed a Geiger counter on an Aerobee – a rocket that could fly above the atmosphere for about a quarter of an hour. The Geiger counter functioned as an X-ray detector and Giacconi and his team found a clear X-ray signal. Their source was somewhere in deep

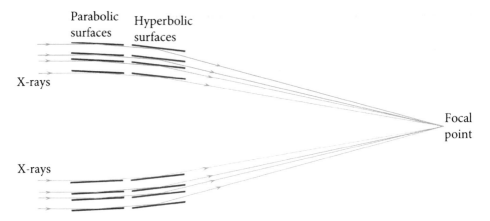

Figure 5.2 *An X-ray telescope works by shepherding X-rays that hit parabolic mirrors at a grazing incidence onto hyperbolic mirrors. The hyperbolic mirrors in turn bring the X-rays to a focus.*

space and it was emitting a hundred million times more energy in X-rays than does the Sun. Indeed, their source was emitting a hundred times more energy in X-rays then the Sun emits at *all* wavelengths. This was a completely new type of object. Giacconi and his team called it sco x-1, which means it was the first X-ray source to be found in the constellation of Scorpius.

An X-ray detector was all very well, but what astronomers really wanted of course was a telescope that could take X-ray images. Taking an X-ray image is difficult, though, because X-rays can be hundreds or thousands of times more energetic than visible light: the techniques employed in building an optical telescope – which typically involves reflecting light with a mirror – simply won't work. X-rays don't bounce off a mirror; they smash into it and keep on moving until they are absorbed. However, there is a method for bringing X-rays to a focus, and it's familiar to anyone who has ever skimmed a stone on a pond. Throw a stone at too steep an angle and it goes straight into the water; throw it at a small angle, though, and the stone skips off the water's surface. The same idea can be used in an X-ray telescope. If you use a mirror made out of a material that doesn't absorb high-energy photons, typically gold or iridium, and catch an X-ray at the so-called **grazing incidence** then the photon skips off the mirror rather than penetrating it. A pair of mirrors of precisely the right shape – a parabolic mirror followed by a hyperbolic mirror – can shepherd X-rays to a focal point. You can take an image. See figure 5.2. The German physicist Hans Wolter first came up

Figure 5.3 *An artist's impression of material being sucked away from the blue super-giant star* HDE 226868 *by the black hole Cygnus* X-1. (ESA/Hubble & NASA)

with this idea for an X-ray telescope, and such an instrument is now called a **Wolter telescope** in his honor. An X-ray telescope perhaps resembles a bucket more than a traditional optical telescope. However, as it stands, the bucket would not be particularly useful: it would catch only those X-rays arriving almost parallel to the mirrors and thus grazing their surfaces. X-rays from the same source arriving in the middle of the bucket would just pass right through without hitting the mirrors at a grazing incidence. The solution? Use a matryoshka doll type of setup, with a series of mirrors nested one inside another. This is the design behind modern X-ray telescopes.

Giacconi and his team made the first imaging X-ray telescope in 1963. Their telescope lacked the sophisticated design of modern instruments, but it was able to take images of hot spots in the upper atmosphere of the Sun. These were crude images but nevertheless useful ones. Giacconi, then, could be said to have initiated X-ray astronomy – an achievement that was recognized by his award of a share of the 2002 Nobel prize in physics.

The field burgeoned in the 1970s, when NASA launched several Earth-orbiting missions to survey the X-ray sky: **Uhuru**, a small telescope, launched at the start of the decade; **Einstein**, a much larger instrument, worked all the way through to 1981. Uhuru discovered 339 X-ray sources, ranging from supernova remnants to galaxy clusters. It was Einstein, however, that really changed our view of the X-ray sky: it cataloged thousands of X-ray sources including the outer regions of stars and the central regions of galaxies. After Einstein, X-ray observatories became a priority. And, as we know, one of the things they could look for was black holes.

One of the first X-ray sources to be discovered turned out to be the first widely accepted solar-mass black hole candidate. In 1964, instruments on board Aerobee rockets found Cygnus X-1. After intense study astronomers now know that Cygnus X-1 is a blue supergiant star with a compact companion having a mass of perhaps nine times that of the Sun: that's well above the neutron star limit, and the companion is thus almost certainly a black hole. See figure 5.3. Another X-ray source in the same constellation, V404 Cygni, also turns out to be an excellent candidate for black hole status: in this case a star not too dissimilar to the Sun is orbiting a compact companion having ten to fifteen solar masses. It's difficult to think of what this compact companion could possibly be if it isn't a black hole. To date a further dozen X-ray binaries have joined Cygnus X-1 and V404 Cygni as prime stellar-mass black hole candidates. The accretion disks in these systems are all different in character. The emission from Cygnus X-1 is intense but erratic; that from V404 Cygni is generally quiet, implying that the influx of material onto the disk is a slow trickle rather than a deluge. The common theme is that each of these disks surely surrounds a black hole.

Following on from the success of Uhuru and Einstein, several other X-ray satellites were launched. For example, **Ginga** (the Japanese word for 'galaxy') studied the X-ray sky between 1987 and 1991. The German **Röntgensatellit** (ROSAT) operated between throughout the 1990s. In 1995, NASA launched the **Rossi X-ray Timing Explorer** (RXTE), an observatory that was developed in order to measure how cosmic X-ray sources vary with time. RXTE was extremely successful, and sent back data until it was decommissioned in early 2012. One of its many discoveries is that of the smallest black hole candidate: the X-ray binary XTE J1650-500 almost certainly contains a black hole with a mass just 3.8 times that of the Sun. It was in 1999, however, that a revolution took place in X-ray astronomy; it's a revolution that's still transforming our knowledge of the Universe.

In 1999, two large orbiting X-ray observatories were launched, one by NASA and one by ESA. The **Chandra X-ray Observatory** became NASA's third 'Great Observatory'. Chandra was named in honor of Subrahmanyan

Figure 5.4 *An artist's impression of* xmm-*Newton in orbit. The observatory's three X-ray telescopes are clearly visible. Each follows the design shown in Figure 4.3, with 58 wafer-thin gold-covered mirror shells bringing X-rays to a focus. (*esa/D. Ducros*)

Chandrasekhar, a fitting name since one of his many achievements was the determination of the critical mass beyond which a white dwarf inevitably becomes a neutron star. The **X-ray Multimirror Mission–Newton** (xmm-Newton) was one of esa's 'cornerstone' missions in its Horizon 2000 Science Programme. See figure 5.4. The mission was named in honor, of course, of Sir Isaac. Although Newton lived a long time before X-rays were discovered, he so influenced our understanding of gravity, optics and spectroscopy – all of which are relevant to xmm – that the name is surely appropriate. The two observatories complement each other: xmm-Newton has the larger collecting area, Chandra the superior resolution. At the time of writing, both xmm-Newton and Chandra are still collecting data and returning it to Earth. They have exquisite sensitivity (they are about a hundred million times more sensitive than Giacconi's first X-ray telescope) and the capacity to take long exposures of the sky (so after the initial discovery of sco x-1, astronomers

Figure 5.5 *An artist's impression of the ring of dust that typically surrounds a super-massive black hole and its accretion disk. If we view the ring edge-on then light from the accretion disk is blocked by the dust. An X-ray telescope, however, can see through the dust and study the accretion disk directly. (ESA/V. Beckmann, NASA-GSFC)*

with these telescopes have been able to catalog more than a hundred thousand X-ray sources, the light from the most distant of which took 13 billion years to reach us). Together, these two observatories have investigated the environs of more than just stellar mass black holes: as we shall now see, they've also studied much bigger beasts.

Hearts of darkness

One of the most surprising astronomical discoveries in the latter half of the last century was that of quasi-stellar objects – **quasars**. These objects are incredibly bright, with luminosities that are often hundreds of times greater than the Galaxy's luminosity. On the other hand, they can vary in brightness on timescales as short as a few days, which implies that all this energy

Figure 5.6 *The giant elliptical galaxy M87 lies in the constellation of Virgo. A jet of fast-moving electrons and other charged particles shoots out from the galaxy's centre. As the particles accelerate through the twisted magnetic field lines in the jet, synchrotron radiation is produced. The jet is about 5000 light years long in visible light; at radio wavelengths the jet is much longer. Astronomers have observed the jet in X-rays as well as the longer wavelengths; Chandra, for example, discovered that the X-ray jet has a 'knotty' structure. The engine that's powering all this is a supermassive black hole that's billions of times more massive than our Sun. A similar mechanism is believed to power quasars. (NASA and the Hubble Heritage Team)*

emanates from a volume billions of times smaller than a typical galaxy. What engine could possibly power such a beast? As astronomers looked for more quasars to study they uncovered a further mystery: nearly all quasars possess large redshifts and are thus at large distances from us. Now, looking deep into space with a telescope is like looking through a time machine into the past: the higher an object's redshift the further back in time the light we observe left that object. By studying the redshifts of quasars astronomers came to realize that the number of quasars peaked about 2.5 billion years after the Big Bang and has been declining ever since. Compared with the number of quasars that lived about 11 billion years ago, only one in 500 is around today. Where did all the quasars go?

The most plausible hypothesis anyone came up with to explain their huge luminosity and compact nature was that quasars are **supermassive black holes** lying at the heart of young galaxies. 'Supermassive' in this case means millions or even billions of solar masses. Note that if astronomers detect such a black hole candidate they can estimate its mass by measuring the speeds of objects orbiting close to the event horizon. The idea is similar to that discussed in chapter 3, where we looked at Keplerian motion in the Solar System. In the case of a supermassive black hole, the more massive the hole the faster objects close to it will orbit. As a supermassive black hole sucks in surrounding gas, squeezing and compressing it, the accretion disk becomes extremely hot – and radiation, including high-energy X-rays, is emitted. Another common feature of accretion disks – **relativistic jets** – could also play a role in the emission. In a process that is still not entirely clear, but that probably involves intense magnetic fields, a highly energetic plasma can get carried away in directions perpendicular to the accretion disk. The emission appears as a pair of fast-moving jets. The jets can also emit X-rays. Thus X-ray telescopes such as Chandra and XMM-Newton are ideal tools for studying supermassive black holes.

The high-energy photons and particles inside the fast-moving jets can travel outward from the black hole for many thousands of light years. On the other hand, even the largest black holes have small volumes – measured better on stellar scales than galactic scales. For example the event horizon of a black hole possessing, say, a billion solar masses would fit into the orbit of Uranus. This explains why quasars can vary on such short timescales: the black hole that powers a quasar is small. Furthermore, this idea provides an answer to the mystery of where the quasars went. In order to explain the exceptional luminosity of a quasar, the black hole engine would have to be gobbling the equivalent of twice Earth's mass every second, which equates to a diet of 200 solar masses every year. Such a rate of consumption can't last forever: there simply isn't that much surrounding matter to be eaten. It follows that a bright quasar must be a temporary stage in a galaxy's existence.

The picture gradually came about that *every* galaxy has a supermassive black hole at its heart. When the Universe was just a couple of billion years

Figure 5.7 *The 'death star' galaxy has been imaged in X-rays by Chandra (shown as purple in this composite image), in the optical and ultraviolet regions by Hubble (shown as red and orange) and in the radio by the Very Large Array and* MERLIN *(shown as blue). Taken together, these images show how a jet powered by a supermassive black hole at the center of the primary galaxy (lower left) hits a companion galaxy that's 20 000 light years distant (upper right) and then moves on into space. (X-ray:* NASA/CXC/D.Evans et al.; *Optical/UV:* NASA/STSCI; *Radio:* NSF/VLA/D.Evans et al., STFC/JBO/MERLIN)*

old, galaxies interacted with each other much more frequently than they do now; after all, the Universe was a much smaller place back then. Full-on mergers of galaxies, or even just close encounters, would see tidal forces increase the rate at which gas was funneled into the central black holes – and the holes would feed voraciously. So back then the Universe was home to lots of quasars. Perhaps *every* galaxy once shone as brightly as a quasar. Those supermassive black holes are still around – and indeed are more massive than ever, since they've had billions of years to feed – but the rate at which they accrete gas is much less than it once was. Nowadays, those black holes at the hearts of galaxies are quiet.

Quasars were just one of several types of bright, compact object that astronomers discovered in the last century. In 1943, twenty years before Maarten Schmidt identified the first quasar, Carl Seyfert discovered a class of galaxy that possesses a particularly bright nucleus. Such a **Seyfert galaxy** can be bright at wavelengths from infrared all the way up to X-ray; a few are bright at radio wavelengths too. And in 1978 Edward Spiegel coined the term **blazar** to describe a particular type of compact object – a 'blazing' quasar – that was violently energetic. One of the triumphs of modern astronomy has been the realization that all these seemingly different types of object – quasars, blazars, Seyfert galaxies along with various subtypes – are all essentially the same. They are all examples of **active galactic nuclei** (AGN). In each case the object is a galaxy whose central supermassive black hole is gorging on gas. The differences we observe are mainly dependent on our line-of-sight view. If the emission from the accretion disk is pointed in our general direction we see a quasar; if the emission is pointing roughly perpendicular to our line of sight we might identify a Seyfert galaxy; and if we look directly down a jet that's heading towards us we might see a blazar.

This unified model, the notion that such disparate objects can all be explained in terms of active galactic nuclei, was hugely attractive. For the model to become fully accepted, however, astronomers needed to observe the central supermassive black hole in an active galactic nucleus. (Well, they wouldn't be able to observe it directly of course. As we noted above, the event horizon around even a gigantic black hole is tiny; the accretion disk around the hole is larger, but still too small even for a telescope as powerful as Hubble to resolve it. Nevertheless, it should be possible to observe the *effects* of these black holes.) In the 1990s, astronomers began to search in earnest for these monsters. X-ray telescopes were in the forefront of the search.

A Japanese mission called the **Advanced Satellite for Cosmology and Astrophysics** (ASCA) was launched in 1993 and, two years later, astronomers pointed its X-ray telescope towards a spiral galaxy with the unprepossessing name of MCG-6-30-15. Although the galaxy lies about 100 million light years away its nucleus is exceptionally bright – it's an AGN (of the Seyfert class, as it happens). When astronomers studied the galaxy's X-ray spectrum they found that it was dominated by an emission line due to iron. That in itself was not unusual; line emissions simply denote the presence of very hot gas. However, such a line emission should be symmetric and very narrow – in other words very much like, well, a line. Instead they saw a broad feature, more like a hump, which was heavily skewed. What could explain the width and asymmetry? Well, an emission line due to iron would look *precisely* like this due to the relativistic effects that occur as material is whipped around by a supermassive black hole. So that's one galaxy with a monster at its heart.

After the initial discovery by ASCA, astronomers investigated MCG-6-30-15 in increasing detail. The findings are wonderfully strange. For example, astronomers using XMM-Newton have gone looking for the presence of an iron emission line in the very hottest part of the accretion disk, close to the event horizon. They found a line so broad that it was almost horizontal. This observation is at the limit of what is technically possible with the telescope, but if it is correct then the width of the emission line implies that the matter is whizzing around the black hole at almost the speed of light. What's more, the intensity of the X-ray glow implies that the material is so hot that the conventional explanation for its temperature – namely that the particles in the disk heat up through collisions as they get closer to the event horizon – does not work. Another mechanism is required to explain these high energies. Fortunately, theoreticians have a couple of ideas of how energy can be extracted from black holes. One ready-made explanation for the observation by XMM-Newton is the so-called **Blandford–Znajek mechanism**, developed by Cambridge physicists Roger Blandford and Roman Znajek in the 1970s.

The idea behind the Blandford–Znajek mechanism uses the fact that a galaxy's central supermassive black hole is likely to be spinning. The hundred million solar mass black hole at the center of MCG-6-30-15, for example, will be rotating once every couple of hours and its event horizon will be moving at close to light speed. What's also important in this mechanism for extracting energy from a black hole is the fact that electrically charged black holes act very much like an electrical conductor. So how does a spinning conductor, such as those we find at the centers of galaxies, pump out energy? Well, the gas in the accretion disks is likely to be threaded with magnetic field lines – and *any* spinning conductor in a magnetic field develops a voltage difference between equator and poles. In other words, a spinning supermassive black hole at the center of a galaxy will behave rather like a car battery. The difference is one of scale: a car battery develops $12\,V$ whereas a supermassive black hole develops $10^{15}\,V$. To extract energy from a battery requires a circuit. In a car the circuit consists of wires; in an active galactic nucleus the circuit is the ionized gas that surrounds the black hole, linking pole to equator. In the Blandford–Znajek picture, then, a spinning black hole acts as a generator that develops a potential difference large enough to fling electrically charged particles into space along its poles with *tremendous* energy.

Figure 5.8 *This is a view of Sgr A*, the supermassive black hole at the center of our Milky Way galaxy, taken by Chandra. The high-energy X-rays detected by Chandra are colored blue in this image; intermediate-energy X-rays are yellowish-green; low-energy X-rays are red. (NASA/CXC/MIT/F.K. Baganoff et al)*

The black hole is thus converting its energy of rotation into the acceleration of particles. The Blandford–Znajek process is extraordinarily efficient, and so could provide the explanation of the relativistic jets that can be seen emanating from some active galactic nuclei. And it may be that XMM-Newton has seen it in action in MCG-6-30-15.

These observations are at the limit of what's possible with the current generation of X-ray telescopes. Even more sophisticated instruments are needed to study the process further. It's worth pointing out, though, that it's not just X-ray telescopes that can be used to search for supermassive black holes. For example, astronomers have used radio and optical observations of how objects orbit a galactic nucleus, and compared them with an expectation of how they *should* move under Kepler's laws, to deduce the presence of supermassive black holes. So radio observations of a water maser near the nucleus of the nearby spiral M106, a type of Seyfert galaxy, implied the presence of a central object with a mass 40 million times greater than the Sun. And optical observations by Hubble enabled astronomers to deduce the presence of a supermassive black hole at the center of a couple of dozen galaxies. One of the objects that Hubble detected, the dense central object in the giant elliptical galaxy M87, held the record for the most massive black hole then found: 3.5 billion solar masses. (That's puny compared to the current record holder, the black hole at the heart of an object known as OJ 287: that black hole is believed to have a whopping 18 billion solar masses.)

Astronomers have even confirmed the presence of a supermassive black hole in our own Galaxy. Two groups, one led by Andrea Ghez and the other Reinhard Genzel, have spent several years making accurate observations of stars near **Sagittarius A*** (Sgr A*), a radio source close to the center of the Galaxy that was first discovered in 1974. See figure 5.8. It has long been known that Sgr A* is small – about the size of the orbit of Saturn. And now, thanks to the observations of Ghez and Genzel, we know that Sgr A* has a mass of about 4 million solar masses. Such a large mass in such a small volume can be nothing other than a black hole. As supermassive black holes go this is rather small, and the low luminosity of the emission lines in the radio and infrared suggest that Sgr A* is not a particularly vigorous eater; indeed, it's almost anorexic. But since a supermassive black hole can be found in the Milky Way, which is a rather gentle and inactive sort of galaxy, it does rather suggest that they can be found in all galaxies.

In 2000, when looking at all the supermassive black holes that had been found, two independent groups of astronomers (the so-called 'Nuker' team of astronomers interested in studying galactic nuclei; and Laura Ferrarese and David Merritt) noted a tight relationship between the mass of a galaxy's central supermassive black hole and the motion of stars near the center. Stars swarm around a galaxy's central regions much like bees around a honeypot. By making Doppler measurements of these stars astronomers can quite easily determine the velocity dispersion of the stars; in essence, the velocity dispersion just measures the typical velocity of stars as they move through a given region. It turns out that the mass of the black hole is proportional to the fourth power of the velocity dispersion. The relationship hints at a deep relationship between the central black hole and the galaxy that hosts it.

This relationship is important for many reasons. One reason is that observing the effects of a supermassive black hole is rather difficult but measuring a stellar velocity dispersion is, relatively speaking, easy. So it becomes relatively easy to estimate the mass of a central black hole. By applying this relationship to large numbers of galaxies, astronomers were able to clarify several interesting points.

First, there is now little doubt that a supermassive black hole is indeed a common feature of massive galaxies. Second, about one thousandth of the luminous mass of a galaxy (that's the mass we *see*, not mass in the form of dark matter) belongs to its central supermassive black hole. In other words, the more massive the galaxy the more massive the black hole. Why should there be this correlation? Well, the most likely reason seems to be that the central black hole regulates the amount of gas that's available for star formation in its host galaxy. In other words, black holes profoundly influence the evolution of galaxies. Bear in mind that the size of a central black hole compared to the size of its host galaxy is the same as the size of you, the reader, compared to the size of Earth. This demonstrates the huge gravitational potential energies involved with black holes. Finally, the number of supermassive black holes in the nearby Universe is similar to what astronomers would expect from observations of distant quasars. So now we know for sure where all those quasars went: the quasars didn't die, they just faded away.

Enigmas large and small

In the summer of 1981, as I was lazing away the months before leaving for university, I came across an article in that estimable magazine *Sky & Telescope*. The article was called 'ss 433 – Enigma of the Century', a title intriguing enough to capture anyone's attention, and it described an object that was baffling astronomers: a faint star in the constellation Aquila, about 18 000 light years away, called ss 433. It's worth noting that the object might have had a different name: the astronomer Larry Krumenaker published details of the star in 1975 and called it κ16. Krumenaker had been looking for stars with extra emission lines in their spectra – lines that denote the existence of particular chemical elements in or around the star – and he found that κ16 possessed a H-alpha line, a red spectral line that occurs when the electron in a hydrogen atom falls from the third to the second lowest energy level. That discovery itself was not noteworthy; it merely indicated that a cloud of glowing hydrogen gas surrounded the star. What was striking about κ16 was that the H-alpha line was about four times wider than normal. Unfortunately, with the crude technology available to him, Krumenaker was unable to come up with an explanation for his observation. In 1977, the astronomers

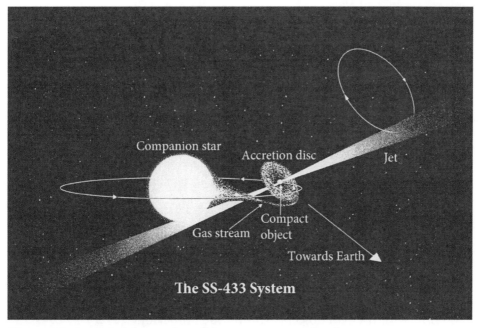

Figure 5.9 An artist's impression of the ss 433 system. (ESA)

Bruce Stephenson and Nicholas Sanduleak published a catalog of objects with emission lines in their spectra and, because this was the 433rd object in their catalog, it became known as ss 433 – a name that stuck.

Real interest in ss 433 took off in 1978, when Peter Martin, Paul Murdin and David Clark took a more detailed spectrum of the object and found that some of the emission lines could not readily be identified with any known element. This was not just striking, it was a mystery. Just a few months later, a team led by Bruce Margon found that those mysterious emission lines were in fact *pairs* of emission lines – one line of each pair was redshifted and the other line was blueshifted. Furthermore, the lines were not static: over a period of about five months they moved and crossed each other. A redshifted or blueshifted emission line is not unusual, of course: an emission line shifted towards the red implies that the line's source is moving away from us, a line shifted towards the blue implies motion of the source towards us, and the size of the shift is a direct measure of the velocity. It's just the Doppler effect. The *pairs* of Doppler-shifted lines found by Margon's team were an oddity, however, and Margon's team, like Martin's team, were unable to identify the chemical elements giving rise to the lines – so not only were they in the dark about its chemical make-up, they were unable to determine how fast ss 433 was moving. Cue *huge* interest in the object.

The following year astronomers, having pored over old photographs of ss 433 dating back to the 1920s, showed that it was a variable star: it varied in brightness with a period of 162 days – the same period exhibited by the emission line pairs. And soon after astronomers discovered that the line pairs were in fact line *triplets*. That was the clue they needed. Finally, they twigged what those emission lines were: they were just normal H-alpha lines. What was unique about ss 433 was that the central line stayed put, while the two shifting lines either side travelled at speeds up to a quarter of the speed of light, before appearing to slow and then change direction. People called ss 433 the 'coming and going star'. Soon after, David Crampton, Anne Cowley and John Hutchings found that even the central emission line of the triplet moved slightly, this time with a period of 13 days.

By the time the *Sky & Telescope* article hit the stands, astronomers had a rough idea of the sort of object ss 433 must be. It had to be *two* objects rather than one, with the two objects orbiting each other once every 13 days. The system had to be spewing out hydrogen atoms in a pair of jets moving in opposite directions. The gas in the jets had to be moving at speeds up to a quarter of the speed of light. Finally, those jets had to rotate around the center in a period of 162 days. It was clear that ss 433 was unlike anything seen before, hence the title of that magazine article.

There was an element of hype in the magazine's description of ss 433, of course. Two decades or so earlier and the magazine could equally have well run a piece entitled '3C 273 – Enigma of the Century' since back then the most puzzling type of object was the quasar. Quasars such as 3C 273 were controversial at first because their spectra contained unidentified emission lines, but eventually astronomers realized that those lines were simply hydrogen lines that had been Doppler shifted a long way to the red end of the spectrum. Later still, as we have seen, astronomers came to understand how supermassive black holes provided the power source for quasars. The stories behind the distant enigmatic 3C 273 and the nearby enigmatic ss 433 are thus not too dissimilar. Indeed, it now seems that objects such as 3C 273 and objects such as ss 433 are essentially the same type of phenomenon. The difference between them is purely one of scale.

Astronomers now believe ss 433 consists of a spinning black hole (with a mass perhaps eight times that of the Sun) which, every 13 days, orbits a white supergiant star (with a mass perhaps 27 times that of the Sun). The

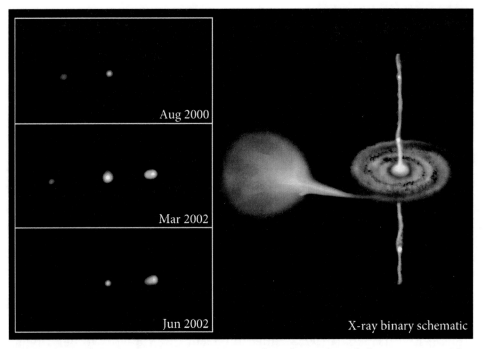

Figure 5.10 *The microquasar* XTE J1550-564 *lies about 17 000 light years away. It's a binary system in which an old, orange–red star is being eaten by a black hole companion that has a mass greater than ten times the mass of the Sun. The object is usually extremely faint at X-ray wavelengths, but in 1998 it became the brightest X-ray source in the sky and it was studied by* RXTE. *As the illustration on the right shows, the gas being pulled from the star forms an accretion disk around the black hole and fast-moving jets of particles are expelled at the poles. In August 2000, Chandra observed the left-moving jet, which is tilted towards Earth. In March 2002, Chandra observed the opposing right-moving jet, which is tilted away from Earth. A few months later, the left-moving jet had faded. Between the time the microquasar brightened to the time the jet faded, the particles in the jet had travelled more than three light years. A microquasar such as this is an ideal laboratory for learning more about supermassive black holes, where the timescales and distances involved are much larger. (Left: X-ray - NASA/CXC; right: illustration - CXC/M. Weiss)*

black hole sucks gas from the star and as this material swirls towards the hole it forms an accretion disk. As always, the disk gets extremely hot and it is the disk that produces most of the visible light that we see from the system. It was the accretion disk, for example, that Krumenaker identified as being a 'star' on his photographic plates. The disk wobbles with a period of 162 days. And, to complete the picture, an opposing pair of jets shoot out at right angles to the disk and at speeds that are a significant fraction of lightspeed.

The description of ss 433 is thus reminiscent of the description of a quasar. The differences that exist depend on the mass of the black hole – a solar-mass black hole in the case of ss 433, a supermassive black hole in the case of a quasar. In ss 433 the accretion disk has an average temperature of millions of degrees, in a quasar just a few thousand degrees; in ss 433 the accretion disk is perhaps a thousand kilometers in size and radiates mainly in the X-ray region, in a quasar the disk is a million times larger and radiates mainly in the ultraviolet and optical regions; in ss 433 the particles in the jets travel for a few light years, whereas particles in quasar jets can travel for millions of light years. But these are purely differences in scale, not differences in kind. Thus Felix Mirabel, in 1992, coined the term **microquasar** to describe an object such as ss 433. Since the name 'quasar' initially meant a quasi-stellar object it would have made more sense to call ss 433, which consists of a star and a stellar-mass black hole, a 'quasar' and reserve the term 'megaquasar' for its supermassive cousins. Unfortunately, these objects weren't discovered in the right order and we're stuck with the names we have.

In the years since ss 433 was first recognized as being worthy of close attention, astronomers have found more microquasars. Cygnus x-1, the first black hole candidate identified by astronomers, is a microquasar. The closest known microquasar to Earth (it's also the closest known black hole) is v4641 Sagittarii, which lies about 1600 light-years away. The microquasar LS 5039 is about 10 000 light years distant; it has jets that reach out 2.5 billion kilometers into space. The microquasar GRS 1915+105 is about 40 000 light-years away; its jets have been observed to eject material at 92% of lightspeed. In fact, astronomers now have more than a dozen microquasars to study – and have used Chandra and XMM-Newton to do so. The importance of microquasars is that they can act as laboratories for studying quasars: events that happen on a timescale of days in the case of microquasars might take years in the case of quasars. Astronomers can thus use microquasars to understand more about relativistic jets and the extreme spacetime conditions that happen near the event horizon of a spinning black hole.

Now available: black holes in size M

In 2000, Chandra observed M82 – the Cigar galaxy – in detail. This relatively nearby object is a **starburst galaxy**, one that has a much higher rate of new star formation than is typical. In the case of M82 this rapid star formation is probably caused by the gravitational influence of its much larger neighbor, the beautiful spiral M81: when the Cigar galaxy gets too close to its neighbor, tidal forces cause gas to channel into the central regions and stars can then form in large numbers. It's this intense activity, which is associated with extremely high temperatures, that makes M82 a prime target for X-ray observatories.

Chandra found that M82 is host to many point X-ray sources. The brightest of those sources, M82 X-1, is so bright that it is hard to account for – unless it is a red giant star providing food for a black hole with a mass perhaps 200–5000 times that of the Sun. This, then, is an example of a candidate object called a mid-mass black hole or an **intermediate-mass black hole** – an object that's much larger than a stellar-mass black hole yet puny when compared with the supermassive black holes at the center of galaxies.

The obvious question to ask is: how could such a mid-mass black hole come into existence since, in the present-day Universe, we see no stars with a mass greater than about 150 times that of the Sun?

Well, astronomers came up with two plausible scenarios to explain the object in M82 X-1. One idea was that it formed through the merger of several pre-existing stellar-mass black holes. The other idea was that a gigantic star was indeed involved, and at the end of its brief life it underwent gravitational collapse to a black hole. Such a gigantic star might be able to form in M82 because it contains regions where lots of stars are packed into a small volume; for example, M82 X-1 lies in a cluster where a million stars are packed into a volume that has a radius of just 50 light years. To give some idea of what this density of stars means, consider that there are only a dozen stars within a 10 light year radius of Earth – and that number includes the Sun. Where such large numbers of stars exist in such crowded conditions, it's possible that collisions between stars might create a short-lived, incredibly massive 'superstar' that could collapse into a black hole. (There was a third possibility: perhaps M82 X-1 wasn't a black hole at all, and astronomers just needed to be more imaginative in thinking up explanations.)

While M82 X-1 was the only possible example, there was too little information to understand mid-mass black holes. Gradually, however, Chandra

Figure 5.11 An artist's impression of the incredibly luminous X-ray source HLX-1 (represented by the bright object to the top left of the galactic bulge) in the periphery of the edge-on spiral galaxy ESO 243-49. This X-ray source, first observed by XMM-Newton, is strong evidence for the existence of mid-mass black holes. (CNRS/INSU/ Illustration Heidi Sagerud)

and XMM-Newton found several other candidate mid-mass black holes. And in 2009, a team of astronomers working with XMM-Newton announced the discovery of an object that seemingly just can't be explained – unless it's a black hole that's about 500 times more massive than the Sun. The team studied an object that they called the Hyper Luminous X-ray Source 1 (HLX-1), which lies in the outskirts of a galaxy called ESO 243-49. See figure 5.11.

At first the team members thought they were looking at a typical X-ray binary, with a normal star orbiting a black hole with a mass of a few times that of the Sun. However, when they looked more closely with large optical telescopes it became apparent that the source was extremely distant – about 290 million light years away. For it to have the apparent brightness it does, despite lying at such a distance, its intrinsic brightness must extreme. In fact, it must be the most luminous compact X-ray source ever seen – hence the choice of 'hyper luminous' in the name. Only one model can explain all its features: here is a 500 solar mass black hole gorging on the material in its environment.

It seems, then, that mid-mass black holes really do exist. And they may prove to be the key to understanding how supermassive black holes form: it may well be that it's the merger of many mid-mass black holes that creates the monster that lies at the heart of all galaxies.

Chandra and XMM-Newton have demonstrated that black holes come in size S (stellar-size objects with a mass of between about 3–30 solar masses), size M (mid-mass objects with a mass of between about 500–50 000 solar masses) and size XXL (supermassive objects with a mass that can be millions or billions times that of the Sun). And these two venerable observatories have revealed much about the events that can take place around the various types of black hole. But so many questions about black holes remain and neither XMM-Newton nor Chandra will be around much longer to give answers. They are already operating long past their original mission lifetimes: engineers designed the Chandra mission to last five years and the XMM-Newton mission to last just two years. So is the future for further observations of black holes … well, black?

X-ray observatories: the next generation

The number of open questions in science seems to grow exponentially. Take supermassive black holes, for example. Where exactly does the correlation between black hole mass and stellar velocity dispersion come from? No one knows for sure. The correlation implies the existence of a close link between the formation and evolution of black holes and that of their host galaxies. But did black holes cause galaxies to form, or did galaxies form first and then generate black holes, or was some other mechanism at work? What role did dark matter play in all this? And why are all these supermassive black holes so peaceful nowadays? There may not be the same amount of fuel available as when these objects shone as quasars, but there should be enough gas and dust around for them to shine more brightly than we observe. Then there are questions about the effects of spinning black holes on the surrounding spacetime, and questions about the mechanism by which black holes accelerate particles to such high energies. To help answer these and other questions several X-ray telescopes have been planned, successors to RXTE, Chandra and XMM-Newton.

The first telescope to see action will be a NASA mission called the **Nuclear Spectroscopic Telescope Array** (NUSTAR), which has a planned launch date in mid-2012. NUSTAR will be the first space-based mission with the capacity

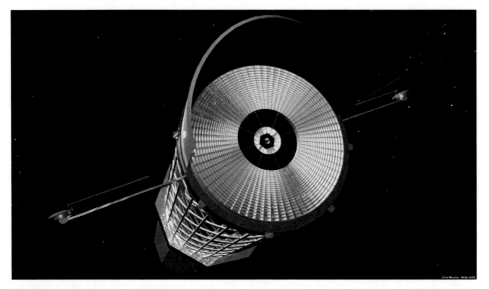

Figure 5.12   *An artist's depiction of the proposed International X-ray Observatory. Funding difficulties mean that this particular design will not be taken further.* (NASA)

to focus hard X-rays; that is, X-rays with high energy – comparable to those used in medicine. Recent advances in mirror technology have made a hard X-ray mission such as NUSTAR possible – and the result should be the ability to peer into regions that have so far been hidden from view. In particular, NUSTAR should be able to see many more supermassive black holes and tell us more about the power sources of active galactic nuclei.

A year or so after NUSTAR launches, scientists hope that a German X-ray telescope called **Extended Roentgen Survey with an Imaging Telescope Array** (EROSITA) will be launched on a Russian rocket. EROSITA will be the first mission to image the whole sky in the medium-energy X-ray range. One of the outcomes of this mission should be a catalog of all the feeding black holes in nearby galaxies, even those obscured from direct view, along with the detection of more than 170 000 new active galactic nuclei.

In 2014, the Japanese space agency hopes to start operating an X-ray satellite called **Astro-H**. The satellite will contain a number of instruments for studying both hard and soft X-rays. Another instrument that aims to study hard X-rays is the **Hard X-ray Modulation Telescope** (HXMT), whose four-year mission will begin when it is launched in 2015. After completing an initial all-sky survey, HXMT will focus on supermassive black holes.

Waiting in the wings, however, is an observatory that would provide an unprecedented view of the X-ray sky. At least, that is what X-ray astronomers were hoping for; whether it happens is uncertain. For ESA, the successor to XMM-Newton was planned to be an observatory called **X-ray Evolving Universe Spectroscopy** (XEUS). For NASA, the successor to Chandra was an observatory called **Constellation-X**. The hope was for both these missions to launch some time around 2017 or 2018. In 2008, however, ESA, NASA and JAXA (the Japan Aerospace Exploration Agency) agreed to study the possibility of a joint mission, a merger of XEUS and Constellation-X. The result was a proposal called the **International X-ray Observatory** (IXO).

The IXO proposal was for a densely packed shell of 15 000 thin glass mirror segments to form a Wolter telescope with an effective collecting area at certain energies of three square meters – much larger than Chandra and XMM-Newton combined. Indeed, mission scientists claimed that going from Chandra to IXO would be like going from the famous 5 m Hale telescope at Palomar to a telescope with a 22 m mirror. In addition to the telescope, the plan was for IXO to carry a host of other instruments that would allow astrophysicists to study black holes in unprecedented detail. Two high-resolution spectrometers, instruments sensitive enough to detect black hole driven winds, would probe conditions near supermassive black holes. A high timing resolution spectrometer would study how the X-ray output of accretion disks varies with time, thus providing information about the size of disks and how they evolve. A wide field imager and a hard X-ray imager would together scan large areas of the sky looking for the most energetic X-rays and thus those black holes that are currently obscured from view. Finally, an X-ray polarimeter would study the polarization of these most energetic photons, thus supplying detailed data about the nature of the twisted space-time around spinning black holes. Taken together, the various techniques employed by IXO would supply many of the missing pieces of the black hole puzzle. A mission such as IXO would tell astronomers not only how black holes form and grow, but also how they influence much larger structures in the Universe.

IXO was one of three candidate missions competing for launch in 2020 (the others being a space-based gravitational wave observatory, which is discussed in chapter 9, and a mission to Jupiter). These were to be joint missions between space agencies. In 2011, though, it became clear that NASA would be unable to fund any of the mission concepts and so ESA withdrew from the partnership. IXO will not fly.

The future for X-ray astronomy is not entirely bleak. Astronomers will still have NUSTAR and EROSITA. Furthermore, ESA have announced a new mission concept called the **Advanced Telescope for High Energy Astrophysics** (ATHENA), which will look at how to deliver some of the IXO science at a reduced cost. From Constellation-X and XEUS to IXO to ATHENA: the next generation of X-ray telescope has travelled a rocky road. But such an instrument opens up too many opportunities for it not to happen. One day.

Darkness visible

We can't, of course, see a black hole. But it's not just X-ray telescopes that will search for the *effects* that black holes have on their surroundings. For example, gamma-ray telescopes (which are the subject of chapter 8) can find hints of black holes feeding: in 2011, an orbiting gamma-ray telescope saw the flash of gamma-rays from a star being ripped apart as it fell into a supermassive black hole. Optical telescopes, working at much lower energies in the electromagnetic spectrum, will study the accretion disks around black holes. The new generation of extremely large optical telescopes (see chapter 12) will be able to resolve the disks around supermassive black holes in galaxies as far away as the Virgo cluster. Meanwhile, infrared telescopes (see chapter 11) will allow astronomers to see accretion disks that are otherwise blocked from view by dust clouds. And telescopes working at various wavelengths may be able to make use of the gravitational lensing properties of a black hole (gravitational lenses were discussed in chapter 4).

The extreme conditions close to the event horizon of a supermassive black hole can give rise to energetic particles as well as energetic photons. Cosmic ray telescopes (which are the focus of the following chapter) and neutrino telescopes (the chapter after next) may be able to probe the relativistic jets that emanate from black holes.

The digestive processes that occur when a black hole devours a star are likely to generate gravitational waves, as are the processes that lead to black hole formation in the first place. Gravitational wave observatories (which are the focus of chapter 9) are hoping to detect these signals.

6

The 'Oh my God' particles

Every so often a particle from space hit's Earth's atmosphere with an energy exceeding anything that a manmade accelerator can achieve. Where do these particles come from? And how do they get to be so energetic? These are two of the oldest unsolved questions in astronomy – but the new generation of cosmic ray detectors promises to provide the answers.

A cosmic rain – Power showers – If the 'Oh my God's' are bullets, what's the gun? – LOFAR so good – Catching bullets

S. Webb, *New Eyes on the Universe: Twelve Cosmic Mysteries and the Tools We Need to Solve Them*,
Springer Praxis Books, DOI 10.1007/978-1-4614-2194-8_6, © Springer Science+Business Media, LLC 2012

On the morning of 10 September 2008 the first proton beam travelled around the world's largest particle accelerator, the Large Hadron Collider (LHC). The LHC accelerates particles via a ring of superconducting magnets inside a circular tunnel that is 27 km in circumference. It was built so that physicists can smash together opposing beams of protons at high energies and then study the debris. One of the hopes for the LHC is that the Higgs boson, the so-called 'God particle', the missing piece in the standard model of particle physics, will be lurking in the fragments when the protons collide.

Figure 6.1 *A man stands in front of the* ATLAS *detector, one of six detectors attached to the Large Hadron Collider. (*ATLAS *Experiment,* CERN*)*

When this monster of a machine operates at its full capacity both of the proton beams will possess an energy of 7 TeV – about a millionth of a joule. Put like that, this may not seem worth getting excited about. One joule is the energy released when a small apple falls one meter to the ground; a *millionth* of a joule is tiny in our everyday world. But to put all that energy into subatomic particles … well, on that scale the energy is immense. On the subatomic scale energies are usually described in terms of electronVolts (eV): the energy of a photon of visible light is between 1.5–3.5 eV; a mol-

Figure 6.2 *The Fly's Eye telescope, an experiment led by George Cassiday, consisted of 67 modules in corrugated steel barrels located within the US Army Dugway Proving Ground in Utah. Each module contained a mirror and a dozen photomultiplier tubes that searched for flashes of light that might indicate the arrival of a cosmic ray. Each tube produced a view of a hexagonal patch of sky; the coverage of the sky into hexagonal sections is similar to the compound eye of a fly – hence the name of the experiment. (University of Utah Department of Physics and Astronomy)*

ecule of salt will dissociate into its sodium and chlorine ions if 4.2 eV of energy is supplied; when you receive a medical X-ray the photons involved possess an energy of about 200 000 eV. So to accelerate a subatomic particle to an energy of 7 TeV – that's 7 000 000 000 000 eV – requires engineering and scientific prowess of the highest order. That's why the construction of the LHC required funding by a group of countries acting in concert and the collaboration of more than ten thousand scientists, engineers and technicians.

It's easy to wax lyrical about the LHC, which is undoubtedly one of the greatest technological achievements of humankind, but it's always best to keep a sense of humility. Nature, it turns out, has particle accelerators that dwarf anything we can ever hope to construct. On the dark, moonless night

Figure 6.3 *A Fly's Eye module: a mirror and photomultiplier tubes were housed in a corrugated steel barrel. When the detector was not observing, the barrels were pointed down and away from sunlight. When observing, on clear moonless nights, they pointed upwards. (University of Utah Department of Physics and Astronomy)*

of 15 October 1991, an array of telescopes operated by the University of Utah recorded the blue flashes that result when a very high-energy cosmic ray smashes into Earth's atmosphere. That evening the telescope array, called the **Fly's Eye**, detected an incoming particle with an energy of 3.2×10^{20} eV. That's enough to run one of the energy-efficient lightbulbs in my house for about 3.5 seconds. Again this may not seem a great deal, but this energy was being carried by a *single subatomic particle*: the Fly's Eye saw a macroscopic amount of energy, an amount of energy we meet in everyday life, being carried by a particle that was tens of thousands of times smaller than an atom.

It's difficult to comprehend how a subatomic particle came to possess such an energy. Consider how fast it must have been moving. The particle detected by Fly's Eye that October night is generally assumed to have been a proton –

but it carried almost 50 million times as much energy as the most energetic protons the LHC will produce. If it was a proton it must have been travelling essentially at the speed of light – in fact, just one and a half femtometers per second slower than lightspeed. If the Fly's Eye proton and a photon were in a race then, after one year, the photon would be ahead by just 46 nm – about the size of a virus. It seems safe to say that engineers will never be able to accelerate protons to that speed, no matter how big they make their accelerators. As astrophysicists at the time declared, if the Higgs boson is the 'God particle' then the Fly's Eye proton is an 'Oh my God' particle.

The discovery of such an extraordinary entity was a shock because astrophysicists thought it could not exist. Finding the source of such ultra-high-energy cosmic rays is one of the key quests in modern astronomy. (As with many other entities in this book, an acronym is involved. However, the acronym for ultra-high-energy cosmic rays is UHECR. This is so horrible – it reminds me of my father's morning expectorations before he quit smoking – that I'll write it out in full if you don't mind.)

A cosmic rain

When astronomers learned in the 1930s that cosmic rays are affected by Earth's magnetic field they knew the rays must be electrically charged particles. What precisely, though, were these particles that rain down on us? Well, one early finding regarding cosmic rays was that they exhibit a wide variety of energies. Astronomers have found that at low energies about 90% of cosmic rays are protons; a further 9% are alpha particles, the nuclei of helium atoms; the remainder are electrons plus the nuclei of heavier elements.

What about the composition of cosmic rays at higher energies? That's more difficult to ascertain because the more energetic the particles the fewer of them there are. The general assumption, perhaps, is that they are mainly protons.

Even more vexing than the question of cosmic ray composition is the question of their source. One difficulty arises because the particles are electrically charged so their paths through space are twisted and bent by magnetic fields. The paths of low-energy cosmic rays are particularly affected. Nevertheless, it's known that some cosmic rays with small and moderate energies originate from the Sun. The source of cosmic rays possessing moderate to high energies is more difficult to discern but it's thought that many of them originate from supernovae that exploded in the Galaxy. The supernova explosion itself does not necessarily generate high-energy cosmic rays. Rather, as Enrico Fermi first proposed, charged particles such as protons get caught in the magnetic fields of the expanding gas clouds that result from the explosion. These particles can become trapped between reflecting magnetic fields, bouncing to and fro for thousands of years and receiving an energy boost with each bounce. Eventually they have so much speed that the supernova remnant cannot hold them. When they escape, they become cosmic rays. This so-called magnetic shock acceleration model can explain cosmic ray energies up to about 1000 TeV – far higher energies than the LHC can generate. What about the really high-energy particles, the ones with energies more than 1000 TeV? Neither the Sun nor supernova remnants can accelerate particles to these energies. Something else must be going on.

It seems likely that cosmic rays with energies up to about three million TeV originate within the Galaxy. Even higher-energy cosmic rays, however, probably originate from outside the Milky Way. The reason for this belief is that the highest-energy cosmic rays are moving so fast that their paths are hardly deflected by the Galaxy's magnetic field and so should point back to their origin. The fact that we don't see a preponderance of them arriving from the disk of the Milky Way or from its center suggests they have an extragalactic origin. However, when astrophysicists track back the path of ultra-high-energy cosmic rays they don't seem to find a possible source.

The puzzle gets even thornier.

In 1966, Kenneth Greisen and, independently, Georgiy Zatsepin and Vadim Kuzmin, argued that cosmic ray protons will interact with the photons in the cosmic microwave background. Any protons possessing an energy above about 5×10^{19} eV will interact with the photons and produce particles called pions; this process will continue until the proton energy drops below 5×10^{19} eV (the so-called GZK cutoff). Now, physicists understand this interaction extremely well and they can calculate how far on average a proton can travel before its energy drops below the cutoff value: it turns out that we should detect very few cosmic ray protons with an energy greater than the GZK cutoff from any source more distant than about 160 million light years. In effect, the GZK cutoff acts as a filter against cosmic rays from the distant Universe. Earth might be hit by cosmic rays with ultra-high-energies, but such particles should be extremely rare.

Well, astronomers *do* detect cosmic rays with energies greater than 5×10^{19} eV. A cosmic ray with an energy greater than the cutoff was observed as long ago as 1962, and since the Fly's Eye record-breaker in 1991 astronomers have seen more than a dozen comic rays with an energy in excess of the cutoff. That raises an obvious question: are we seeing more ultra-high-energy cosmic rays than we expect? And of course there is a related question: the GZK suppression suggests that these cosmic rays must come from the local neighborhood (astronomically speaking), but what sources within 160 million light years of Earth could possibly produce such ultra-energetic particles? No one knows for sure. Could it be that ultra-high-energy cosmic rays aren't protons after all (the GZK suppression would not be so pronounced if cosmic rays were massive atomic nuclei)? Or is something completely unexpected happening?

An ultra-high-energy cosmic ray is like a bullet. And just as detectives need to know certain pieces of information about any bullets they come across – what's it made of, how was it fired, where did it come from – so too astrophysicists need to understand cosmic rays. To gain that understanding, and solve several long-standing mysteries, they first need an efficient way to catch those cosmic bullets.

All the observatories described in this book study the Universe using electrically neutral messengers – photons, neutrinos, gravitational waves. All observatories, that is, with the exception of cosmic ray observatories: these observe the Universe using electrically charged particles. If astrophysicists can catch these cosmic bullets and discern their origin then they will have initiated a completely new type of astronomy: charged particle astronomy.

Power showers

Ground-based cosmic ray detectors don't observe cosmic rays as such. If you want to catch a cosmic ray particle *directly* you must place your detector above most of Earth's atmosphere, which lets you detect the particle before it has had chance to interact with air molecules. Balloon-borne experiments have a long history, of course, starting with Hess in 1912. Such experiments continue to this day and, as we shall at the end of the chapter, space-based cosmic ray detectors fly even higher. Interestingly, the vitreous humor in Apollo astronauts' eyeballs may have formed the first space-based cosmic ray detectors. Neil Armstrong and Buzz Aldrin both reported seeing flashes of white light, at the rate of about one every three minutes, even when they had their eyes closed. Experiments on subsequent Apollo flights led credence to the idea that the flashes were the result of cosmic rays passing through the mens' eyeballs. One possible mechanism for this is that the vitreous humor was acting as a Cerenkov detector, with the cosmic rays passing through the vitreous humor faster than light itself can travel through that medium. (Cerenkov detectors are described in more detail later.) Nevertheless, the detectors that astronomers can deploy on a balloon or a space-based experiment are necessarily smaller and less durable than can be built on Earth's surface. Most cosmic ray research takes place with ground-based detectors.

So how does a ground-based observatory detect a cosmic ray? Well, as implied above, a high-energy cosmic ray particle that enters Earth's atmosphere will soon collide with an atomic nucleus in an air molecule. This collision will produce a large number of secondary particles, which collectively share the energy of the primary cosmic ray. These secondary particles will then soon collide with other atomic nuclei in the atmosphere, creating tertiary particles, which in turn collide with further atomic nuclei . . . the resulting cascade of particles is called an **extensive air shower**. This air shower can contain billions of particles and, depending on the energy of the primary, by the time it hits Earth's surface it may cover an area as large a 16 square kilometers. It's this shower of particles that astrophysicists attempt to detect.

There's no way of knowing in advance when or where those air shower are going to occur, but it would help to know the *rate* at which showers are generated. That at least is something astrophysicists have a handle on; they know quite accurately how many cosmic rays of any particular energy hit Earth in any given interval. It turns out that the rain of low-energy cosmic rays – with energies of a few million electron volts – is, in the terminology of TV weather forecasters, persistent. About two hundred such cosmic ray particles hit every square meter of Earth *every second*. It was that steady, persistent rain that Hess discovered a century or so ago. The high-energy cosmic rays – more hailstones than rain – are not so common. On average, cosmic ray particles with energy greater than 10^{18} eV hit an area of one square kilometer each week; for energies above 10^{19} eV the arrival rate is only one particle per square kilometer per year. If a cosmic ray particle has an energy of 10^{20} eV it can be classed as an ultra-high-energy cosmic ray – and only one such particle strikes within a square kilometer every *century*.

Figure 6.4 *An artist's view of a cosmic ray hitting Earth's atmosphere above Uluru. When the cosmic ray strikes an atomic nucleus high in the atmosphere the collision triggers a shower of particles. By the time the shower reaches Earth's surface it can contain billions of particles. (Australian Nuclear Science and Technology Organisation)*

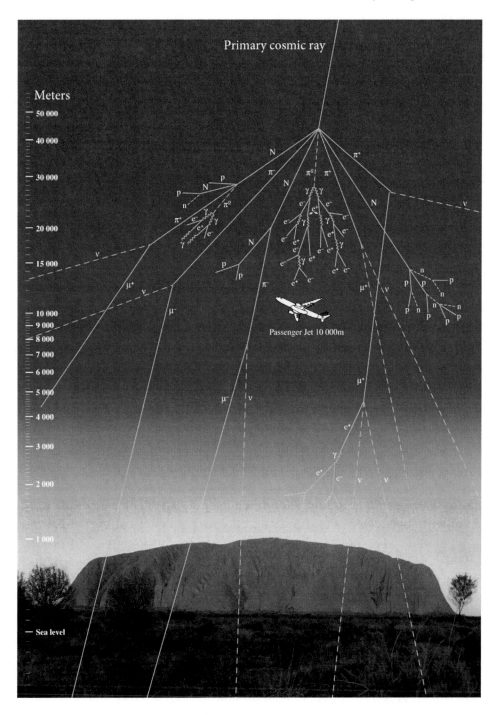

Primary cosmic ray

Passenger Jet 10 000m

Thus in order to detect a reasonable number of ultra-high-energy cosmic rays, and thereby try to learn something about their origin, astrophysicists must either run their experiments for a long time, much longer than the span of a typical scientific career, or else build detectors that cover a large area. Not surprisingly, astrophysicists have tended to pursue the latter option. The Fly's Eye was a typical example of the approach.

The Fly's Eye team designed and constructed a follow-up to their original experiment at the Dugway Proving Ground. Conditions at this site, incidentally, are ideal for cosmic ray studies: the desert floor lies more than a kilometer above sea level and it's a rare day there when there's precipitation. Furthermore, towns are few and far between and, in any case, light pollution is generally hidden by mountains. The team's follow-up telescope, the **HiRes Fly's Eye,** watched the skies from 1997 until 2006. The new instrument (actually two sets of telescopes separated by about 12 km) had a much better resolution and sensitivity than the original – hence the term 'hi-res'. As with the original, the HiRes Fly's Eye detected cosmic rays by making use of a process called **fluorescence.** When the charged particles in an air shower collide with molecules of atmospheric nitrogen, the nitrogen can be excited – raised to a higher energy level. This excitation energy can then be emitted in the form of a faint ultraviolet glow. The clusters of light sensors in a fluorescence telescope can measure this faint glow. At least, they can on a dark, clear, moonless night; they thus tend to operate on only a few days each month. To a human eye an air shower is invisible; to the Fly's Eye an air shower looks like a bright ultraviolet streak across the sky, increasing in brightness as it increases in size until it vanishes when it strikes the ground. By measuring the amount of fluorescent light, scientists can estimate the number of particles in the shower and thence the energy of the primary.

The HiRes Fly's Eye was not the only instrument looking for ultra-high-energy cosmic rays during that period. In Japan, for example, scientists constructed a ground-based detector called the **Akeno Giant Air Shower Array** (AGASA). Rather than use fluorescence, AGASA employed a different method. With really high-energy events, enough particles in the shower reach the ground for an array of scintillation detectors to catch them. (As we saw in chapter 3, several experiments are trying to detect dark matter using scintillation detectors; the underlying idea is the same.) The shower hits the ground in the form of a thin 'plate' of particles travelling at close to lightspeed; by measuring the arrival time of this plate at different detector stations astronomers can determine the direction of the primary cosmic ray.

In addition to surface detectors it's possible to employ underground particle detectors, which are useful for detecting muons; this combination of detectors allows astronomers to determine the composition of the primary cosmic ray as well as its direction. The AGASA instrument used 111 scintillation detectors on the surface and 27 particle detectors buried beneath Earth's surface, with everything connected by fiber optics. The entire array covered about 100 square kilometers. AGASA searched for ultra-high-energy cosmic rays between 1991 and 2004 and during that time it detected about a thousand particles with an energy greater than 10^{19} eV; on 3 December 1993 it detected a particle with the second-highest energy ever observed – 2×10^{20} eV (another OMG particle).

Between them the two observatories, AGASA and HiRes, produced results that generated a conundrum. Although AGASA was four times less sensitive than HiRes it detected 11 ultra-high-energy cosmic rays (in other words, 11 particles with an energy above 10^{20} eV). If the AGASA results were correct then the GZK cutoff did not exist. The astronomers on the HiRes team, however, calculated that if the GZK cutoff didn't exist then during the period 1997 to 2006 they should have observed 43 ultra-high-energy cosmic rays; they detected only 13. *That* result strongly suggested the GZK cutoff is real. The discrepancy is at the heart of the high-energy cosmic ray enigma: the AGASA result implies that ultra-high-energy cosmic rays come from relatively nearby sources while the HiRes result suggests that the sources are distant. In both cases, the precise nature of the sources is unknown.

The two results were mutually exclusive. So, which team was right? HiRes or AGASA? To decide between them, even bigger and better cosmic ray detectors are required. Fortunately, astronomers now have two such detectors.

The **Telescope Array** (TA) project is a cooperative venture between the HiRes and AGASA science teams, and as such is rather unusual: it's not often that competing groups join forces to resolve differences. In addition to the Japanese and American teams, institutions from China, Taiwan, Russia and South Korea are involved. As with HiRes, TA is located in the Utah desert, although the telescope had to be moved away from the Dugway Proving Ground: the increased security levels put in place after the events of 9/11 made it difficult to operate that site.

In an attempt to understand the differences between the HiRes and AGASA observations, TA uses both types of detector – fluorescence and ground-based scintillation – at the same site. The array of scintillation detectors is similar to AGASA, but on a much larger scale: the 576 surface detectors, set about 1.2 km apart from each other, cover an area nine times larger than AGASA. Observing the sky above the scintillation detectors are three fluorescence detectors, which are set about 30 km apart from each other and in the form of an equilateral triangle. The fluorescence detectors consist of 12 sets of mirrors, each with 16 segments, housed in a building with garage-type doors on the roof; the doors open when night sky is dark and clear – allowing the mirrors to collect the faint flashes of an air shower. Although Japan provided much of the funding for TA, the decision to combine ground-based detectors with fluorescence detectors ruled out Japan as a site for the project: the air in Japan is too humid for such detectors to work.

The TA project, which is the biggest cosmic ray detector in the northern hemisphere, began taking data in 2008. But by the time TA started operations a much larger detector in the southern hemisphere had already been working for four years.

The **Pierre Auger Observatory** (PAO) – one of the largest observatories in the world – was the brainchild of Alan Watson and the 1980 physics Nobel laureate James Cronin. In 1992, they proposed the construction of a cosmic ray detector covering an area greater than the size of Luxembourg. The hope was that an instrument covering such a large area might be able to detect about 30 ultra-high-energy cosmic rays each year, which would give scientists enough data to begin to pinpoint the sources of these particles and thus narrow the range of possible production mechanisms. It took several years for the hundreds of scientists (and 17 countries) involved in researching the detector to agree a design and venue. They settled on an Argentinian plain called Pampa Amarilla as the site and in 1999 a ground-breaking ceremony took place in the Mendoza province of Argentina, just east of the Andes. In 2008, more than nine years after construction began, the Pierre Auger Observatory was inaugurated. It's currently the world's most powerful instrument for studying cosmic rays. (Pierre Auger, incidentally, was a French nuclear physicist. It was Auger who, in 1939, discovered that extensive air showers are generated when high-energy cosmic rays collide with Earth's atmosphere.)

Figure 6.5 Top: one of the 1600 water-filled tanks that go to form the PAO. The solar panel on top of the tank produces power for the electronics; the communication mast houses a radio antenna and a GPS receiver. Bottom: One of the four PAO fluorescence detectors overlooking another of the tanks. (Pierre Auger Observatory)

As with the TA project, the PAO uses two different techniques to capture ultra-high-energy cosmic rays. The first uses an array of **Cerenkov radiation** detectors. These detectors are simply polyethylene tanks, each containing 12 000 liters of pure water. The tanks are closed and opaque to light, and so are totally dark inside, but the particles generated in an air shower can pass through the polyethylene. The key to this technique is that those fast-moving charged particles travel faster through the water than light itself can move in water. Whenever charged particles outpace light in a particular medium a shock wave is created, the optical equivalent of a sonic boom. This Cerenkov radiation has a characteristic deep blue hue, and it can be detected with relative ease; the PAO measures the radiation using photomultipliers mounted on the tanks. Background sources can also generate Cerenkov radiation, but this can be easily eliminated since background radiation is unlikely to cause simultaneous triggering of independent detectors. So the idea is simply to connect the tanks to a data center. If the data center registers simultaneous bursts of Cerenkov light in several tanks then you know there's an air shower taking place. Well, PAO has more than 'several' such water tanks: it has 1600 of them arranged in a regular grid covering about 3000 square kilometers of the Pampa.

By combining measurements of the amount of Cerenkov light detected, it's possible to estimate the energy of the primary cosmic ray particle that initiated the shower; the greater the energy of the primary the larger the number of detectors that will register the shower. And by noting slight differences in the detection times at various tank locations, it's possible to estimate the incoming trajectory of the primary. The estimates are not perfect, since only a fraction of the shower is sampled – the tanks see only particles that reach the ground. But the estimates can be improved when the data are compared with results from PAO's second method for studying cosmic rays: the familiar fluorescence technique. The PAO has four fluorescence detectors, each employing six telescopes, situated on hills around the periphery of the ground array. If two of the fluorescence detectors observe the same shower then it's possible to calculate with accuracy the direction of the incoming primary cosmic ray particle.

So the PAO looks for air showers in two ways: the vast array of water tanks, working all the time, detects lots of events while the smaller array of fluorescence detectors, working on a few dark nights each month, detects fewer events but with more accuracy. Since the PAO is much bigger than any other cosmic ray observatory it is perhaps best placed to solve the mystery of these high-energy bullets from space. And in a few years the observatory may become even more potent: scientists hope to build Auger North, a northern-hemisphere sister observatory in Colorado near the city of Lamar, which will cover an area on 10 000 square kilometers. That's about half the area of Wales. If this were built, there'd be all-sky coverage of those mysterious ultra-high-energy cosmic ray events.

If the 'Oh my God's' are bullets, what's the gun?

As already mentioned, one of the key questions in astrophysics is: what mechanism is Nature using to accelerate particles to the 'Oh my God' sort of energy? The list of proposed mechanisms is a long one. Perhaps the most favored explanation involves the supermassive black holes that, as we saw in the previous chapter, lie at the center of every active galactic nucleus – it's at least plausible that such black holes could provide the acceleration mechanism. Other astrophysicists have proposed that the shock waves generated when galaxies collide might be large enough to accelerate protons to the energies we occasionally see. The galaxies NGC 4038 and NGC 4039, for example, have been ploughing into each other for the past 100 million years; perhaps galaxies such as these are the culprits. Yet other astrophysicists suggest that a **magnetar** – a neutron star possessing an incredibly large magnetic field – would be capable of accelerating cosmic rays to the energies observed. Some cosmologists suggest that cosmic rays aren't *accelerated* at all. Perhaps, instead, cosmic rays are generated by the collapse of topological defects – hypothetical objects such as cosmic strings that would be a relic from the Big Bang. A few physicists even suggest that ultra-high-energy cosmic rays might be the result of completely new physics; perhaps the 'Oh my God' object came from the decay of a hypothetical supermassive particle known as a wimpzilla. (The wimpzilla, assuming it exists, is a dark matter candidate – as discussed in chapter 3. It's such a good name that one can only hope this is the correct explanation.)

Well, one would hope that an observatory that covers 3000 square kilometers would be able, finally, to put astrophysicists out of their misery and reveal the source of ultra-high-energy cosmic rays. Even before its inauguration the PAO was observing air showers; in the three years after it began taking data, the instruments observed about a million showers. Nearly all of the showers, of course, came from the incessant background rain of moderate-energy cosmic rays but some came from their ultra-high-energy counterparts. It turned out that 77 of the showers came from cosmic rays that had energies above 4×10^{19} eV – just under the GZK threshold. But only two of them came from cosmic rays that had energies above 10^{20} eV. These numbers led the Auger team to an interesting conclusion.

Recall that the AGASA and HiRes observations led to a controversy regarding the existence of the GZK cutoff. These results seemed to settle the controversy. If there were no suppression of cosmic rays with energies greater than the GZK limit then the PAO should have seen about thirty events with an energy greater than 10^{20} eV – significantly more than the pair it actually saw. This was clear evidence that HiRes was right and the GZK cutoff exists. This in turn lends weight to the idea that the cosmic ray sources are distributed throughout the Universe; the 'drag' felt by particles as they travel through the cosmic microwave background means that they cannot reach us from the most distant reaches of the cosmos with the ultra-high-energies they began with. The ultra-high-energy cosmic rays we *do* observe must come from within the GZK sphere, which as we have seen has a radius about 160 million light years.

The PAO team went on to analyze the 27 most energetic cosmic rays in the sample, those with energies exceeding 5.7×10^{19} eV. When a cosmic ray possesses such a high energy it is only slightly deflected by galactic and intergalactic magnetic fields, so by determining the direction at which it hit Earth's atmosphere it should be possible to track its path back to a source. In 2007, the PAO team revealed that 20 of these highest-energy particles could be traced back to parts of the sky lying within 3.1° of nearby active galactic nuclei. Four of the particles could be traced to within 3.1° of Centaurus A, which at a distance of 13.7 million light-years is the closest active galaxy. The Auger scientists argued that if the rays came randomly from all directions, only five or six would correlate with the known locations of the active galaxies. There was less than a 1% chance that the correlation between ultra-high-energy cosmic rays and active galactic nuclei was random rather than real.

So was the mystery solved at last? Were ultra-high-energy cosmic rays accelerated by active galactic nuclei in a mechanism that remained unknown but presumably somehow involved the central supermassive black hole?

Not everyone was convinced. Some critics argued that the active galaxies claimed to be associated with cosmic rays were really rather ordinary. If activity were measured on a scale of 1–10, these would be a 1 or a 2. How could such relatively feeble objects shoot out ultra-high-energy particles? Besides, weak active galaxies are so common that you might expect to find

Figure 6.6 *The active galaxy Centaurus A is the closest galaxy with a jet powered by a supermassive black hole. This photograph from Chandra clearly shows opposing jets of fast-moving particles (the one pointing to the upper-left is 13 000 light years long). The highest-energy X-rays are colored blue in this image; intermediate-energy X-rays are colored green; and low-energy X-rays are colored red. Could active galactic nuclei like this one be the source of cosmic rays? (NASA/CXC/R. Kraft et al)*

a few of them within 3° of any random direction in the sky. For the critics, then, the correlation found by PAO was likely to be nothing more than a coincidence. This suspicion was strengthened when it became clear that the HiRes team found no such correlation between cosmic rays and active galaxies in their data.

By early 2010, the PAO team had more ultra-high-energy events they could analyze. The correlation they had seen between the directions of cosmic rays and active galactic nuclei was much weaker: the directions were not completely random across the sky, but fewer than 40% of the events seemed to be associated with active galaxies. Furthermore, the new analysis of those ultra-high-energy events threw up a fresh problem: the showers seen by the PAO seem to indicate that the cosmic rays that initiate them are iron nuclei – rather than the protons everyone assumed them to be. This is a puzzle because not only are iron nuclei much less common than protons, they surely would be prone to disintegration in the violent processes that accelerate the rays. And there's another part to the puzzle: the HiRes results, admittedly based on a smaller number of events, suggest that ultra-high-energy cosmic rays not only come from random directions in the sky but they are, indeed, protons.

The mystery of those cosmic ray guns remains unsolved.

LOFAR so good

The cosmic ray puzzle is clearly a difficult nut to crack. Perhaps, then, other techniques should be employed to supplement the approach taken by the Pierre Auger Observatory and the Telescope Array project?

Recall that ground-based observatories such as PAO and the TA project do not detect primary cosmic rays directly. Rather, they detect electromagnetic radiation from a shower – visible light in the form of Cerenkov radiation, ultraviolet light in the form of fluorescence – which is used to reconstruct the event that initiated the shower. The electromagnetic spectrum is broad, however. Can astronomers 'see' ultra-high-energy cosmic rays using other forms of electromagnetic radiation?

It turns out they can. They can use radio waves.

As mentioned in the introductory chapter, Karl Jansky's prototype radio telescopes, back in the 1930s, received waves at the relatively low frequency of 20.5 MHz, corresponding to a wavelength of about 14.4 m. The subsequent development of radio astronomy generally neglected these low frequencies because they produced images that were often thousands of times less sharp than optical images of the same object. It didn't help, either, that Earth's ionosphere tends to disturb radio images at these frequencies.

There are ways around the first of these difficulties. For example, the resolution of a radio telescope gets better the larger it is. This fact led to the familiar giant radio dishes at sites such as Jodrell Bank in the UK or Arecibo in Puerto Rico, but even these monsters don't work well at low frequencies. However, as we shall see in more detail in chapter 13, it's possible to link widely separated radio telescopes together. If they are separated by, say, a few hundred kilometers then the effect is the same as having a single telescope with an aperture that's a few hundred kilometers across. This allows astronomers to take radio images of the sky that have a resolution comparable to visible images. Recently astronomers have pursued this idea further, but instead of linking together complicated radio telescopes they have linked together widely separated arrays of simple dipole antennae – these are just wires that work in the same way as FM receivers. Thus radio astronomers can construct a large, and therefore sensitive, telescope at relatively small cost. Indeed, most of the cost comes not from the telescope itself, but in the electronics and software that's required to combine the signals from the individual antennae so that the output of a single telescope can be emulated.

Overcoming the disturbances caused by the ionosphere is more tricky, however. The complicated structure of the ionosphere varies over time and, just as the atmosphere causes stars to twinkle, so the ionosphere causes jittering in radio images. Until recently, it was this particular difficulty that tended to dampen astronomers' enthusiasm for studying the Universe at low radio frequencies. But technology advances with the same inexorability as Omar Khayyam's moving finger. The astounding increase in available computing power means it is now possible to process images of wide areas of the sky on such short timescales that ionospheric jitter can be corrected for. Antenna design has improved, too, so that a low-frequency antenna can now monitor several different regions of the sky at the same time. These improvements in electronics, computing and software allow a radio telescope array with a baseline of a few hundred kilometers to operate productively at a frequency of down to about 10 MHz – an even lower frequency than Jansky worked at.

A new telescope called the **Low Frequency Radio Array** (LOFAR) employs all these technological developments in order to survey the far Universe at frequencies between about 10–240 MHz, corresponding to wavelengths of between about 1.5–30 m. This frequency range is one of the few windows of the electromagnetic spectrum still largely unexplored. LOFAR consists of an array of small antennae – 15 000 of them in the first phase of the project, 25 000 of them in the full design – collected in 36 stations spread out over an area in Drenthe, a province in the north-east of the Netherlands (it's primarily a Dutch telescope) and five stations in nearby parts of Germany. The signals are sent to the University of Gronigen, which possesses what is currently one of the fastest computers in the world, for analysis. For the first part of the project the telescopes were located within an area having a radius of about 100 km^2, but in 2011 a LOFAR station sited in Chilbolton in England became a full member of the array and stations have been funded in France

Figure 6.7   The plan is eventually for LOFAR to consist of stations all across Europe. Once complete, the resulting array will form a telescope that will allow astronomers to observe the low-frequency Universe with unprecedented accuracy and sensitivity.   (Emde-Grafik)

and Sweden. Further stations are planned in Poland, Austria and Italy. From being a Dutch telescope, the project is turning into a wider European telescope – E-LOFAR. See figure 6.7. As LOFAR grows, it will become increasingly effective.

A telescope such as LOFAR can be pressed into action on many fronts; we shall see one application in chapter 13. What concerns us here, however, is how it can help solve the mystery of ultra-high-energy cosmic rays.

When a primary cosmic ray smashes into the atmosphere to produce an extensive air shower, many of the secondary particles quickly create electron–positron pairs. As these electrons and positrons move down through the atmosphere they get deflected by Earth's magnetic field. And whenever an electron (or a positron) changes direction in this way it emits synchrotron radiation. The great speeds involved with ultra-high-energy cosmic rays mean that this radiation will have a frequency of less than 200 MHz. In other words, LOFAR should be able to detect the synchrotron radiation associated with an air shower. By measuring the arrival times of the radio pulse at the different antennae, along with the intensity of the pulse, it should be possible to determine the direction of the primary particle with great accuracy. The technique is known to work: in 2005, a test instrument for LOFAR called the **LOFAR Prototype Station** (LOPES), based in Karlsruhe in Germany, detected

Figure 6.8 *Some of the most interesting astronomical observations over the next few years will come not from space-based telescopes but from nondescript grassland, such as this muddy field in the Netherlands! The* LOFAR *telescope consists of a European-wide collection of such fields, each containing an arrays of dipole antennae. Because the design is so simple, and because there are no moving parts, the costs of construction and subsequent maintenance are low. Most of the cost involves the electronics and software that's needed to bring the signals from the antennae to a central computer for analysis. (Hanny van Arkel – www.hannysvoorwerp.com)*

radio pulses that were coincident in both time and direction with extensive air showers. A further benefit of the LOFAR approach is that, unlike fluorescence detectors, a radio telescope does not require dark skies; rather than working for just a few days each month, LOFAR can work essentially nonstop.

Calculations by the LOFAR team suggest that the synchrotron radiation pulse coming from a cosmic ray having an energy in the range 10^{17} eV to 10^{19} eV will be bright enough for a single antenna to pick up. Various trigger mechanisms would ensure that signals from surrounding antennae are saved and analyzed. The *really* high-energy events, however, occur so rarely that LOFAR is unlikely to see one. That is, LOFAR is unlikely to see any ultra-high-energy cosmic rays smashing into a few hundred square kilometers of Earth's atmosphere. However, it *is* likely to be able to see the effects of ultra-high-energy cosmic rays smashing into the ten million square kilometers of the Moon's apparent surface. When an ultra-high-energy cosmic ray hits the Moon's regolith – relatively uniform, dense soil – a flash of bright, 300 MHz radiation would be emitted. And LOFAR would be able to detect those flashes. Using the Moon in this way creates an observatory with a vast surface area. With luck, LOFAR will help solve the cosmic ray mystery.

Catching bullets

Observatories such as PAO, TA and LOFAR are ground-based. They work by looking upwards, trying to observe the debris left behind when cosmic rays smash into something. Are there any alternatives to this? Well, astronomers could attempt to get above the atmosphere and look down on the debris. They could even try to catch cosmic rays directly.

The first dedicated space-based cosmic ray experiment is the **Payload for Antimatter Matter Exploration and Light-nuclei Astrophysics** (PAMELA). This instrument, which was developed by an Italian-led collaboration, was launched in 2006. It aims to study cosmic rays directly, particularly those that consist of antimatter (in the form of positrons and antiprotons). One of the reasons behind searching for positrons is that they may come from the annihilation of WIMP dark matter (see chapter 3). PAMELA has indeed seen large numbers of positrons – a finding that may be explained by dark matter (though of course there may be other explanations for the observations). Although the experiment is insensitive to high-energy cosmic rays, let alone the ultra-high-energy monsters we have discussed above, results from PAMELA are of interest for cosmic ray researchers: the more knowledge that can be gleaned about cosmic rays in general, the more that can be inferred about the nature those 'Oh my Gods'.

The **Alpha Magnetic Spectrometer** (AMS-02) experiment, which in 2011 was placed on the **International Space Station** by the penultimate mission in the Space Shuttle program, has similar goals to PAMELA. The experiment is the brainchild of Sam Ting, a particle physicist who won the 1976 Nobel prize. The advanced particle detectors on AMS-02 are studying the flux of cosmic rays at low to medium energies and for nuclei ranging from protons up to iron. The detectors are currently capturing cosmic rays at the rate of

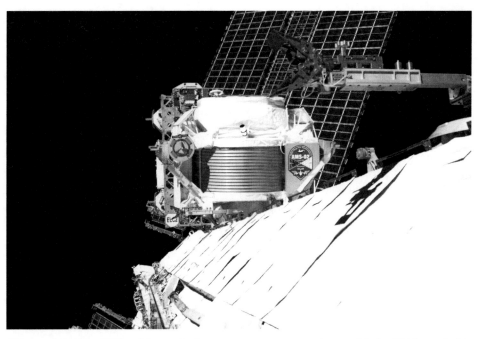

Figure 6.9 *The Alpha Magnetic Spectrometer was put into orbit in 2011, and this small detector is capturing cosmic rays at a prodigious rate – tens of millions of them every day. By analyzing so many particles, the experiment should illuminate some of the puzzles surrounding cosmic rays. (NASA)*

about 45 million per day. As with PAMELA, the hope is that the experiment will shed light on dark matter – but it will also help astronomers understand in detail the nature of cosmic rays.

Orbiting gamma-ray telescopes (which are the focus of chapter 8) can also be used to investigate cosmic rays. When cosmic rays smash into interstellar gas clouds they can produce gamma rays that may be detected. An analysis of such gamma rays has already backed up Fermi's long-standing proposal that supernova remnants can act as vast particle accelerators.

And, in a nice twist, it's possible for a space-based cosmic ray observatory to perform by looking *down* at Earth. The idea would be to use all of Earth's atmosphere as a cosmic ray detector: a satellite with a suitably large field of view would be able to look down at the extensive air shower created by a high-energy particle barrelling into air molecules. A satellite could track the fluorescence streak from a shower as it happened, and study precisely how the shower developed. A suitably equipped satellite could effectively be hundreds of times 'bigger' even than the Pierre Auger Observatory. Such a satellite was investigated by ESA; it was called the **Extreme Universe Space Observatory** (EUSO). Mission studies were discontinued by ESA in 2004, but Japanese scientists carried on with the design. The observatory is now called JEM-EUSO (with JEM standing for 'Japanese Experiment Module', a science module on board the International Space Station, which is where the observatory would be based). The hope is for a 2013 launch for JEM-EUSO.

In the coming years, then, results from LOFAR will be combined with results from PAO, TA, PAMELA, AMS-02 and JEM-EUSO. When that is done surely astrophysicists will be able to answer one of the oldest questions in their science: where the heck do those 'Oh my God' particles get their energy?

7

Deep sea, deep snow ... deep space

We know that ultra-high-energy protons can travel on average only about 160 million light years before interacting with the cosmic microwave background and losing energy. Ultra-high-energy photons can't even travel that far before being absorbed by the background. Does that mean it's impossible to investigate the distant, high-energy Universe? Fortunately not. There's another messenger that can bring information on distant high-energy sources: the neutrino.

Just passing through – Looking inside the Sun – Looking inside a supernova – Nets for catching ghosts – Improving the view

S. Webb, *New Eyes on the Universe: Twelve Cosmic Mysteries and the Tools We Need to Solve Them,* 159
Springer Praxis Books, DOI 10.1007/978-1-4614-2194-8_7, © Springer Science+Business Media, LLC 2012

A question often asked is: what's the most beautiful image ever captured by a telescope? Any answer will be subjective, of course, but for my money the award goes to Hubble's Ultra Deep Field. Between September 2003 and January 2004, the Hubble Space Telescope looked at a region of sky in the direction of the constellation Fornax. It was a small region of sky – about the size of the letter x here if you hold the book at arm's length – but that small patch of the heavens contains more than ten thousand galaxies, some of them nearby, most of them at an almost unimaginable distance. It's a truly beautiful image.

A question less often asked is: what's the most boring view of the sky captured by a telescope? The answer this time is more straightforward. It has to be the view afforded by a neutrino telescope. With a neutrino telescope you can detect the Sun. Oh, and in 1987 three neutrino telescopes detected a supernova, a source that quickly dimmed. Much less spectacular.

That vista is set to change, however. The neutrino, a ghostly particle that's not too far removed from a definition of 'nothingness', may provide astronomers with their only way of directly investigating the high-energy Universe beyond the immediate cosmic neighborhood – and their only way of peering directly into hot, dense regions that are otherwise hidden from view: the central regions of stars, supernovae and active galactic nuclei.

Just passing through

The neutrino is a type of elementary particle – a *fundamental* particle, in the same way as the electron is fundamental. Its existence was first postulated by Wolfgang Pauli in 1930. He introduced it to solve a puzzle, namely the peculiar way in which a neutron decayed into a proton and electron. Physicists had studied the phenomenon in detail and it seemed that neither energy, momentum nor angular momentum were necessarily conserved when a neutron decayed. And yet the conservation laws – of energy, of momentum and of angular momentum – were a cornerstone of physics. Pauli suggested that another particle, hitherto undetected, must be involved in the neutron decay process. That mysterious particle would carry away the observed difference between the energy, momentum and angular momentum of the initial and final particles.

It was clear from the start that neutrinos would be difficult to detect: they would have a zero or minuscule mass and thus travel at or close to the speed of light; and they would have zero electric charge and thus pass through

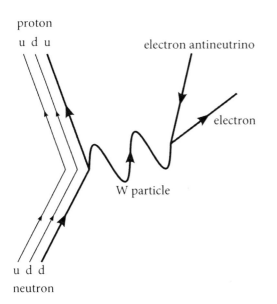

proton

u d u

electron antineutrino

electron

W particle

u d d

neutron

Figure 7.1 A neutron consists of three fundamental particles called quarks: to be precise, two d quarks and one u quark. One of the d quarks can turn into a u quark by emitting a particle called a W boson. The three quarks that remain – two u quarks and one d quark – constitute a proton. The W then almost immediately decays into an electron plus an electron antineutrino. Thus the net result is that a neutron has decayed into a proton, an electron and an electron antineutrino.

matter undisturbed. It turned out that they interact so weakly with matter that a neutrino could typically shoot through several million kilometers of lead without being stopped. In the time it has taken you to read this paragraph, trillions of neutrinos emanating from nuclear reactions taking place in the heart of the Sun will have passed straight through you; your body is unlikely to have stopped any of them. Indeed, over the course of your entire life, on average perhaps only one neutrino will interact with the atoms in your body. Detecting neutrinos was always going to be a tough job.

In 1956, more than a quarter of a century after Pauli first postulated the particle, Clyde Cowan and Frederick Reines confirmed the neutrino's existence in experiment. (To be pedantic, they in fact detected the antineutrino – the neutrino's antiparticle.) The neutrino was, in the words of Reines, 'the most tiny quantity of reality ever imagined by a human being'. Nevertheless, the two physicists reasoned that they could detect these particles near a nuclear reactor. Neutrinos are created in many nuclear reactions, both fusion processes (where light nuclei fuse together to form heavier nuclei, as takes place in stars) and fission processes (where massive nuclei break down into less massive nuclei, as takes place in a reactor). Thus a nuclear reactor is a neutrino production facility: the reactions taking place therein shoot out 50 trillion neutrinos per square centimeter *every second*. With so many of them around, Reines and Cowan believed it should be possible to stop a few:

place water-filled tanks around the reactor and then surely, once in a while, a 'tiny quantity of reality' would strike one of the protons in the water.

If a neutrino does interact with a proton, one of two things can happen. Either the neutrino carries on as a neutrino after the collision, which is a so-called 'neutral current interaction', or else it converts into an electron at the same time that some other conversion takes place in order to ensure conservation of electric charge, in which case the collision is called a 'charge current interaction'. The experiment devised by Cowan and Reines involved a charge current interaction: if an incoming antineutrino came sufficiently close to a proton then the proton would convert into a neutron with the emission of a positron (an antielectron). The two newly created particles would not long survive: the positron would quickly meet an electron and annihilate into two gamma rays, while the neutron would soon be captured by an atomic nucleus, with the release of a gamma ray. The coincidence of these two events – annihilation of a positron and capture of a neutron – gives a unique signature of a neutrino interaction. (Well, an antineutrino interaction.) Cowan and Reines were able to detect gamma rays in the water tanks by using scintillators – material that emits light flashes in response to gamma radiation. In turn, that light was detected by photomultiplier tubes.

Cowan and Reines ran their experiment for months and found that their setup detected about three neutrino interaction events each hour. This was undoubtedly an experiment worthy of the Nobel prize, and sure enough the prize was given. Unfortunately, the award was delayed. For his discovery of the neutrino, Reines was awarded the Nobel prize in 1995; Cowan, though, had died in 1974.

In 1962, Leon Lederman, Melvin Schwartz and Jack Steinberger demonstrated the existence of a new type or 'flavor' of neutrino – the muon neutrino. The **muon**, a fundamental particle, is essentially a more massive version of the electron. And just as the 'ordinary' neutrino is a fundamental partner to the electron so the muon neutrino is the equivalent partner to the muon. Thus in a charge current interaction involving the muon neutrino, a muon is created – in the same way that an electron neutrino gives rise to an electron. It was this property that let Lederman, Schwartz and Steinberger to discover the muon neutrino. The three men shared the 1998 Nobel prize.

And in 2000, scientists at Fermilab announced the first direct evidence for the tau neutrino. This third flavor of neutrino is the partner to the **tau** particle which, like the muon, is a more massive version of the electron. The tau is about 17 times more massive than the muon, and 3477 times more massive than the electron. The tau neutrino is another of the fundamental entities in the standard model of particle physics. As with the other two flavors of neutrino, it was the existence of the charge current interaction – in which the tau neutrino gives rise to a tau – that enabled the discovery to take place. In due course, those involved may well receive the Nobel prize.

Looking inside the Sun

In the early 1960s, the astrophysicist John Bahcall used a standard model of solar physics to calculate how many neutrinos must be emitted by the nuclear fusion processes taking place in the Sun's core. The Sun shines by 'burning' hydrogen: four hydrogen nuclei are burned to form a single helium nucleus, and in the process energy is emitted. This reaction also emits two positrons and two neutrinos. Since a vast number of such reactions are taking place, it turns out that a vast number of neutrinos stream out into space – about 10^{38} of them every second. Knowing how many neutrinos are being emitted by the Sun, Bahcall was able to calculate how many neutrinos any particular experimental setup should detect. His calculations made clear the scale of the challenge: although a hundred billion solar neutrinos pass through your thumbnail every second, a detector the size of an Olympic swimming pool would stop only a few neutrinos every week.

Raymond Davis Jr, was sure he could meet the challenge and test Bahcall's prediction. Davis set up an experiment using a 100 000 gallon tank and filled it with perchloroethylene, a substance rich in chlorine and commonly used as a dry-cleaning fluid. He reasoned that every time a neutrino hit a chlorine nucleus in the fluid, the chlorine atom would transform into a radioactive isotope of argon. By collecting the argon atoms every month or so he would know how many neutrinos his tank had captured. It's worth noting at this point that Davis built his detector more than a kilometer underground in the Homestake gold mine in South Dakota. His reasons for doing so, of course, were the same reasons that lead dark matter researchers to construct their observatories underground. He had to protect the chlorine target from cosmic rays and other radiation that might contaminate his results.

Davis got the first experimental result in 1968. He then ran the experiment continuously for almost a quarter of a century, from 1970 to 1994. Time and again, whenever Davis counted the argon atoms he found that his setup had detected only one third of the neutrinos that Bahcall's calculations suggested should be found. Bahcall rechecked his calculations and found no errors; other theoreticians performed similar calculations and reached the same conclusions as Bahcall; Davis checked and rechecked his experiment and could find nothing wrong. The discrepancy between the predicted rate of neutrino detection and the measured rate of detection became known as the solar neutrino problem.

One proposed solution to the problem was that the solar model used by Bahcall was wrong. Perhaps, some astronomers suggested, the Sun had 'shut down' – and the number of nuclear reactions taking place in our star's core had reduced. If that were the case then the number of neutrinos streaming out from the Sun would be reduced. We would not necessarily notice such go-slow behavior on the part of the Sun: it takes many thousands of years for heat energy to move from the Sun's core to the surface. This solution turned out to be wrong, but the mere fact that such a suggestion could be made implied something important: *a neutrino telescope provides a way to look deep inside the Sun.*

The correct solution to the solar neutrino problem turned out to reveal something important about fundamental physics: the three flavors of neutrino have mass; not much mass, but a non-zero mass nonetheless. And because they possess mass, the three different neutrino types can oscillate between each other and, as it were, 'change personalities'. An electron neutrino produced in a solar nuclear reaction can, en route to Earth, turn into a muon neutrino and the muon neutrino can turn into a tau neutrino. The Davis experiment was sensitive only to electron neutrinos, and so could detect only one third of the neutrinos produced in the Sun's core. The **Sudbury Neutrino Observatory** (sno), located more than two kilometers underground in a nickel mine in Ontario, was developed so that it would be sensitive to all three neutrino flavors. In 2001, the sno team published the first clear evidence that solar neutrinos do in fact oscillate.

The painstaking work that Davis carried out laid the foundations for neutrino astronomy. For this reason he shared the award of the 2002 Nobel prize in physics. At 88, he was the oldest ever recipient of the prize.

Looking inside a supernova

The Mozumi zinc mine, near the city of Hida in Japan, is the underground host of the **Kamioka Observatory**. Kamioka has a long history of hosting important experiments in particle physics, many of them involving neutrinos. Consider, for example, the Kamiokande-II experiment.

Figure 7.2 *This photograph, taken in 1966, shows the construction of the tank used in Ray Davis' experiment in the Homestake gold mine. The tank, which was about 6 m in diameter and about 14.5 m long, was located in a cavern about 1.5 km underground. Davis Cavern, as it is now known, is currently being used in dark matter searches; see figure 3.6. (Brookhaven National Laboratory)*

Kamiokande-II was a neutrino telescope designed to check the results of the Davis solar neutrino experiment. Like the Davis experiment, a great deal of rock helped to shield the Kamiokande-II detector from cosmic rays: the detector was 1 km underground. Unlike the Davis experiment, which used a radiochemical technique to determine the presence of neutrinos, Kamiokande-II used a Cerenkov detector. The idea was that if two quite unrelated techniques measured the same number of neutrinos then clearly the Davis result would stand.

The Kamiokande-II detector consisted of 3000 tons of extremely pure water. Occasionally – *very* occasionally – a neutrino would collide with the nucleus of an oxygen atom in the water and in doing so convert into a muon.

Two things ensure that such a muon is relatively easy to spot. First, muons are electrically charged. Second, any muon produced in such a collision will be moving through the water at close to the speed of light; in fact, it will be moving faster than light itself can move through water. And as discussed in the previous chapter, an electrically charged particle moving through a substance faster than light itself can travel in that substance generates the optical equivalent of a sonic boom – Cerenkov radiation. The characteristic blueish–purplish glow of Cerenkov radiation is rather feeble, but is readily observable in a transparent medium such as water. By using photomultipliers to detect the weak Cerenkov light, the Kamiokande-II scientists could track the path of the muon through the water; and since the direction of travel of the muon would be pretty much the same as the neutrino that created it (any deviation would be half a degree or less) they could determine the direction of the incoming neutrino.

Well, Kamiokande-II started detailed testing in 1985 and scientists expended a great deal of effort ensuring that particles produced from various sources of background radiation would not contaminate the results. In 1987 Kamiokande-II began taking data in earnest. It quickly became apparent that Kamiokande-II was detecting more neutrinos coming from the Sun's direction compared to other directions in the sky. This information on the *direction* of neutrinos showed for the first time that true neutrino astronomy was possible. Furthermore, it was the first *direct* demonstration that the Sun is a source of neutrinos – though in a sense that was just a confirmation of the obvious.

Once a few years' worth of data had been compiled it became clear that Kamiokande-II was seeing about half of the neutrinos that the standard solar model predicted. This was at first glance in conflict with the Davis experiment – but the same resolution of solar neutrino problem neatly explained both the Davis and Kamiokande-II results: the Davis experiment was completely insensitive to muon and tau neutrinos whereas Kamiokande-II had some sensitivity to them. Thus Davis saw only one third of the solar neutrinos while Kamiokande-II saw a few more.

These were important results obtained at the Kamioka Observatory. Another key result was the first evidence in 1998 that neutrino flavor could oscillate, which was three years before SNO showed explicitly that solar neutrinos oscillate. Perhaps the most interesting observation at Kamioka, however, came on 23 February 1987 – soon after Kamiokande-II had started taking data.

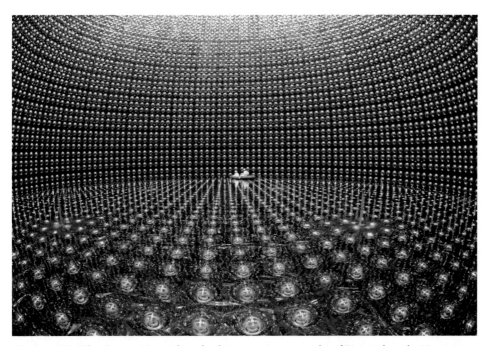

Figure 7.3 *The Super-Kamiokande detector, an upgrade of Kamiokande-II, consists of 50 000 tons of pure water surrounded by 12 200 photomultiplier tubes. The structure is 41.4 m tall and 39.3 m wide. Super-Kamiokande started data collection in 1996. Note the two men in the dinghy at the edge of the detector. (Kamioka Observatory, ICRR (Institute for Cosmic Ray Research), The University of Tokyo)*

Astronomers in February 1987 were ecstatic because Sanduleak −69° 202a, a star in the neighboring Large Magellanic Cloud, underwent a supernova explosion. It was the closest supernova since 1604: SN1987A, as it was named, was just 160 000 light years away. For the very first time, astronomers could observe a nearby supernova using telescopes. This particular supernova occurred because the core of the progenitor star collapsed. Now, in the process of core collapse lots of neutrinos are emitted. In fact, although a core-collapse supernova looks incredibly bright, most of the energy is radiated away by neutrinos rather than photons. It is estimated that SN1987A must have generated 10^{58} neutrinos – a vast number. The directors of neutrino telescope observatories immediately asked themselves whether their detectors might have picked up any of those neutrinos.

Well, the Kamiokande-II detector saw 11 neutrinos from SN1987A. An interesting finding was that the neutrinos reached Earth before light from the explosion. (The neutrinos didn't move faster then light, despite claims in late 2011 by CERN's OPERA experiment that such behavior might be possible. Rather, it's just that they fled the scene of the collapse immediately, whereas photons from deep inside the star's core took hours to struggle to the surface: the neutrinos had a head start.) Kamiokande-II wasn't the only detector to register the explosion. The Irvine–Michigan–Brookhaven detector, which was deep underground in a salt mine on the shore of Lake Erie, saw 8 neutrinos; the Baksan Neutrino Observatory, which was buried beneath 3.5 kilometers of rock in the Caucasus mountains, spotted 5 neutrinos. Thus astronomers found 24 neutrinos in total. Compared to the 10^{58} neutrinos that SN1987A generated, 24 events might seem like a paltry sum. Nevertheless, it is a staggering thought that astronomers here on Earth could detect particles that came from the heart of an exploding star. Not only could neutrino telescopes look inside the Sun, they could look inside a supernova too. For the detection of these astrophysical neutrinos, and the various other discoveries made by Kamiokande-II, the director of the experiment, Masatoshi Koshiba, shared the 2002 Nobel prize with Davis.

Neutrino telescopes thus have huge potential. They can reach parts of the Universe other telescopes can't. Even so, only two neutrino sources have been found in space: the Sun and SN1987A. Are there other sources to study?

Solar neutrinos and supernova neutrinos are produced in nuclear fusion reactions. These reactions have a typical energy scale associated with them and, compared to certain other astrophysical processes, that energy scale is not particularly large. Thus the neutrinos found by the Davis experiment, Kamiokande II and similar detectors are all low-energy neutrinos. However, some astrophysical processes take place at much higher energy scales. We know that active galactic nuclei, for example, emit photons with extremely high energies and they could also be the source of ultra-high-energy cosmic rays. And as we shall see in the next chapter, a phenomenon known as a gamma-ray burst is the most powerful event in the Universe. It's possible that active galactic nuclei, gamma-ray bursts – or some other objects – could generate high-energy neutrinos. And these neutrinos would be unaffected by magnetic fields; unlike cosmic rays the neutrinos, if they could be detected, would point straight back to their source. They would give astronomers a new window on the Universe, a new way of studying some of the most energetic processes taking place in the Universe.

But detecting low-energy neutrinos from the nearby Sun is hard enough. Is it possible to detect high-energy ghosts from the distant cosmos?

Nets for catching ghosts

Imagine a high-energy neutrino from the deep cosmos colliding with an atom here on Earth. How could astronomers detect it?

Well, the neutrino collision could generate a high-energy electrically charged particle – electron, muon or tau, depending upon the neutrino flavor. If the charged particle were to move faster than light does in some medium (water, for example) then astronomers could study the event by detecting the resulting Cerenkov radiation. In practice, muons penetrate matter much more easily than electrons or tau particles and so muons would produce the longest tracks in a Cerenkov detector. Thus such a detector is likely to be more sensitive to muon neutrinos than the other two flavors.

The problem with constructing a Cerenkov detector for neutrinos on Earth's surface is that it would be swamped with muons: there are vastly more muons produced in cosmic ray collisions than are produced in high-energy neutrino collisions. "No problem," you say. "Build the detector at the bottom of a mine. A kilometer of rock will shield against most of the unwanted muons." Well, that was the logic for building Kamiokande-II, the Irvine–Michigan–Brookhaven detector and the Baksan Neutrino Observatory deep underground. But even that much rock doesn't shield against *all* unwanted muons; ten kilometers of rock would be better, a hundred better still. Carrying that logic to its extreme, the best shielding for a high-energy neutrino detector would be the bulk of the Earth itself. Suppose scientists built a neutrino detector at the South Pole, say, and used Cerenkov radiation to distinguish between those muons travelling 'up' and those travelling 'down'. They would know with near certainty that a muon travelling 'up' had been produced by a high-energy neutrino colliding within a few kilometers of the detector: only neutrinos can penetrate Earth's bulk. Even with this setup many of the detected neutrinos would be generated when high-energy cosmic rays hit the far side of the Earth – so-called atmospheric neutrinos. Nevertheless, there would be some neutrinos that came from the depths of space. The key would be to study both the energy and the arrival directions of the neutrinos; a statistical analysis would determine which neutrinos had their origin in space. Atmospheric neutrinos would arrive randomly from all directions underneath the detector. An excess of neutrinos from a particular region of the sky would be the sign of an extraterrestrial source.

A neutrino detector at the South Pole would be a strange telescope indeed. It would be looking 'down' not up; it's 'dark sky' would be beneath it not above. But it would be a great way of catching such ghostly particles. That's the idea behind the **IceCube Neutrino Observatory**, a neutrino telescope built by a collaboration of American, German and Swedish institutions.

Few people have to endure the inhospitable working conditions suffered by the scientists and engineers who developed IceCube. Simply getting to the project was an adventure: a commercial flight to Auckland in New Zealand, followed by a transfer to Christchurch; from Christchurch, via military plane, to the McMurdo station on the Antarctic coast; and finally a flight to the South Pole. Once there, the environment facing them was harsh in the extreme. The Amundsen–Scott South Pole station is almost three kilometers above sea level, so altitude sickness can be a problem until the body becomes acclimatized. The cold, of course, is intense: during winter the average temperature is −65 °C, although it has dropped as low as −82.8 °C. Finally, the work took place on the windiest and driest place on Earth. Nevertheless, the science that IceCube permits more than counterbalances the difficulties faced in constructing it.

Work began in 2004 and was completed in 2010. The construction process, which could only take place during the Antarctic summer, consisted of blasting scalding hot water into the ice at a pressure of 68 atmospheres. This 'hot-water drill' was used to melt 80 vertical holes evenly spaced over a square kilometer, so that 125 m separated one hole from its nearest neighbor. Each hole was 2.45 km deep. While the hot water was still circulating in the hole, a 1 km-long string was lowered into it until the bottom of the string reached the bottom of the hole. See figure 7.5. Attached to each string, like pearls on a necklace, were 60 optical sensors. These sensors, each one separated from its neighbor by 17 m, are so exquisitely sensitive that they will respond to just a single photon.

Any photons that the sensors detect are amplified a hundred million times and translated into a digital electrical signal, which is sent to the central data acquisition system in the station above. Each sensor also has its own ultra-precise clock, which can pinpoint the arrival time of a photon to within 5 nanoseconds. Once the string is in place, the ice refreezes around it – and there it will stay until the ice has migrated to Antarctica's coast, whereupon the string will fall into the ocean. Unless global warming hastens this process, the strings will be in ice for about 25 000 years. The detector itself, though, is likely to last a much shorter time – about 25 years, or so.

Now that the construction work is complete, IceCube is the biggest particle detector in the world: its array of sensors covers a cubic kilometer. And

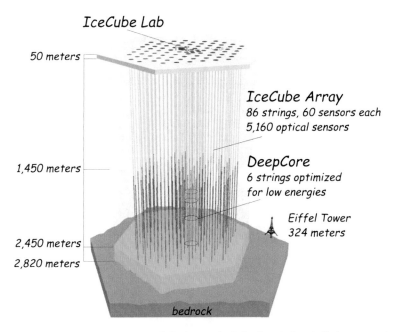

Figure 7.4 *A schematic diagram of the IceCube lab. (IceCube Collaboration/*NSF*)*

although it is in an uninviting location, IceCube is in almost the perfect place for a neutrino detector. First, the detector material needs to be transparent in order for the Cerenkov effect to work. Well, the subterranean Antarctic ice is in its purest form here, so distortions from bubbles and air pockets are minimal. Furthermore, if one hopes to detect a reasonable number of neutrino collisions then one needs lots of material. Well, IceCube has a volume of 1 teraliter (1 000 000 000 000 liters) – *much* larger than previous detectors. It turns out that the effective size of the detector is even larger: at the high particle energies that IceCube hopes to study (it will be sensitive to neutrinos in the energy range 10^{11} eV to 10^{21} eV) a muon can travel up to 15 km before being absorbed. It follows that IceCube can detect a neutrino even if it collided kilometers away from the sensor itself. IceCube should be able to detect a hundred high-energy neutrinos every day. An upgrade to IceCube called DeepCore will allow the detection of lower-energy neutrinos, of the sort that supernovae will generate. Finally, the 1.45 km of ice covering the detector does a fair job of shielding the sensors from naturally occurring radiation while, as already mentioned, Earth's bulk acts as a shield against upward-moving muons from other sources. Icetop, which is a detector on

Figure 7.5 *The photograph shows a digital optical module (DOM) – a sophisticated optical sensor – being lowered into one of the holes that the IceCube team drilled into Antarctic ice. Now that IceCube is complete, there are 5160 DOMs buried in the ice, patiently looking for flashes of light that might signal the arrival of a high-energy neutrino. (IceCube Collaboration/ NSF)*

the Antarctic surface, looks for air showers generated by cosmic rays; this helps IceCube distinguish between events that might cause the emission of Cerenkov radiation.

Even before engineers completed the construction of IceCube it was clear that the telescope worked. For example, the partially-constructed telescope observed the muon shadow of the Moon: about nine million muons a month rain down on IceCube from above, but the detector sees far fewer when the Moon is overhead and blocking a certain fraction of those particles. More importantly, perhaps, IceCube has already detected neutrinos – about ten thousand of them. None of these neutrinos came space (they were created when charged cosmic rays hit particles in the Earth's atmosphere) so at the time of writing the neutrino sky remains empty as far as IceCube is concerned. But that situation will surely change soon – perhaps even as this book is published.

Figure 7.6 *An artist's impression of the array of optical sensors, buried in the ice of Antarctica, that form the IceCube telescope. If a high-energy neutrino interacts with an oxygen atom in the ice, a charged particle can be produced that will be moving through the ice faster than light can travel. A cone of Cerenkov radiation, with its characteristic blue hue, will be produced. It's this Cerenkov radiation that the sensors will detect. (IceCube Collaboration/NSF)*

All in all, then, the South Pole was a good choice.

There's a problem with IceCube, though: since it's looking *down* it's looking only at the northern hemisphere. To study the neutrino sky in the southern hemisphere astronomers need a detector based in northern latitudes. This is where European scientists have proven to be much smarter than their American colleagues: they are building a neutrino detector on the French Riviera, where conditions are rather more temperate.

Rather than use ice as the neutrino detecting medium, a collaboration called **Astronomy with a Neutrino Telescope and Abyss Environmental Research** (ANTARES) uses water – courtesy of the Mediterranean Ocean. ANTARES is situated about 40 km off the shores of Toulon, France. This location was, of course, not chosen because of its climate. For various reasons the North Pole is an unsuitable venue for a neutrino detector, so an antipodean version of IceCube is not feasible. Instead, astronomers chose to use a volume of ocean rather than a volume of ice as the detector medium. Now, any ocean-based Cerenkov detector must be below the 1 km depth to which daylight can penetrate, and it really needs to be deeper so that the water can shield it against background muons. Well, the Mediterranean is up to 5 km deep in places. Better yet, some of those deep sites are close to shore, which makes it easier from a logistical point of view to build the telescope. Such an ocean location also makes it possible to maintain the telescope, by replacing damaged detectors for example, something that is impossible in an ice-based telescope such as IceCube. Furthermore, water is in some ways a better medium than ice for a Cerenkov detector: light scatters less in water than it does in ice so that ANTARES has a much greater pointing accuracy than IceCube. There are trade-offs, however: the ANTARES scientists have to understand how the ever-changing volumes of water might affect their signals; they even have to make allowances for the light emitted by bioluminescent bacteria.

The idea behind ANTARES was to anchor a collection of twelve 450-meter long strings into the bed of the Mediterranean, with each string containing 25 'storeys' – groups of photomultipliers and sensors embedded in a protective glass sphere. The active part of the strings – the 350 meters that contain the storeys – are under 2.5 km of sea water, so the detector's 875 photomultipliers are shielded from those cosmic ray-induced muons. In the inky blackness the detectors then search for brief flashes of Cerenkov radiation that might denote the presence of muons generated by neutrino collisions. The experiment will be particularly sensitive to neutrinos with energy in the range 10^{11} eV to 10^{14} eV – so, like IceCube, it will be looking for high-energy phenomena.

The ANTARES collaboration began in 1996, but it was ten years before the real work started and the project's first string was installed. In 2008, two years after that first string, the final line was powered up and the ANTARES detector was complete. It is now detecting neutrinos, and at the time of writing has seen more than a thousand of them. However, that's not yet enough neutrinos to tell whether any came from an extraterrestrial source – astronomers will need to see many more before they can determine whether neutrinos are arriving in large numbers from particular regions of the sky. One interesting discovery, though, was that of deep-diving sperm whales in the

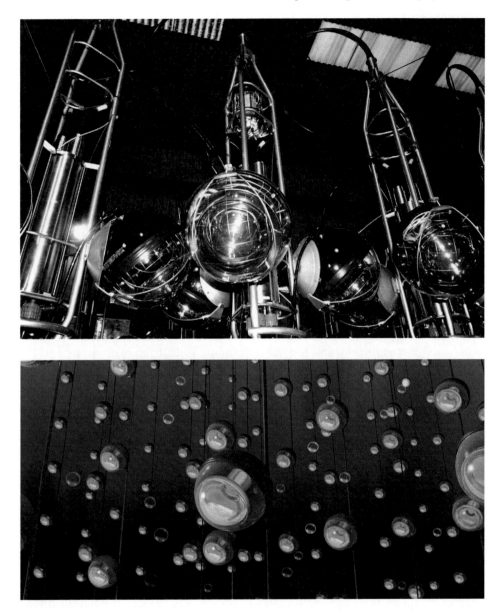

Figure 7.7 Top: A string of ANTARES eyes. These glass spheres contain photomultipliers that look for Cerenkov radiation. The strings had to be installed on the Mediterranean seabed by a special craft. Bottom: An artist's impression of the underwater environment in which detectors for KM3NET, a proposed follow-on to the ANTARES experiment, will take place. The KM3NET experiment will make use of a cubic kilometer of Mediterranean seawater. (Top: L. Fabre/CEA; Bottom: Aspera/G. Toma/ A. Saftoiu)

Mediterranean. Physicists on the ANTARES project investigated whether it would be possible to detect neutrinos from the sound waves generated by the particle shower coming from a neutrino interaction. It turns out that the frequency of such sound waves is in the same range as the frequency of the clicks emitted by whales.

The ANTARES detector covers a surface area of 0.1 km^2 and its photomultipliers are arranged over a height of about 0.35 km. In other words, although it is the largest neutrino detector in the northern hemisphere, it is a significantly smaller detector than the cubic kilometer IceCube. But experiments such as ANTARES, and two smaller detectors currently working in the Mediterranean, are a step towards construction of the **km^3 Neutrino Telescope** (KM3NET) – a neutrino detector whose medium will be a cubic kilometer of the Mediterranean Sea.

The KM3NET project is currently in the design phase, with representatives from 40 universities and institutes in ten countries studying the best way to take such an ambitious initiative forward. If the political, legal and financial aspects of the project can be agreed, and KM3NET is completed, then Earth will house two cubic-kilometer sized neutrino detectors: one situated in the northern hemisphere to study the southern sky and one in the southern hemisphere to study the northern sky. Astronomers will have a net for catching ghosts.

Improving the view

Look at the sky through a neutrino telescope and the vista remains uninteresting. Will telescopes such as Icecube (and KM3NET if it's built) *really* improve the view? There's every hope that they will.

As Yogi Berra said, "You can see a lot just by observing". A problem many astronomers face is that they can't directly observe many of the interesting regions they'd like to – hot, dense regions of the Universe, such as the cores of supernovae, or the highest-energy parts of active galactic nuclei and microquasars, or (as we shall describe in the next chapter) gamma-ray bursts. As mentioned earlier, the hurdle is that for the most part neither protons nor high-energy photons can reach Earth from those distant, hot, dense regions where all the interesting action is taking place. But neutrinos *can* reach us from those places. Neutrinos stream out from the cores of supernovae or the central regions of AGN as if there were nothing in their way. Neutrinos can travel the length of the Universe without hindrance from the microwave background. So neutrinos, if they can be caught, can illuminate what happens in regions that are presently hidden from view. And, since they point straight back to their source, neutrinos may help solve the mystery of the origin of ultra-high-energy cosmic rays (as we discussed in the previous chapter).

Neutrino observatories may be able to tell us even more. Throughout the history of the Universe there must have been billions of core-collapse supernovae, each one of which will have poured out vast numbers of neutrinos. Space must thus be bathed in neutrinos – the so-called **diffuse supernova neutrino background**, an accumulation of all those relic neutrinos from all the past supernovae that have occurred. At the time of writing this background has not been observed; it's at the edge of observability for existing neutrino experiments such as that at Kamiokande. However, European physicists are currently investigating suitable locations for the next generation of neutrino experiment: the **Large Apparatus studying Grand Unification and Neutrino Astrophysics** (LAGUNA). Such as instrument, which could contain up to one million cubic meters of liquid, would certainly have the sensitivity to observe the neutrino background. The study of this background would allow astronomers to investigate events that happened at cosmological distances.

Our view of the Sun may also change when astronomers start looking at it through more powerful neutrino telescopes. As we saw in chapter 3, it has been suggested that gravity will cause dark matter particles to accumulate in the center of massive objects such as the Sun. A high density of such particles will lead to an increased chance that they will meet and annihilate, which will lead to a large number of high-energy neutrinos. The details depend on the precise nature of dark matter, but if it turns out that dark matter is primarily a WIMP with a relatively large mass then IceCube and KM3NET should be able to detect the neutrinos.

It's not just astronomers who'll be looking forward to the new view of the sky: physicists can learn much by studying neutrinos from the cosmos. A neutrino is an insubstantial beast, but that means it should be sensitive to new physics. For example, by investigating whether different neutrino flavors interact differently with a gravitational field, physicists will be able to put Einstein's theory of general relativity to a stringent test. They will be able to search for possible signatures of a grand unification of the standard model of particle physics. If such things exist, they might see cosmic strings or magnetic monopoles.

The preceding chapters in this book have concluded with a discussion of how telescopes operating at different wavelengths could help answer the question under discussion. But neutrino telescopes will be opening up a completely *new* window on the high-energy Universe, a view that is hidden from other instruments. Perhaps the only thing to expect with this new view is the unexpected.

8

Far as human eye can see

They are the most luminous phenomena in the known Universe. But what exactly are gamma-ray bursts? What causes these phenomenal explosions? For a long while astronomers had few clues to help them answer these questions. The mystery of gamma-ray bursts is not yet solved – but a succession of orbiting gamma-ray telescopes is pointing the way towards a solution.

A mystery begins – Swift: catching bursts on the fly – A mystery solved? – A mystery remains… – One less thing to worry about – Fermi flies – Back to Earth – Bursts emit more than just gamma-rays

S. Webb, *New Eyes on the Universe: Twelve Cosmic Mysteries and the Tools We Need to Solve Them*, Springer Praxis Books, DOI 10.1007/978-1-4614-2194-8_8, © Springer Science+Business Media, LLC 2012

What's the farthest you can see with the naked eye? With eyesight like mine the answer is about an arm's length. Under *perfect* observing conditions, however, and for those with the very keenest eyesight, the answer would seem to be about 12 million light years. We know this because a small number of amateur astronomers claim to have seen the spiral galaxy M81 without optical aid. This beautiful object lies 11.8 million light years away in the direction of the constellation of Ursa Major, the Great Bear. It contains the mass of about 250 billion suns. Once the blazing light from these billions of stars has reached Earth, the apparent magnitude of M81 has dimmed to the edge of visibility, even for those with exceptional eyesight.

However, if you were looking up at the northern sky in the direction of the Boötes constellation on 19 March 2008, it's just possible that you might have seen the light from something more distant than M81. Much, *much* more distant. That Wednesday, an hour or so before the morning rush-hour began in Britain, an object became visible and stayed so for about half a minute. At its peak, the object was as bright as the planet Uranus. The visually challenged among us would never have noticed it, but a person with keen eyes could – in principle – have seen it.

And how far away was this object that appeared so briefly in Boötes? Well, the light hitting Earth that morning came from an explosion that happened about 7.5 billion years ago – long before Earth had come into existence, and more than half the time since the Big Bang. In rough language, the explosion occurred more than halfway across the Universe.

Just think for a moment of how luminous this object must have been: it outshone an intrinsically bright galaxy such as M81, with its billions of stars, even though it was much further away. The energy output of the event was truly stupendous, many times greater than the brightest supernova ever seen. Astronomy is a science full of superlatives – biggest this, most massive that – but the explosion recorded that March day was truly staggering: one of the most intrinsically bright objects that humans have ever observed.

This remarkable object, which goes by the unprepossessing name of GRB 080319B, was a gamma-ray burst. It was discovered by the orbiting Swift observatory, which keeps a constant watch for such bursts. Swift identifies a gamma-ray burst on average just under once every four days, so such events are not rare. For a time, though, gamma-ray bursts were the most enigmatic and puzzling phenomena in astronomy – and the mysterious case of the gamma-ray bursts is still not fully resolved.

Figure 8.1 *The beautiful spiral galaxy Messier 81 lies 11.6 million light years away. Hubble's sharp vision allows it to resolve individual stars in M81, and the galaxy can be seen easily with binoculars. Remarkably, if you possess perfect eyesight and have perfect observing conditions then you might even be able to see it with the unaided eye. The image shown here is a combination of Hubble images taken at blue, yellow and infrared wavelengths. (NASA, ESA & the Hubble Heritage Team STSCI/AURA; A. Zezas and J. Huchra)*

A mystery begins

Gamma-ray photons possess higher energies than those from the rest of the electromagnetic spectrum: a gamma-ray generated in radioactive decay might typically have an energy of a few hundred keV; a gamma-ray produced in an extreme astronomical event might have an energy as high as 50 TeV. So the problem involved in trying to bring gamma-rays to a focus is even worse than it is with X-rays: a gamma-ray will just shoot straight through a lens or a mirror. It follows that if astronomers want to study the gamma-ray sky they need a type of instrument that can contain the energy of these bullets. Well, physicists have been building gamma-ray detectors for decades so it's certainly possible to construct an instrument for use in astronomy.

However, before looking at the details, consider the environment in which a gamma-ray 'telescope' would have to work. The instrument would likely be sensitive to charged particles (cosmic-ray protons, for example) as well as gamma-rays, so some method would be needed to distinguish gamma-rays from background noise. Furthermore, Earth's atmosphere is an effective shield against cosmic gamma-rays, which is just as well since gamma-rays cause serious damage when absorbed by living cells; in sufficiently large doses they kill. Thus the instrument would have to be sufficiently compact and robust to be put on a balloon or launched into orbit. It's not an easy task.

Note that it's not impossible to do gamma-ray astronomy from the ground, particularly at the highest energies. The highest-frequency photons from space act like cosmic rays, creating showers of fast-moving particles when they strike atoms in the atmosphere, so some of the techniques outlined in chapter 6 apply here too. For example, in New Mexico between 2001–2008 the **Milagro** experiment (which name was blessedly not the result of a hunt for a catchy acronym) searched for traces of Cerenkov radiation in a covered pool of water. The idea was that fast-moving particles from a gamma-ray-initiated shower could be detected by hundreds of photomultipliers inside the pool. Several other gamma-ray observatories work on a different principle, each making use of instruments known as IACTs (imaging atmospheric Cerenkov telescopes). An IACT is a combination of several optical reflectors that capture the faint Cerenkov light from extensive air showers, and ultra-fast cameras that record images, and it allows astronomers to observe the high-energy sky from the ground. In Woomera, the **Collaboration of Australia and Nippon for a Gamma Ray Observatory in the Outback** (CANGAROO) uses the IACT technique to study the southern sky. In Namibia, the **High Energy Stereoscopic System** (HESS) does something similar. The **Major Atmospheric Gamma-ray Imaging Cerenkov Telescope** (MAGIC), which is based in the Canaries, and the **Very Energetic Radiation Imaging Telescope Array System** (VERITAS), which is based in Arizona, are two more large ground-based gamma-ray instruments.

Whichever technique it uses to detect gamma-rays, a ground-based detector spots only high-energy gamma-rays: typically, it will be sensitive to photons with an energy in the range 50 GeV to 50 TeV. To catch lower-energy gamma-rays really requires a detector that's above the atmosphere.

Given the difficulties involved in building a space-based gamma-ray telescope it's not surprising that early progress was slow. The **Explorer 11** satellite launched in 1961, along with balloons sent up in the same period, were

Figure 8.2 *Four* IACTs. *(a) One of the two* MAGIC *mirrors on La Palma. (b) The four 13 m telescopes in the* HESS *array in Namibia. (c) One of the* CANGAROO *telescopes, located in the desert about 500 km north of Adelaide. (d) The* VERITAS *array in Arizona. ((a)* MAGIC *Collaboration/Aspera; (b)* HESS *Collaboration/Aspera; (c) M. Mori; (d) Steve Criswell)*

able to detect radiation that hinted at the presence of cosmic gamma-rays. But it was not until the launch in 1967 of the third in NASA's **Orbiting Solar Observatory** program (OSO-3) that cosmic gamma-rays were definitely detected; OSO-3 saw, in total, 621 gamma-ray photons from space. Gamma-ray astronomy had truly arrived, even if the view it gave was still patchy. The new science then received a boost from an unexpected source.

In 1963, the Limited Test Ban Treaty came into force, prohibiting the test detonations of nuclear weapons except underground. That year, the United States Air Force launched the first of the VELA spy satellites in order to monitor the effectiveness of the treaty. The VELA satellites, hunting in pairs, watched for signs of nuclear explosions. They did this by carrying an array of detectors sensitive to neutrons, X-rays and gamma-rays – stuff that is emitted in copious amounts in a nuclear explosion.

The original detectors were rather poor, but technology improves continuously and with each launch the efficiency of the VELA detectors increased. In 1969, Ray Klebesadel was head of a team developing the detectors for VELA5. As part of this work, Klebesadel analyzed the data from the two VELA4 satellites. He discovered that the detectors recorded a gamma-ray event on 2 July 1967. He checked further, and found that this event also triggered the still operational VELA3 satellites.

At first the US intelligence community was worried. Perhaps the gamma-ray had come from a bomb detonated in space, or perhaps the Soviet Union or China were deliberately jamming the VELA detectors as a means of hiding the tell-tale signs of a nuclear weapons test. However, when VELA5 found more of these gamma-ray bursts – roughly one burst every fortnight – the intelligence community lost interest: even they understood enough to realize that their Cold War enemies lacked the technology to test nuclear weapons every other week. The puzzle was declassified and passed on to astronomers.

News of gamma-ray bursts became public in 1973, when Klebesadel and two of his colleagues at the Los Alamos Scientific Laboratory, Ian Strong and Roy Olson, published an analysis of 16 events observed by VELA5 and VELA6 between July 1969 and July 1972. Astronomers soon observed more bursts with gamma-ray detectors on board satellites whose job was to study the Sun – satellites such as OSO-7 – and they found that the bursts occurred randomly, and seemed to come from all parts of the sky. With this began a controversy: what, precisely, was causing these bursts?

This was a difficult question to answer because, in each and every event that was detected, the burst had vanished before astronomers could locate its position on the sky and subject it to follow-up observations. From the sparse data provided by these early satellites it proved impossible for astronomers to find a counterpart or host, such as a star or a galaxy, to any of the bursts. And without that information it proved impossible to pin down the distance scale associated with what became known as a **gamma-ray burst**. Did the bursts come from the outskirts of our Solar System (perhaps the bursts were the result of colliding comets)? Did they come from the Galaxy (perhaps magnetic flares on the surface of stars held the clue)? Did they come from remote galaxies (perhaps a mechanism involving black holes could solve the problem)? Scientists proposed more than a hundred different mechanisms to explain the origin of the bursts – so at one time there were fewer observed gamma-ray bursts than there were theories to explain them.

Progress in understanding the nature of the bursts picked up in 1991 when NASA launched the second in its 'Great Observatory' series – the **Compton Gamma-ray Observatory**, which was named after the Nobel prize winning physicist Arthur Compton. See figure 8.3. Among its payload was an extremely sensitive gamma-ray detector called the burst and transient source experiment (BATSE), which could identify gamma-rays with energies between 20 keV up to 2 MeV. See figure 8.4. The instrument detected 2704 gamma-ray bursts in nine years. With so many events now to refer to, a naming convention had to be adopted; gamma-ray bursts were given the name GRB followed by six figures specifying the year/month/day on which the event was observed. If more than one burst was detected in one day, a letter was assigned to the burst in order of occurrence.

At last, astronomers had meaningful amounts of data to analyze. BATSE showed that the bursts are intense, not particularly long-lived (with a duration between about 0.1 s to 100 s) and pop off at the rate of about one per day somewhere in the Universe (though the true rate might be larger if we see only a fraction of the bursts). Further study of the BATSE data showed that there are two main classes of burst: those shorter than 2 s and those longer than 2 s. The main difference between the two classes is that the short bursts release more of their energy in very energetic gamma-rays. The short bursts are therefore classed as being 'short and hard' while the others are 'long and soft'. A further difference is that long bursts are more luminous than short bursts. Perhaps the key result coming out of the BATSE data, though, was the confirmation that gamma-ray bursts are uniformly and randomly distributed across the sky; there is no direction from which we detect more bursts. This finding ruled out most of the models put forward to explain the bursts.

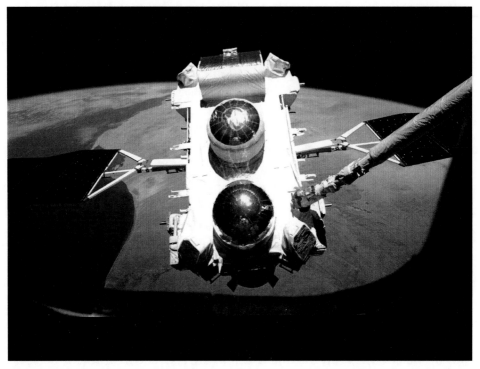

Figure 8.3 *The Compton Gamma-ray Observatory being deployed from the Space Shuttle Atlantis payload bay. (NASA)*

For example, since the bursts were not concentrated along the plane of the Milky Way they could not have an origin within the Galaxy. Indeed, by the mid-1990s only two possibilities seemed tenable: either the bursts originate from a large, diffuse halo surrounding the Galaxy or they originate from distant galaxies far beyond our own Local Group.

But how to distinguish between the two possibilities?

The most direct method of learning more about gamma-ray bursts would be to study their afterglows. It would be strange indeed if the bursts emitted only gamma-rays; most energy sources emit several forms of energy. A flame, for example, does not just gives off visible light – it also emits lots of infrared radiation. Although the initial explosion that generated a burst might last for only a few seconds, the afterglow could linger for weeks. The afterglow would follow a path down the electromagnetic spectrum, firstly emitting mostly in the gamma-ray region, then peaking at X-rays, and so on all the way down to radio waves until the afterglow faded completely from

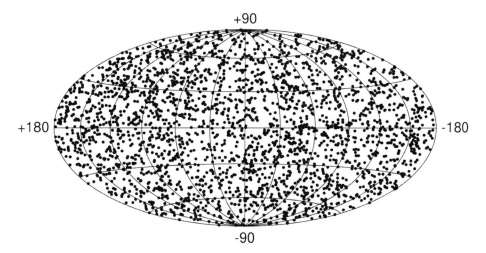

Figure 8.4 *The* BATSE *instrument on board Compton discovered 2704 gamma-ray bursts. As the diagram here shows, the bursts came from all directions in the sky. This result was a clear demonstration that gamma-ray bursts are distant.* (NASA)

view. If astronomers could detect the afterglow of a gamma-ray burst they would be able to study it with variety of telescopes. Finally, they might be able to learn something of the nature of those enigmatic gamma-ray bursts.

In 1996, the Italian Space Agency, in collaboration with the Netherlands Agency for Aerospace Programs, launched the **Bepposax** satellite. The satellite was named in honor of the Italian physicist Giuseppe Occhialini – Beppo was his nickname. sax was the Italian acronym of 'satellite for X-ray astronomy'. In addition to a detector that looked for gamma-rays with an energy up to 300 keV, Bepposax contained four other instruments. Once a gamma-ray burst had been detected, the idea was to use the other instruments to locate its position as accurately as possible; this information would be relayed immediately to powerful optical telescopes on the ground and to Hubble, which would then have a fighting chance of finding the afterglow.

Within ten months of its launch, astronomers succeeded in using Bepposax to detect the X-ray afterglow of the gamma-ray burst GRB 970228. Astronomers immediately began to search near the X-ray source, and they succeeded in obtaining the first optical images of a gamma-ray burst object. The object seemed to be associated with a faint, fuzzy patch of light – presumably a 'host' galaxy from within which some cataclysmic event had led to the burst. The case for supposing gamma-ray bursts come from far away was strengthened: but astronomers still needed to measure the distance to a host galaxy to be sure.

BeppoSAX soon identified and located another burst – GRB 970508. Within four hours astronomers had the world's largest optical telescopes trained on the event. They were able to measure a spectrum, and found a redshift of 0.835 – the burst lay several billion light years from Earth. That settled it: gamma-ray bursts lay at cosmological distances.

For us to observe a gamma-ray burst at those distances, the bursts must be extremely luminous. Indeed, the energies involved are so huge that it's difficult to comprehend the figures – which is the main reason that many astronomers initially believed the bursts must be nearby. A typical burst is one hundred quadrillion times more luminous than the Sun; it releases as much energy in a few seconds as a thousand stars like the Sun emit over their entire lifetime. For a fleeting but glorious moment, a burst outshines the rest of the Universe in gamma-ray light.

It became a priority for astronomers to understand the origin of these extreme events. The success of BATSE and BeppoSAX demonstrated that it was possible to study gamma-ray bursts in detail. What was needed was an observatory that could spot large numbers of bursts and, whenever it saw one, quickly train a variety of instruments upon it. What astronomers needed was a spacecraft like **Swift**.

Swift: catching bursts on the fly

Before we look at Swift in more detail, it's worth describing in general terms how instruments can study the gamma-ray sky if they don't bring those high-energy photons to a focus. The key is to catch the photon and study how it interacts with matter.

When a gamma-ray strikes matter it can do only a limited number of things. It can, for example, collide with an electron and bounce off it – in much the same way as two billiard balls can bounce off each other. The electron gains energy while the photon loses energy and thus decreases its frequency. This is called **Compton scattering**. Or the gamma-ray can be absorbed by an atom at the same time as an electron is ejected from the atom – a process known as **photoabsorption**. Or the gamma-ray can interact with an atomic nucleus and transform itself directly into matter by creating an electron and positron pair – a process called **pair production**. All of these interactions cause electrons to move in some way, which means that an elec-

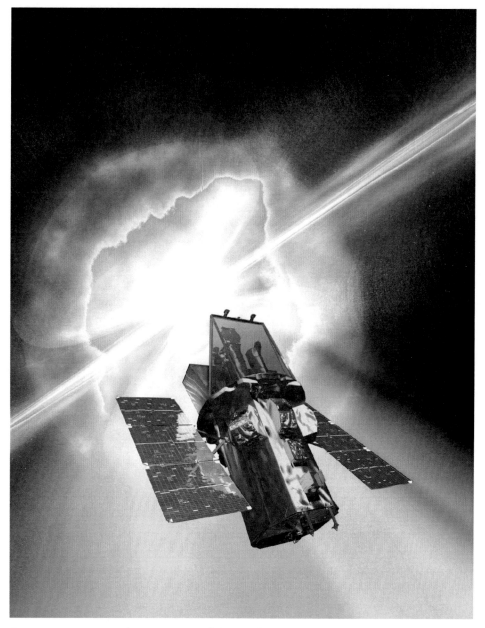

Figure 8.5 *An artist's impression of the Swift observatory, with a gamma-ray burst as a backdrop. Swift was launched in November 2004, and has already exceeded its planned six-year mission lifetime. The spacecraft orbits at an altitude of 600 km and has an orbital period of 90 minutes. On average, Swift detects a gamma-ray burst from somewhere in the Universe at the rate of about one every four days. (Spectrum and* NASA, *Sonoma State University, Aurore Simonnet)*

tric current has been created. By amplifying and measuring these currents one can estimate the energy of the original gamma-ray.

So astronomers build what is essentially a 'light bucket'. Such an instrument needs to be fixed on a particular region of sky and then collect as many gamma-ray photons as possible. A basic detector would use the scintillation technique to transform the gamma-ray into optical signals, which could then be recorded and analyzed. We saw in the previous chapter how scintillators – materials that emit photons in the visible range when struck by a high-energy charged particle – are used in neutrino detection. Scintillators don't detect neutrinos or photons directly, of course, because neither particle possesses a charge. But since a gamma-ray strike creates high-energy electrons or positrons – through Compton scattering, photoabsorption or pair production – the effects of the strike in the scintillator will eventually cause the emission of low-energy photons. Photomultiplier tubes that surround the scintillator then collect the photons and, by adding up all the energy collected, the energy of the initial gamma-ray can be determined.

Most scintillators used in gamma-ray detectors consist of inorganic materials, usually a salt such as sodium iodide or caesium iodide. Added to these salts are impurities known as activators; these impurities are there to enhance the efficiency of light emission and to ensure that the emitted photon is of the appropriate wavelength. Thallium and sodium are commonly used activators. The detector used in BATSE, for example, was a sodium iodide crystal with a thallium activator; other common detectors have used a caesium iodide crystal with a sodium activator.

More recently, astronomers have started to use semiconductors for this type of detector. In these devices an incoming gamma-ray creates an electron/hole pair in the semiconductor. Such solid-state devices are more precise than scintillators at measuring the energy of the gamma-ray, and they can better determine the area in space from which the ray came. Such improvements come at a price: these devices are expensive and must be cooled to achieve optimum operating temperatures.

In order to create a gamma-ray image astronomers can use a technique called **coded aperture imaging**. A mask, made out of thousands of small lead tiles, is mounted in front of a ray of gamma-ray detectors. The tiles are arranged in a random pattern, and cover about half of the available area. When gamma-rays come through the mask, the lead tiles absorb some photons while the open areas let others pass through to the detectors. The mask thus casts a shadow on the detectors. From the position of the shadow, combined with knowledge of the tiles' positions, computers can calculate the direction of the gamma-ray source.

Thus an orbiting observatory armed with solid-state scintillation detectors and coded aperture imaging techniques would be able to make detailed observations of the gamma-ray sky. That's precisely what the designers of Swift had in mind. The Swift observatory was developed by an American–British–Italian consortium, and launched into a circular, low-Earth orbit in 2004. The observatory contains an instrument called the burst alert telescope (BAT) that can be used to study gamma-ray bursts. It detected its first burst within two months of the launch date. In addition to BAT, Swift is also home to an X-ray telescope (XRT) and an ultraviolet/optical telescope (UVOT).

The largest instrument on-board Swift, BAT, is the satellite's watchdog. It can view approximately one sixth of the entire sky with a single glance as it looks for gamma-rays possessing an energy between 15 keV to 150 keV. Within 20 s of detecting a burst, BAT relays an estimate of the burst's position to observatories on the ground. The estimate is accurate to between 1 to 4 minutes of arc. Such accuracy is rather poor by the usual standards of optical astronomy, but it's impressive for gamma-ray astronomy: 1 minute of arc is about one thirtieth the diameter of the full Moon as seen from Earth. The spacecraft also begins to reposition itself in space, so as to point the XRT and UVOT at the burst. Swift uses a system of six on-board momentum wheels – flywheels, essentially – to rotate into the correct direction. It does this so quickly – it can rotate through 50° in just over a minute – that to poetic-minded astronomers it moves like a swift in the sky. It's refreshing to learn that not all astronomy missions have ugly, acronym-based names. This mission is named after a beautiful, agile bird.

Once Swift is pointing in the correct direction, the XRT comes into play and provides a more accurate estimate for the burst's location – an estimate accurate to between 3 to 5 *seconds* of arc. UVOT then images the field and, within a few minutes of the burst being detected, Swift transmits a finding chart to all interested observatories on the ground. Finally, XRT and UVOT settle down to study the gamma-ray burst's afterglow.

Swift thus enables astronomers to observe gamma-ray bursts during the crucial first few minutes and hours. Some of the bursts spotted by Swift have turned out to be particularly interesting. For example, light from the burst GRB 090429B set out on its journey to us when the Universe was just 520 million years old. At the time of writing, this burst is the most distant object ever seen. In 2010, just over five years after launch, Swift detected its five hundredth burst. And with 500 bursts to study in detail, astronomers have been able to learn much more about these events.

A mystery solved?

So, what causes a gamma-ray burst?

As mentioned earlier, there are two different types of gamma-ray burst ('long' and 'short') and presumably, therefore, two different types of progenitor. Information on the short bursts remains relatively scant, but astronomers are beginning to agree on a model that explains the long bursts.

Astrophysicists such as Stanford Woosley have proposed that long bursts are created by a **collapsar**, in an event that is similar to a type II, or core collapse, supernova. The precursor to a collapsar is a rapidly rotating **Wolf–Rayet star** nearing the end of its life. These stars, which are named after the astronomers Charles Wolf and Georges Rayet, possess between 20 to 30 times the mass of the Sun. Once a Wolf–Rayet star has used up its supply of nuclear fuel it has no outward radiation pressure to counterbalance the gravitational pull of its mass so the core starts to collapse; and if the core is sufficiently massive, nothing can prevent it from collapsing into a black hole. See figure 8.6. The black hole immediately begins to pull in surrounding stellar material, and an orbiting doughnut-shaped envelope of material – an accretion torus – quickly forms. The part of the torus closest to the black hole orbits at speeds close to that of light while the outermost parts move less quickly. Now, as described in chapter 5, such a rotating accretion torus will be threaded by magnetic field lines; and those field lines will be violently twisted because the innermost part of the torus is rotating more quickly than the outermost part. This causes two opposing jets of material – electrons, positrons and protons – to shoot outward along the star's axis of rotation in a direction perpendicular to the torus, where the matter density is much less than it is within the disk. By the time the jet punches through what remains of the stellar surface, it's travelling at close to light speed.

There are two key differences between the picture of a collapsar sketched here and an 'ordinary' type II supernova. First, a collapsar requires a rapidly rotating star: only then does a rapidly rotating accretion torus form with the capacity to launch the jet. Second, the progenitor star must have formed in a galaxy that contains only small amounts of elements heavier than helium; in technical terms, the star must have low **metallicity**. The presence of heavier elements inhibits the creation of a gamma-ray burst in two ways. First, these

Figure 8.6 *This artist's impression shows how a massive star 'burns' progressively heavier elements. Once the star has used up its supply of nuclear fuel it collapses into a black hole. Vast amounts of energy can be released in two opposing jets along the axis of rotation. (National Science Foundation)*

heavy elements smother the jets and make it difficult for them to pierce the stellar surface. Second, the magnetic field generated by these elements acts as a brake on the star's spin, slowing it below the level needed to generate a gamma-ray burst. High metallicity means no burst. These constraints mean that only a small fraction of supernovae have the potential to create a burst.

The collapsar is only the first step in making a gamma-ray burst. The next step is the so-called relativistic fireball, a model developed by astrophysicists such as Martin Rees and Peter Mészáros. Imagine the starting line of a chaotic Formula 1 or IndyCar race in which the cars all move at different speeds: as the race starts, collisions occur as the faster-moving cars pile into the slower-moving vehicles. So it is with the relativistic jet created by the collapsar: different parts of the jet will be at different pressures, densities and temperatures, and faster-moving clumps of matter will plough into slower-moving clumps. The result of these collisions is a series of internal shock waves within the jet, and the production of gamma-rays. When matter is at a high temperature, photons cannot travel far before being absorbed. As soon as the jet cools enough to become transparent, however, the gamma-rays begin to travel at light speed in the same direction as the jet. If the jet happens to be pointing in our direction then those gamma-rays are what our detectors pick up: we see a gamma-ray burst. As for the afterglow – well, the fast-moving jet of particles chases after the gamma-rays and eventually ploughs into the material of the interstellar medium. The initial collisions create energetic X-rays; with each collision, though, the material loses kinetic energy and subsequent collisions create radiation possessing less energy and thus longer wavelengths – ultraviolet light, visible light and, eventually, radio waves.

The collapsar–fireball model thus explains many of the properties of long gamma-ray bursts. Furthermore, two pieces of circumstantial evidence support the notion that collapsars are the progenitors of bursts. First, long gamma-ray bursts are seen in regions where there has been lots of recent star formation, such as the arms of spiral galaxies and irregular galaxies. Massive stars, of the type required to form a collapsar, are only ever found in such regions. Second, several cases are on record where a gamma-ray burst was followed within a few days by the appearance of a supernova – and in each case the supernova was of a type consistent with core collapse. Note that the association between gamma-ray bursts and supernovae will only be apparent for relatively close events; most bursts are so distant that a supernova would be too faint to see. Other explanations have been posited to explain at least some of the long bursts. For example, Swift has seen some bursts that might best be explained if they involved not a black hole but a magnetar – a neutron star possessing a magnetic field many trillions of times greater than Earth's. Such a huge magnetic field can slow the star's rotation rate – it's like slamming on the brakes of your car – and the energy of the rapid spin-down can continue to power the burst. It may indeed be that a single model is insufficient to explain every long burst that has occurred. Nevertheless, the collapsar–fireball model is generally successful.

So much for the long gamma-ray bursts. What about the short bursts?

Astronomers have found it much more difficult to study short gamma-ray bursts, and have been able to localize only a few of them to definite galactic hosts. Nevertheless, it's clear that these bursts differ markedly from the long bursts in more than just duration. For a start, even though all the identified hosts have been nearby and thus amenable to detailed follow-up study, in no case has a supernova been associated with a short gamma-ray burst.

Furthermore, the short bursts are not restricted to star-forming regions: several have been found in the outer halo of large elliptical galaxies, where star formation has essentially ceased.

To date there is no universally agreed model for the progenitor of a short gamma-ray burst, but the model that finds most favor is the merger of two neutron stars (or the merger of a neutron star with a black hole). Such binary systems are likely to be rare, but astronomers know of neutron star–neutron star binaries in the Galaxy and there is reason to suppose that neutron star–black hole binaries also exist. The point is that, according to general relativity, such binary systems will inevitably lose energy through gravitational radiation, and as that happens the two compact objects will spiral closer and closer together. Eventually, tidal forces will rip apart the neutron stars (or the single neutron star in the case of a neutron star–black hole binary) and vast quantities of energy will be liberated before the matter is lost from view as it falls over an event horizon. Computer models suggest that the entire process takes only a few seconds, which would account for the short duration of the bursts. And, since the stars involved have already finished their evolution, there is no supernova.

It might appear to be a hopeless task: how can astronomers prove that short bursts originate from the merger of a neutron star with another compact object? How can we possibly observe the merger of such small, dark objects.? Well, a merger would produce a clear signature that astronomers should be able to observe – but that's a subject for the next chapter.

A mystery remains…

Our understanding of gamma-ray bursts has come a long way since scientists first realized they had a puzzle on their hands. Astronomers now possess a wealth of observational data on the bursts and their afterglows, and although the theoretical models can't yet explain every detail of every event they are certainly successful in explaining the main features. It's always the case in science, though, that as more is learned the more there is to learn.

In 2006, Swift discovered a relatively nearby gamma-ray burst called GRB 060614. The burst lasted 102 s, so it was clearly a long burst. Its relative proximity meant that astronomers could train their ground-based telescopes on it in order to study the supernova that should emerge. They saw nothing. Hubble then observed GRB 060614 in exquisite detail. No supernova. Furthermore, Hubble discovered that the event occurred in quite the wrong sort of neighborhood for a long gamma-ray burst: it occurred in a small galaxy with few massive stars, and at quite a distance away from where even those few massive stars reside.

The observation of GRB 060614 would appear to suggest one of three things. If the progenitor was a massive star, as should have been the case according to our understanding of long gamma-ray bursts, then it must have been an odd sort of massive star: one that resided in a quite different sort of stellar neighborhood to what we would expect and that exploded in a quite different fashion, without forming a supernova. No one yet knows what that sort of star could be. So maybe GRB 060614 was a drawn-out type of short burst, as indicated by the lack of a supernova and the type of galaxy in which it occurred? However, if the progenitor was a pair of inspiralling neutron stars, or a neutron star and black hole merger, then all the theoretical models predict that the event would have radiated for a maximum of only a couple of seconds. GRB 060614 lasted fifty times longer than that. The models would need a complete revamp. The third possibility, of course, is that GRB 060614 was something completely new. One suggestion is that Swift saw a star being swallowed by an intermediate-mass black hole.

However you look at it, the mysterious case of those gamma-ray bursts is not quite yet solved.

One less thing to worry about

You will remember that we began this chapter with a discussion of GRB 080319B. For us to see the explosion 7.5 billion years after it occurred – well, it must have been quite some lightshow for any creatures who saw it at the time. That thought raises a question: what would have happened to life on a planet that was close to GRB 080319B when it exploded? A distant planet that was out of the path of the jet might dodge the bullet, but what if it were in the firing line? Any advanced civilization that happened to be sufficiently close and in the path of the jet would face certain annihilation. In 1955, Arthur C. Clarke wrote a moving story called *The Star*. It told how an extraterrestrial civilization was destroyed when its sun went supernova, and how the light from that destructive event was seen on Earth as the Star of Bethlehem. One is immediately reminded of his story when one thinks of the energy that gamma-ray bursts emit and of the threat a burst would pose to any civilizations in the vicinity. Another poignant fact is that GRB 080319B was observed just a few hours after Clarke's death was announced.

However, it's extremely unlikely that GRB 080319B destroyed any high civilizations. Without getting into the wider debate about the existence of extraterrestrial intelligence, which is the subject of chapter 13, it seems likely that a necessary requirement for the development of civilization is the presence of heavy elements. GRB 080319B happened billions of years ago, when galaxies were largely lacking in such elements. Put simply, GRB 080319B occurred too early in the history of the Universe for it to have affected intelligent life. A gamma-ray burst that occurred nowadays, however, would be a quite different matter. In particular, what if Earth were in the firing line when a long gamma-ray burst went off?

If Earth were sufficiently close to a gamma-ray burst then life – certainly the higher forms of life – would be destroyed. Indeed, it has been suggested that the Ordovician–Silurian mass extinction event, which occurred 443 million years ago, was the result of a long gamma-ray burst that occurred within 6000 light years of Earth. Such a burst could have stripped Earth's atmosphere of half of its ozone, leaving surface life vulnerable to high levels of ultraviolet radiation. If that sort of nearby burst happened now, the result might well be another mass extinction. There is even a candidate star that could be a cause for concern: WR 104 is a Wolf–Rayet star with a rotational axis that is aligned within 16° of Earth. This star is only 8000 light years from Earth; some astronomers believe it's even closer. If it exploded as a gamma-ray burst (this is a big 'if' – astronomers don't yet have the ability to predict such things) then it might cause us problems.

Before you start losing sleep, though, it's worth noting that there's no direct evidence linking the Ordovician–Silurian extinction to a gamma-ray burst; at present it's nothing more than a plausible hypothesis. Furthermore, as we have seen, long gamma-ray burst events are much less likely to occur in a large, metal-rich galaxy like our own than in small, metal-poor galaxies. A long gamma-ray burst is unlikely to affect us. It's one less thing to worry about. (However, although short gamma-ray bursts are less energetic than their longer-lasting cousins, a short burst happening close to us would be just as dire for life on Earth. Let's hope there are no nearby neutron stars orbiting each other in a dance of death.)

Fermi flies

In 2008, NASA launched the **Fermi Gamma-Ray Space Telescope** into low-Earth orbit. From its vantage point 550 km above Earth's surface, the telescope is studying a vast range of frequencies at the high-energy end of the electromagnetic spectrum. Fermi can detect photons with an energy as low as 8 keV and as high as 300 GeV; the high-energy gamma-rays it detects are thousands of times more energetic than those detected by Swift. Over its five-year mission, which its designers hope will be extended to ten years or more, Fermi will address several scientific questions. How are particles accelerated in active galactic nuclei? Is there an excess of gamma radiation from the center of the Milky Way that might signify the presence of dark matter? Do primordial micro black holes exist? One of its key tasks, however, is to study the high-energy behavior of gamma-ray bursts.

Fermi carries two instruments. The large area telescope (LAT), the primary instrument, is many times more sensitive than any previous gamma-ray instrument flown in space. The gamma-ray burst monitor (GBM) complements the LAT at particular energies. Both are capable of detecting a gamma-ray burst and, if a burst is seen, the spacecraft turns to get the best view.

The LAT is a remarkable device, which can take in one-fifth of the entire sky in a single glance. One of the most impressive features of the LAT is the success with which it distinguishes cosmic rays from gamma-rays: for every gamma-ray that enters the LAT up to one million cosmic rays are rejected. If the LAT were unable to make this distinction then its ability to detect gamma-ray bursts would be severely reduced. The GBM complements the LAT rather nicely. First, it works at a lower energy range. Together the two instruments provide one of the widest energy detection ranges of any satellite yet built; if Fermi were a piano it would have a range of about 23 octaves. Second, it scans more of the sky at any instant; the GBM has such a wide field of view that it can detect gamma-ray bursts from over two-thirds of the sky at any particular time – essentially it sees all the sky not hidden by Earth's large disk. When the GBM identifies the location of a burst, the LAT can begin follow-up observations almost immediately.

The two instruments are working well. After just 20 months in orbit, the LAT had detected its 100 billionth event – that's a lot of gamma-rays. And, as expected, Fermi has made discoveries. For example, Fermi has shown that the microquasar Cygnus x-3 emits gamma-rays; it has seen the brightest persistent source in the gamma-ray sky (Fermi has been looking directly down the barrel of a blazar jet, which happens to have flared in recent months); and it has found a new structure centered on the Milky Way – two gamma-ray-emitting 'bubbles' that are each 25 000 light years across and whose nature is not yet fully understood. In 2011, the mission team published the

Figure 8.7 *An artist's impression of Fermi in its low-Earth orbit. Our planet blocks out a lot of the sky; nevertheless, Fermi has such a wide field of view that it can take in much of the rest of the sky in one glance. (NASA & General Dynamics)*

second Fermi catalog, which contains the results of a full two years of observation. The catalog contains 1873 gamma-ray sources, of which just over one thousand are associated with active galactic nuclei. What's much more intriguing is that more than 500 of the sources have no obvious counterpart; they remain unidentified. Discoveries surely await. As for gamma-ray bursts, Fermi has detected several record breakers. For example, the most powerful explosion ever seen was GRB 080916C, which was recorded by both the LAT and the GBM; the progenitor star died 12.2 billion years ago. Material from GRB 090510 was ejected with the fastest speed yet recorded: gas from the explosion moved at 99.99995% of the speed of light. And the most energetic gamma-ray ever detected from a burst came from GRB 090902B; the LAT recorded its energy at 33.4 GeV.

Gamma-ray bursts were once rare, enigmatic events. The deployment of Swift and Fermi means that bursts seem almost commonplace. The details of the explosions may not be fully understood – but, unlike the distant bursts themselves, surely a full understanding lies not too far away.

Back to Earth

This chapter has focused on bursts, but gamma-ray telescopes provide valuable information on many other cosmic phenomena that involve high energies: AGN, microquasars, supernova remnants as well as possible signatures of new physics. It's not surprising, then, that in the future astronomers intend to study the gamma-ray sky with new observatories. What's perhaps more surprising, at first glance, is that many of these planned observatories are based on the ground. It makes sense, however. If you are interested in the *really* high-energy gamma-rays produced by extreme processes then space-based observatories are ineffective: the more energetic the photon the fewer of them there are, so the small collecting areas on board a spacecraft are unlikely to encounter them. By using Earth's atmosphere as the collecting medium, and employing the same sorts of techniques that are used to detect cosmic rays, it's possible to study high-energy gamma-rays – as demonstrated by CANGAROO, HESS, MAGIC, Milagro and VERITAS, all of which we have looked at briefly. The successors to these instruments are already being planned.

For example, the **Cerenkov Telescope Array** (CTA) is a European-led project involving 500 scientists from 25 countries that hopes to study the gamma-ray sky at energies from about 50 GeV up to 100 TeV. The idea is to

Figure 8.8 *An artist's impression of a possible design for the* CTA. *Large central mirrors collect light caused by showers from low-energy gamma-rays. Smaller mirrors, spread over a wide area, study rarer but brighter showers. (Aspera/D. Rouable)*

construct two arrays of dozens of imaging atmospheric Cerenkov telescopes. These IACTs, as you will recall, can detect the extensive air showers that are created when a high-energy gamma-ray smashes into Earth's atmosphere. One array will be in the northern hemisphere and concentrate on studying the lower end of the energy scale, and thus extragalactic objects; the second array will be in the southern hemisphere and study the full range of energies, and thus galactic objects. Note that an American counterpart to CTA, called the **Advanced Gamma-ray Imaging System** (AGIS), planned to use an array of IACTs covering a square kilometer on the ground. However, early in its design phase the AGIS project merged with CTA, and the AGIS work packages have been subsumed by the larger European project.

Yet another proposed ground-based observatory is the American–Mexican collaboration known as the **High Altitude Water Cerenkov Experiment** (HAWC). HAWC will take a slightly different approach to the study of high-energy gamma-rays and build on the success of the earlier Milagro experiment. Milagro consisted of a pool of pure water that was about the size of a football pitch. The pool was shielded from light and surrounded by hundreds of photomultiplier tubes. It was thus similar to the Kamiokande Observatory discussed in the previous chapter. Rather than having one large pool, the HAWC collaboration plans to use an array of 900 water tanks covering an area of 150 m by 115 m. The tanks will be situated much higher than the Milagro pool: HAWC will be sited in Sierra Negra, Mexico, 4.1 km above sea level. If completed, the result will be an observatory that will far exceed Milagro's sensitivity to high-energy gamma-rays.

This new generation of ground-based observatories – complemented by balloon flights and, perhaps, new space-based instruments – will ensure that future astronomers will be able to study the sky at the highest possible energies in the electromagnetic spectrum.

Bursts emit more than just gamma-rays

It's not only gamma-ray telescopes that can observe gamma-ray bursts. We've already seen that the study of a burst's afterglow can provide hugely valuable information on the burst itself, such as its distance, how energetic it was, and how much material surrounded the burst. Well, when a satellite such as Swift detects a gamma-ray burst, this information gets sent to astronomers (by email, text message, phone) so that they can immediately search that part of the sky where the burst occurred. They look for an object that's fading quickly. Once they've found the afterglow they can monitor it using telescopes operating throughout the spectrum – first X-rays (see chapter 5), then ultraviolet (chapter 10), optical (chapter 12), infrared (chapter 11) and finally radio (chapter 13). The observations can go on for weeks, months or even years. The study of afterglows is a truly multi-wavelength endeavor, and it is that which has transformed our knowledge of gamma-ray bursts.

Astronomers have yet other ways of prising out the secrets of bursts. For example, it has been suggested that gamma-ray bursts could be the mechanism responsible for creating ultra-high-energy cosmic rays. Well, the new generation of cosmic ray telescopes (see chapter 6) might test that suggestion. Gamma-ray bursts should certainly create ultra-high-energy neutrinos. Well, the new generation of neutrino telescopes (see chapter 7) should be able to detect these; such a detection would present astrophysicists with an exciting opportunity to learn what takes place deep inside a burst. Finally, the creation of a short burst through the merger of two dense objects would inevitably generate gravitational waves – which happens to be the focus of the next chapter.

9

A new messenger from the cosmos

A short gamma-ray burst is formed, so it's believed, when an orbiting pair of neutron stars merge – or else a neutron star is swallowed by an orbiting black hole companion. But how can astronomers study a process that involves such small, dark objects? The key lies in observing a type of radiation that has to date never been detected.

A clock in space – Gravitational waves – LIGO, VIRGO and GEO – Listening to gravity – ET and friends – An array of cosmic clocks

S. Webb, *New Eyes on the Universe: Twelve Cosmic Mysteries and the Tools We Need to Solve Them*, 203
Springer Praxis Books, DOI 10.1007/978-1-4614-2194-8_9, © Springer Science+Business Media, LLC 2012

In earlier chapters we saw how astronomers are using cosmic ray and neutrino observatories to study certain regions of the Universe, and how eventually they may even study the Universe by using dark matter observatories. These are all *particle* observatories. Nevertheless, the commonest way by far through which astronomers learn about the Universe is by catching waves: electromagnetic waves are humankind's main window on the cosmos.

Electromagnetism is one of four fundamental physical interactions. What of the other three? Is it possible to observe the Universe using messengers that carry those interactions? Well, two of them – the weak and the strong interactions – operate at the subatomic realm. These interactions are both crucial in making the world behave in the way it does but they cannot send waves that travel across Universe in the same way that electromagnetism can. The final fundamental interaction – gravity – is by far the weakest of the four. But gravity *can* produce waves that traverse cosmos. And telescopes are being developed that will catch these waves. Soon, another messenger will be delivering secrets from the cosmos to waiting astronomers.

A clock in space

The existence of **gravitational waves** was predicted by Einstein when he developed general relativity – his theory of gravity. According to Einstein, space and time are not some idealized, independent background entities in which events take place. Rather, space and time – or spacetime – can be thought of as constituting a fabric that interacts with matter and energy. Gravity is the distortion brought about by mass–energy: the presence of mass–energy causes the spacetime fabric to curve, and objects moving in that spacetime take a path that follows the curvature. The American physicist John Wheeler put it best: 'mass–energy tells spacetime how to curve; curved spacetime tells mass–energy how to move'.

Now, Maxwell's theory of electromagnetism says that whenever electric charges accelerate – either because they change speed or they change direction – electromagnetic waves are generated. In the same way, Einstein's theory of gravity tells us that accelerating masses can generate gravitational waves. So your hands, as they move to turn the pages of this book, are generating gravitational waves. Tiny, undetectable waves, of course.

Gravitational waves can be described in terms of the properties that describe any other type of wave. For example, gravitational waves can be

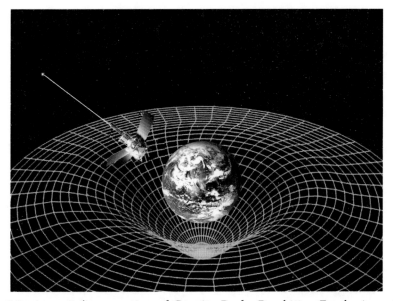

Figure 9.1 *An artist's conception of **Gravity Probe B** orbiting Earth. According to general relativity, Earth's mass causes spacetime to curve. A satellite such as Gravity Probe B, moving freely through space, takes the straightest possible path through spacetime. But that straightest path will follow the curvature. The task for Gravity Probe B is to measure the curvature of spacetime around our planet. (NASA)*

described in terms of their frequency. Although gravitational waves can in principle exist at any frequency they will in general have much lower frequencies than, for example, electromagnetic waves: high-frequency gravitational waves would be generated by large masses undergoing large accelerations, but no known cosmic sources behave in this way. However, lower-frequency waves – in the region, say, 0.1 mHz to 1 kHz – would be generated by processes such as binary stars, stellar remnants spiraling to oblivion in central supermassive black holes, or inspiraling neutron stars. Such frequencies are more typical of audible sound waves than light waves.

Gravitational waves can also be described in terms of wavelength and amplitude. They also have a speed: for a gravitational wave of small amplitude, any particular point on the wave travels at the speed of light. As with electromagnetic waves, a gravitational wave is transverse – in other words, gravitational waves vibrate at right angles to their direction of propagation. And, as with electromagnetic waves, gravitational waves can be polarized (although the polarizations of an electromagnetic wave are at 90° whereas gravitational waves are polarized at 45°).

So far, so similar. However, there are a few differences between electromagnetic and gravitational waves. For example, electromagnetic waves propagate in the framework of spacetime; gravitational waves, on the other hand, are waves *of the spacetime fabric itself*. A gravitational wave thus affects *everything*. If a gravitational wave passes through you as you read this book, your body might get a bit shorter and a bit wider, and then a bit taller and a bit thinner. The exact amount of stretching would depend on various factors, but it's not going to be a large amount. If you were a few light years tall – with your feet on Earth and your head on Alpha Centauri – then the stretching would be of the order of a hair's breadth. Furthermore, electric charges come in two varieties – positive and negative – whereas mass has only one type of 'charge'. One consequence of this is that any process possessing perfect spherical symmetry – such as the spherical pulsation of a spherically symmetric star – will not generate gravitational waves. Nevertheless, there should be thousands of sources of gravitational radiation out there in the cosmos – for example, stars that explode with anything less than perfect spherical symmetry; colliding neutron stars; black holes swallowing the matter in its vicinity. And just as we can observe electromagnetic waves we should, in principle, be able to observe gravitational waves.

'In principle' and 'in practice' are two entirely different things. Calculations show that gravitational waves from even the most violent events will be horrendously difficult to detect. Developing the necessary apparatus and techniques to catch gravitational waves is a huge task and the question must be asked: is it worth embarking on such a gargantuan task? After all, there is a risk that the time, money and effort involved in building the detector might be wasted if it turns out that gravitational waves don't exist: although Einstein's theory predicts the existence of gravitational waves his theory could be wrong. Before astronomers try to build a gravitational wave observatory it would be prudent of them to point to independent evidence for the existence of such waves. Fortunately, astronomers have such evidence. Before describing the new generation of gravitational wave detectors, and discussing their importance for astronomy, it's worth spending a moment to discuss an astronomical object called PRS 1913+16. The study of this object demonstrated the reality of gravitational radiation – and thus that a gravitational wave observatory at least has a chance of success.

As we learned in earlier chapters, a star that's a few times more massive than the Sun is destined to meet a violent fate: once such a star has used up all its nuclear fuel it's destroyed in a supernova explosion and a dense, compact object – a neutron star – is left behind. A neutron star is an extreme object in several ways. Take, for example, its spin. A typical star might spin on its axis once every few weeks. The regions near the Sun's equator, for example, rotate about once every 25 days. As the original star undergoes collapse it will spin faster – just as ice skaters spin faster when they tuck their outstretched arms into their bodies. Well, the fastest neutron stars can spin up to 600 times a second. An even more extreme property of a neutron star is its magnetic field. A typical strength for a star's magnetic field is a few gauss. A randomly chosen place on the Sun, for example, will typically have a magnetic field strength of one gauss. During stellar collapse, the magnetic field strength increases roughly by the ratio of the volumes of the original star to the neutron star. The result is a magnetic field that can be trillions of times stronger than the original star's magnetic field and that is billions of times stronger than anything that can be produced in a laboratory.

So, in the remnants of a supernova explosion we are likely to find an extremely dense, rapidly spinning object with an intense magnetic field. Such an object will emit radio waves in two cone-shaped beams from its magnetic poles and, as it rotates, these beams will sweep round in space like a lighthouse. If one of the beams happens to be pointing towards Earth we will receive a radio pulse; we will, in fact, receive many pulses each second, with a frequency that depends on how fast the neutron star is rotating. Such a beacon would act like a clock in space. In 1967, the radio astronomers Jocelyn Bell and Antony Hewish discovered just such an object – they called it a **pulsar**, a shortened form of 'pulsating star'.

After that discovery, many astronomers began looking for other examples of this new phenomenon. Joseph Taylor and Russell Hulse, for example, used the Arecibo radio telescope to search for new pulsars. They discovered 40 of them but one in particular, discovered in the summer of 1974, was rather different from the rest. The new object, which they named PSR 1913+16 ('PSR' stands for pulsar; '1913+16' gives its position in the sky), emitted pulses that seemed to vary in their timing. The pulses themselves were stable: the time between two sweeps of the beacon was 0.059 03 s, with a precision that rivaled an expensive Swiss timepiece. But, superimposed on that regularity, the pulses 'bunched' with a period of about eight hours: the pulses would arrive sooner than expected, then about four hours later they would arrive slightly later than expected, then in another four hours the pulses would arrive once more ahead of schedule. Hulse and Taylor soon realized that they were observing a pulsar that was orbiting another object

Figure 9.2 *An artist's conception of the binary pulsar. Note that this is not drawn to scale: the purple discs should really be separated by about 200 m! (Michael Kramer, Jodrell Bank Observatory, University of Manchester)*

with an orbital period of 7.75 hours: as the pulsar approached Earth, the Doppler effect meant that the pulses reached us more frequently; and as it receded, the pulses reached us less frequently. A careful analysis of the shift in the pulses demonstrated that the pulsar's companion was the same mass as the pulsar itself: both had a mass about 1.4 times that of the Sun. And since neither object showed up on photographs, Hulse and Taylor concluded that the companion was also a neutron star. The system became known as the 'binary pulsar'. Note that the companion might also be emitting radio beams like its pulsar companion; we would not see them, however, if the beams happened not to sweep across our line of sight.

Over the next few years, Hulse, Taylor and their colleagues studied the pulses from this rare system in exquisite detail. They found that the orbital period was getting shorter: in other words, the two neutron stars were rotating faster about each other in an increasingly tight orbit. The effect was small – each year the orbital period gets smaller by 76.5 millionths of a second, so it will be about 300 million years before the two objects spiral into each other – but by observing the binary pulsar for long enough they were able to measure the effect. And why should the orbital period change in this way? Why should they **inspiral**? Well, this is *exactly* what general relativity predicts will happen in such a system: as these dense objects orbit each other, the system emits energy in the form of gravitational waves. The agreement between what Einstein's theory predicts and what Hulse and Taylor observed is striking. It is beautiful. It is proof that gravitational radiation exists and it was no surprise when Hulse and Taylor won the 1993 Nobel prize for the discovery of this cosmic clock.

Gravitational waves

Einstein predicted the existence of gravitational waves as far back as 1917 and yet, at the time of writing, gravitational waves have still not been observed directly. Why the delay in what would be a key discovery? Well, theorists didn't begin to examine the gravitational wave concept with real thoroughness until the late 1940s. And it wasn't until the late 1960s that experimenters really tried to detect them. The scale of the challenge was perhaps too daunting for everyone concerned.

Let's look at some of the difficulties faced by experimenters when they try to detect gravitational waves. To begin with, in order to detect the effects of a passing gravitational wave they must define some reference against which they can make measurements. In the case of electromagnetism there are no such problems: uncharged particles can be used as a reference against which the relative motion of charged particles can be measured. A gravitational wave, however, interacts with *all* particles it meets. There are no 'uncharged' particles as far as gravity is concerned: gravity affects everything. One solution is to consider a ring of free particles, as in figure 9.3. If a weak gravitational wave passes through the particles in a direction following your line of vision into the book, in other words if the wave travels perpendicular to the plane of the particles, then the ring will be stretched in one direction and squeezed in another. In fact, the particles can oscillate in two different ways, depending upon the polarization of the wave. A so-called **plus-polarized gravitational wave** causes the oscillation pattern shown in the upper part of the figure; a **cross-polarized gravitational wave** causes the oscillation pattern shown in the lower part of the figure. Note that such a wave would cause no motion along its direction of propagation; the particles would stay in their plane, but move to and fro in that plane.

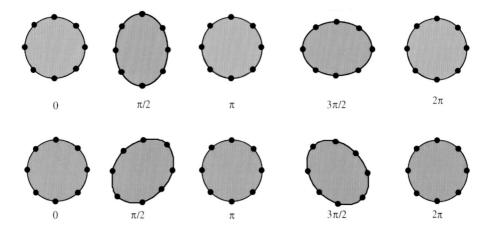

Figure 9.3 *Top: how a ring of particles would be affected if a plus-polarized gravitational wave moved through the ring in a direction perpendicular to the ring (in other words, in a direction following your line of vision into the book). Bottom: how the same ring would be affected if a cross-polarized gravitational wave passed through it. As you can see, a plus-polarized wave causes the ring to move along axes defined by a plus sign; a cross-polarized wave cases the ring to move along axes defined by a cross. The number underneath the ring gives the phase of the oscillation: over one full period, the phase of the wave goes from 0 to 2π.*

So experimenters have a signal they can look for when searching for gravitational waves. But how *big* is the signal? In other words, how much would a ring of free particles be distorted by a passing gravitational wave? This is where things get really difficult. The problem is that although the fabric of spacetime is elastic, and thus permits the existence of gravitational waves, the fabric is really very, very, *very* stiff. The relative change in distances due to a passing gravitational wave is tiny. For example, suppose a supernova occurred somewhere in the Galaxy. Unless by some fluke the explosion was purely spherical, such an event would emit gravitational radiation. If physicists could somehow arrange for test particles to be put into a ring with a radius of about 1 km then the gravitational waves from the event would distort the ring by about one thousandth the size of an atomic nucleus. So *that* is the challenge. It isn't that gravitational radiation from the supernova carries little energy: on the contrary, at its peak the supernova would drench every square meter on Earth with kilojoules of energy each second. The difficulty is that experimenters must be able to measure distances with incredible precision; as hinted earlier, it's like measuring the distance to Alpha Centauri to within the width of a human hair.

It's therefore almost a miracle that the current generation of gravitational detectors have already achieved this sort of sensitivity. And, as we shall soon describe, their sensitivity is set to improve. Even more promisingly, researchers in different laboratories have begun to collaborate and pool their resources. Almost a century after Einstein first proposed the existence of gravitational waves it seems as if the direct detection of gravity waves may soon happen.

LIGO, VIRGO and GEO

The best way to measuring small changes in distance has always been to use an **interferometer** – and to make the interferometer as large and sensitive as possible. Let's look at the sort of interferometer that's required to detect changes caused by a passing gravitational wave.

First, imagine two steel tubes, both of them several kilometers long, arranged so that from above they seem to form an L shape. Light from a laser (let's make it a really powerful laser) enters the interferometer at the corner of the L, where a mirror splits the beam into two and sends one beam down each of the tubes. This is not a problem in itself, but for the interferometer to work the frequency of the light must be *incredibly* stable: over a period of one second the light must not vary by more than about a hundred millionth of a cycle. Second, as the light beams travel down the tubes it's critical that they suffer the smallest possible amount of scattering from stray gas molecules. Evacuating the tubes to a pressure of one trillionth of an atmosphere, and ensuring that the steel is of such high quality that nothing can leak inside, should do the trick. Each tube, in other words, must hold a large, sustained, ultra-high vacuum. Third, when the beams reach the end of their respective tubes they must bounce off a mirror that's suspended there. Since the interferometer must be sensitive to small distance changes it follows that the mirrors at the ends of the tubes, and the mirror at the corner of the L, must be stationary; in other words, all three mirrors must be protected from vibration. Indeed, the mirrors must be so well protected that the random motion of atoms within the mirrors and the fibers from which they are suspended become detectable. Finally, software must continuously monitor the whole system – lasers, mirrors, tubes – to check everything is aligned and in position.

An interferometer set-up such as that described would be a hugely impressive feat of precision technology. And the point of it all?

Well, when the two laser beams recombine at the corner of the L they form an interference pattern. If the paths taken by the two light beams differ by half a wavelength, then the crest of one wave combines with the trough of the other wave and the combining waves completely cancel: destructive interference occurs and no light is seen at the detector. If the paths taken are such that the two beams are always in phase, then constructive interference occurs and one observes alternating bright and dark spots. In the general case, the interference pattern would be some arrangement of bright and dark spots. Now, if nothing moves – if the three mirrors maintain a perfectly constant distance from each other – then the interference pattern remains constant. But suppose that a gravitational wave passes through the interferometer, squeezing and stretching space as it does so. The wave would stretch the distance along one arm while shortening the distance along the other. The distance between the three mirrors would move slightly – *and the interference pattern would change*. If the 'arms' of the L were long enough even a change in length smaller than an atomic nucleus would make itself visible through a changing interference pattern. The changing pattern of light would encode information about the relative change in length between the two arms, and that in turn would give scientists information about what produced the gravitational wave.

The difficulty of making a working gravitational wave interferometer is immense, but in the 1990s a collaboration between two American institutions took on the challenge. In 1992, the Caltech physicists Kip Thorne and Ronald Drever along with the MIT physicist Rainer Weiss founded a project called **Laser Interferometer Gravitational-wave Observatory** (LIGO); in 1999, LIGO had its inauguration ceremony. LIGO actually operates at two sites: at Hanford Observatory near Richland in Washington State, and at Livingston Observatory in Louisiana. Both observatories are home to a set-up similar to the one described above – an L-shaped interferometer with an arm length of 4 km. In fact, by bouncing the laser beam several times off the mirrors the effective optical length of the arms becomes even longer.

Why the need for two observatories in the LIGO experiment? Two reasons. The first is that almost anything can shift those mirrors much more than gravitational waves can: the rumble of a remote earthquake; the murmur of distant road and train traffic; even the thumps from amorous cattle nearby can affect the mirrors. By comparing the shifting interference patterns from

Figure 9.4 Top: an aerial view of the LIGO experiment at Hanford Observatory in Washington. The LIGO experiment has a similar interferometer at Livingston Observatory in Louisiana. Bottom: an aerial view of the VIRGO experiment based at the European Gravitational Observatory near Pisa, on the river Arno plain. (Top: LIGO Laboratory. Bottom: Aspera/CNRS/IN2P3)

two widely separated interferometers there's a chance that much of this random 'noise' can be recognized and thus ignored. The second reason is that, since gravitational waves travel at the speed of light, a gravitational wave will not hit the detectors at the same time. About 3000 km of Earth's bulk separates Richland and Hanford so, depending upon arrival direction, a gravitational wave could take up to one hundredth of a second to travel between the two LIGO observatories. Measuring the time delay between gravitational wave signals at the two observatories would determine where the wave came from in the sky.

If other detectors joined in the hunt then other properties of a gravitational wave could be determined. Well, although LIGO possesses the greatest sensitivity of any detector, others have been built. A joint French–Italian project called **VIRGO** (which name, to the best of my knowledge, is not an acronym) is a laser interferometer similar to LIGO but with slightly smaller arms: each arm is 3 km long. The VIRGO interferometer started its first science run in 2007. A joint German–British project called **GEO600** is another laser interferometer – this time with arms that are each 'only' 600 m long; GEO600 has been operational even longer than VIRGO. (The project was initially called the Gravitational European Observatory, or GEO for short, and was planned to have an arm length of 3 km. Funding cutbacks meant that the project was never built, and instead scientists constructed a smaller interferometer with an arm length of 600 m – hence the name GEO600.) These three observatories, as with all other ground-based interferometers, are sensitive to gravitational waves in the frequency range between about 10 Hz up to about 5 kHz.

The three observatories, LIGO, VIRGO and GEO600, pool their results. The amount of data they collect is staggering. LIGO alone generates so much data that a distributed computing project called **Einstein@home** has been developed; the software uses the idle time of computers belonging to members of the public to crunch through the numbers, in the same way as the popular **SETI@home** project uses a worldwide network of computers to search for radio signals from extraterrestrial civilizations. In addition to pooling their data, each of the observatories continually look to upgrade their equipment. Scientists involved with the LIGO project, for example, will soon replace the initial set of detectors with a new, more sensitive set; the metal wires from which the mirrors are suspended will be replaced with a special fiber made of glass, whose atoms jostle each other less than in metal and thus jiggle the mirrors slightly less; and a new laser will have twenty times more power than previous lasers used in the experiment. These upgrades mean that the Advanced LIGO project, when it starts operation in 2014, will be ten times more sensitive than LIGO itself.

The various interferometers around the world are now surely sensitive enough to detect certain types of gravitational wave and that sensitivity is increasing. So if the present understanding of general relativity is correct then these detectors will soon ring to the sound of a passing gravitational wave. But where would these waves come from?

Listening to gravity

The binary pulsar PSR 1913+16 is radiating gravitational energy as the two neutron stars inspiral – that much is known. The amplitude and frequency of the gravitational waves from the system are essentially unchanging. Perhaps LIGO and the other detectors could observe gravitational waves from this system? The waves would show up as a regular stretching and squeezing of the two arms of the detector, similar to the signal shown in figure 9.5 but at a much lower frequency. It's a nice idea, but unfortunately this won't happen. The binary pulsar is currently emitting gravitational waves of extremely low frequency. The frequency range over which the current crop of interferometers is sensitive – about 10 Hz up to about 5 kHz – is far too high to allow detection of waves from PSR 1913+16.

However, as the inspiral process in a binary system proceeds towards the inevitable collision – with the two objects colliding and merging – both the frequency and the amplitude of the gravitational waves increases. It turns out that for two objects, each with a mass of 1.4 solar masses, the gravitational waves emitted by an inspiraling system run through an interferometer's range of sensitivity in the last half-minute or so before collision. It just so happens that the frequency range to which LIGO and the other detectors are sensitive is in the range of human hearing. If those gravitational waves were represented by sound waves then the final moments of an inspiraling binary system would sound like a bird's chirp. So a **chirp** is the tell-tale signal that LIGO and the others are listening for. If LIGO hears a chirp then it will have heard the death cry of two inspiraling neutron stars – and, presumably, the sound of a short gamma-ray burst, as discussed in the previous chapter. This, then, is how astronomers can study the details of short gamma-ray bursts, a process that involves such small, dark objects: they can watch the process unfold using gravitational waves.

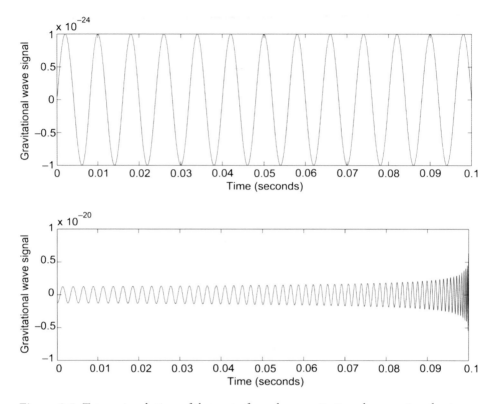

Figure 9.5 Top: a simulation of the sort of regular gravitational-wave signal astronomers expect to originate from a system such as two neutron stars orbiting each other at a large distance. The gravitational waves they produce will be weak and have a nearly constant frequency. In this example, the waves shown have a frequency of about 125 Hz. Bottom: a simulation of the gravitational-wave signal astronomers expect to detect from an inspiraling pair of neutron stars. Note the vertical scale: the signal strength is much larger than in the upper figure. As the neutron stars get closer the gravitational wave amplitude increases. The gravitational wave frequency also increases towards the end: this is the so-called chirp. (A Stuver/LIGO Laboratory)

Unfortunately, in the case of the binary pulsar gravitational wave observatories will have to wait a further 300 million years or so to hear the system's death chirp. Clearly, then, we shan't be hearing any time soon from PSR 1913+16. But the Universe is large. There are lots of galaxies out there, and thus lots of opportunities for neutron star binary systems to form. Since LIGO is sensitive to binary inspirals from as far away as the Virgo cluster, which is 15 Mpc away, there's a good chance that it will hear such a merger;

the Advanced LIGO detector, with its much greater sensitivity, will surely hear these events.

Although, at the time of writing, no gravitational wave observatory has heard a chirp from a short gamma-ray burst, or indeed any of the other gravitational 'sounds' that the Universe might make, even such non-detection can provide information to astronomers. In 2007, Swift and other gamma-ray satellites detected a short burst of radiation seemingly originating in the direction of the nearby galaxy M31 (better known as the Andromeda galaxy). If this burst – GRB 070201 – was indeed a short gamma-ray burst caused by the merger in M31 of two inspiraling neutron stars then LIGO would certainly have detected gravitational waves from the collision. But LIGO heard nothing. That non-detection demonstrates either that the collision occurred much farther away and just happened to be in the line of site with M31, or else the event occurred in M31 but some other, less extreme process was involved.

Already, then, gravitational wave observatories are adding to astronomers' store of knowledge. It may seem strange that the non-detection of a hypothetical entity counts as a discovery. Nevertheless, the orbits of the binary pulsar PSR 1913+16 can *only* be explained in a simple way if gravitational waves exist; and LIGO is now so sensitive – it's perhaps the most sensitive instrument ever built – that it *would* have heard the chirp of gravitational waves from a pair of nearby inspiraling neutron stars. There is no doubt that even now LIGO is making contributions to astronomy. Advanced LIGO – and the other upgraded detectors – will make even deeper contributions. The era of gravitational wave astronomy is almost upon us.

ET and friends

The current generation of gravitational wave observatories will surely soon start to hear the chirp from an inspiraling pair of neutron stars, or a neutron star/black hole binary. But if observatories could be made even more exquisitely sensitive, or if they could be made to hear other gravitational wave frequencies, then astronomers might listen to different phenomena: waves from the first trillionth of a second after the Big Bang; waves from supernovae; perhaps even waves from sources such as cosmic strings, hypothetical structures that might have been generated in the very early Universe, that would be completely new to physics. The possibilities for science are immense – if only astronomers could open this new window on the cosmos.

What, then, are the prospects for the next generation of gravitational wave observatories?

The omens are mixed. On the bright side, the European Commission has given the go-ahead for eight institutions involved in gravitational wave research to undertake a study on how to design a so-called third-generation observatory. The current VIRGO, GEO600 and LIGO projects are all first-generation observatories; the enhancements that all three teams are currently making will lead to a second-generation of observatories. But the European third-generation **Einstein Telescope** (ET), if it's constructed, would listen for gravitational waves with a hundred times the sensitivity of the first-generation detectors.

In 2011, scientists from the 17 countries involved in the ET project released some design details. In order to limit the effects of seismic noise the observatory will be built underground, perhaps as deep as 200 m. There will be three independent detectors, each consisting of two arms with a length of 10 km. The arms will be parallel to Earth's surface and join at an angle of 60°. The mirrors will be cooled to a temperature of 10 K in order to limit the random jiggling caused by thermal effects. A site for ET has yet to be chosen, and a host of technical challenges must still be overcome, but it is hoped that the telescope will be completed by 2017. However, even if it is completed by then it won't start making observations until later – and my guess is that a final go-ahead for the project might require the direct detection of gravitational waves by the second generation of detectors. However, if ET works, the size of the payoff more than matches the size of the technical challenges: the telescope will possess good sensitivity at the relatively low frequencies of 1–10 Hz. An instrument such as ET would enable completely new astronomical observations to take place.

The Europeans are not the only scientific community involved in thinking about new gravitational wave observatories: the Japanese government, for example, has given the go-ahead for the construction of the **Large Scale Cryogenic Gravitational Wave Telescope** (LCGT). The LCGT will be built in the Kamioka mine and will have an arm length of about 3 km.

These new observatories are pushing the boundaries of what's technically possible when building a gravitational wave detector on Earth. But, just as our view of the visible Universe has improved by using space-based telescopes, so our view of the gravitational Universe might improve if we could build a space-based gravitational wave observatory. Is such a development feasible? Here the situation is less promising.

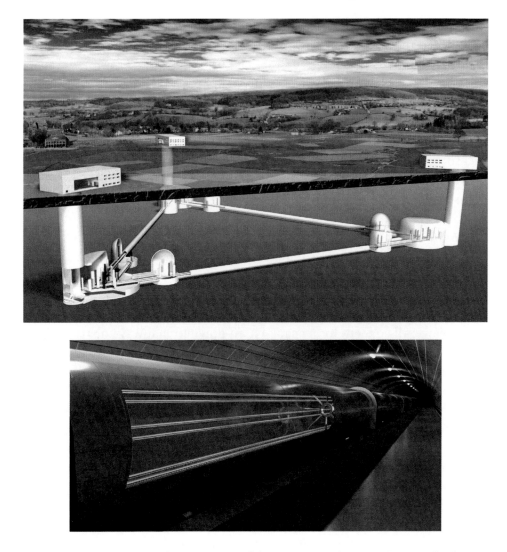

Figure 9.6 Top: an artist's impression of the Einstein Telescope. Current thinking is that the Einstein Telescope will consist of three nested detectors. The interferometer arms, which will be constructed 100–200 m underground, will be 10 km long (so 2.5 times the arm length of LIGO and 3.3 times the arm length of VIRGO). One interferometer will detect low-frequency gravitational waves (in the range 1–40 Hz) while the other will detect higher-frequency gravitational waves. The site has yet to be decided; seismic noise will play a role in determining the best site, but other factors will include the geological characteristics of the soil, human activities, and environmental conditions such as wind and ocean waves. Bottom: an artist's impression of one of the Einstein Telescope interferometer arms through which a laser beam will travel. (Top: CERN; bottom: Aspera/G. Toma/A. Saftoiu)

The **Laser Interferometer Space Antenna** (LISA) was a planned joint project of NASA and ESA. In 2010, the influential decadal report of the US National Research Council recommended that LISA should be one of two large space missions for NASA in the next decade. The hope was that, if a Pathfinder mission were able to test LISA technologies and find them successful, then work on LISA could begin in 2016 with a launch date some time around 2025. LISA would not only have the potential to be even more precise in its measurements than ground-based observatories but, more importantly, it would operate over a completely different frequency range: 0.0001–0.1 Hz, so oscillation periods would vary between a few seconds to a few hours, which is the sort of timescale in which lots of interesting events occur. Just as the Arecibo radio telescope produces a different view of the sky compared to the Keck optical telescope, for example, so LISA would produce a different view of the sky compared to LIGO. Whole new areas of study would open up. That was the hope, anyway.

Scientists had been studying possible designs for LISA for many years, and the overall concept for a space-based gravitational wave observatory is pretty much accepted. Such an observatory would consist of three identical spacecraft arranged in an equilateral triangle, with each arm of the triangle being 5 million kilometers long. The center of this equilateral triangle would orbit the Sun in an Earth-like orbit, but 20° 'behind' Earth; by arranging to have the spacecraft so far away from home, the gravitational effects of the Earth–Moon system – effects that might distort the LISA triangle – would be minimized. The arms of the triangle, of course, would form an interferometer; in fact, they would form two independent interferometers. As a gravitational wave passed through the triangle, squeezing and stretching space as it did so, the distance between the spacecraft would vary – and the interferometers would pick up these characteristic changes in distance.

An experiment like LISA could only succeed if the motions of the three spacecraft were well understood and their relative positions were stable. Scientists developed an ingenious method for ensuring that the LISA constellation would be as stable as possible. The plan was that inside each satellite would be a small cube of gold–platinum alloy – a 2 kg test mass – and it's these cubes that would fall freely through space and respond to gravitational waves. So LISA would measure not the separation between the three satellites themselves but rather the separation between the three freely moving test masses. Now, there are many potential disturbances that could affect the free motion of these test masses; since it's critical that the test masses

Figure 9.7 *An artist's impression of the three* LISA *spacecraft, orbiting the Sun far away from Earth. The spacecraft, 'connected' by lasers, would search for the gravitational waves emitted by phenomena such as gamma-ray bursts.* (ESA)

are disturbed only by gravitational waves, these other disturbances – both external and internal – would have to be recognized and, where possible, eliminated. An example of an external disturbance would be the pressure of light from the Sun. An example of an internal disturbance would be the onboard computer: it would generate an electric field that could act on the test masses. These are minuscule effects but they would swamp the effect of a passing gravitational wave. These problems can be minimized by using the spacecraft itself. For example, the material of the spacecraft could act as a protective shield against the Sun's radiation. And, by ensuring that the separation between test mass and spacecraft remains fixed, disturbances could be prevented from affecting the free motion of the test mass. For this to work the separation between test mass and spacecraft would have to be monitored constantly and, if it changed, sensitive microthrusters would have to fire and gently accelerate the spacecraft back into position. This method of positional control is called **drag-free operation**. The technique has been used in satellites that are in low-Earth orbit, but with nothing approaching the exquisite accuracy demanded by LISA: the position of the spacecraft, relative to the test masses, would have to be controlled to within a hundredth of the wavelength of visible light.

In order to capture the signal of a gravitational wave, LISA would have to measure changes of 5 pm (that's 5 picometers, or just 5 millionths of a millionth of a meter) between test masses. This is possible using laser interferometry, but the techniques required are slightly different to those employed by LIGO and the rest of the current generation of detectors. LIGO, after all, has an arm length of 4 km; the LISA arm length would be 5 million km! The laser beams would spread out over such large distances and it's unreasonable to expect that a beam could be sent out by one spacecraft, reflected by a second, and then picked up again by the first craft. However, there's a way round this. Rather than reflecting it, the second spacecraft could measure the phase of the incoming laser beam; the second spacecraft could then send back its own laser beam at full power, with a phase that was set using information from the first beam. These are relatively new techniques. Can they work? In 2010, LISA scientists tested the new measurement techniques and verified that they do indeed have the required sensitivity.

If LISA's ears were working clearly then it would hear the chirps from inspirals. But it should hear more: bursts from supernovae (which could be correlated with signals from more traditional observatories and also even neutrino detectors) and periodic blips from pulsars. And by studying how the signals change over a year, as the LISA constellation orbits the Sun, it would even be possible for astronomers to determine the rough direction in the sky from which the gravitational waves came. It seems likely, however, that something like LISA would hear a cacophony, a mix of sounds that astronomers have not imagined: after all, if you had never heard a sound before what would you expect to hear if you were suddenly given hearing? That's the exciting situation in which LISA scientists were hoping to find themselves.

Unfortunately, the same NASA funding cuts that in 2011 drove ESA to withdraw from collaboration in the International X-ray Observatory (see chapter 5) also drove them to withdraw from the LISA collaboration. The technologies outlined above are currently under study for a possible European-only mission; but inevitably such a mission would be narrower in scope than the original LISA design.

Japan, meanwhile, is still looking at designs for a space-based mission called the **Deci-hertz Interferometer Gravitational Wave Observatory** (DECIGO). It gets its name because it would be most sensitive to gravitational waves in the frequency range 10 Hz down to 0.1 Hz – and 0.1 Hz is a deci-hertz. The basic design is similar to LISA: there would be three drag-free spacecraft, but the arm length would be only 1000 km. The planned launch is 2027, but at least two pathfinder missions would have to be launched before then, and concluded successfully, before DECIGO itself was launched.

Even if DECIGO and a revised LISA mission (or perhaps a combined LISA–DECIGO mission?) go ahead, one sound from the Universe will be at the very limit of their hearing: the low thunder coming from the Big Bang itself. Another proposed gravitational wave mission aims to detect radiation from the instant the observable Universe came into being. The thinking behind it is as follows.

As described in chapter 2, telescopes that catch photons cannot look back further in time than the recombination epoch about 380 000 years after the Big Bang. The reason for this is that no photons exist from before that time; only after recombination were photons free to stream through the Universe. Neutrino telescopes (which we discussed in chapter 7) could, in principle, do better job of illuminating the nature of the Big Bang; neutrinos were able to move freely from about a second after the Big Bang. If astronomers could observe these relic neutrinos, particles that form a cosmic neutrino background, then they would be looking at particles that came into existence when the observable Universe was one second old. Even if astronomers could observe these relic neutrinos, however, they wouldn't actually be that much closer to the Big Bang itself: the really interesting events, the physics that drove inflation, took place in the first 10^{-35} s of the Universe's evolution. It's impossible to detect photons or neutrinos from those earliest times; such entities did not even exist. But gravitational waves from the birth of the Universe have been travelling since then essentially unimpeded – and in principle it should be possible to observe them.

The **Big Bang Observer** (BBO) is an international proposal to build an instrument with the capacity to observe gravitational waves from the inflationary Universe. (Of course such an observatory would necessarily have the capacity to observe gravitational waves from many other interesting sources, too.) The proposal is at the very earliest stages, with scientists merely investigating the concept; it's a follow-on observatory to LISA and, since the outlook for LISA is far from clear, it's completely unknown when or whether BBO could fly. Nevertheless, the initial thinking is that BBO would consist of three constellations of spacecraft. Two of the constellations would consist of three spacecraft arranged, as with LISA, to form an equilateral triangle; the third constellation would consist of two sets of three spacecraft arranged at the six points of a Star of David. The three constellations would follow Earth's orbit around the Sun, with one constellation being about 20° of longitude away from Earth and the other two constellations being 120° ahead and behind the first. As with the initial LISA design, BBO would measure the distances between freely floating masses inside the spacecraft; but whereas the LISA arm length was planned to be 5 million km, the BBO arm length would be a hundred times shorter – only 50 000 km.

The BBO, if it worked, would look for gravitational waves with a frequency of about 1 Hz with a sensitivity that would enable it to investigate whether those waves came from the earliest moments when the Universe was undergoing inflation. But it's a big 'if'. The BBO would require vast improvements in many areas of the technology that scientists hoped to use in LISA. For example, the accuracy of distance measurements would have to be improved by a factor of a million compared to what LISA required. So while the BBO might give us a glimpse of the Big Bang it certainly won't be doing so anytime soon: perhaps it's a mission that our grandchildren can look forward to with interest.

An array of cosmic clocks

As in the previous chapters, I'd like to conclude with a short section on how other types of observatory could be used to address the question under discussion. In this case, how could other types of telescope study inspiraling neutron stars or the inspiral of a neutron star into a black hole? As we saw in the previous chapter, the short gamma-ray burst to which such an inspiral event gives birth can be studied by telescopes working throughout the electromagnetic spectrum. The trouble, though, is that in the case of short gamma-ray bursts – and in other extreme cases too, such as events that took place in the earliest moments of the Universe – electromagnetic radiation is not the best information carrier. Gravitational radiation carries information to us in a way that electromagnetic radiation cannot. That's the whole point behind the development of gravitational wave observatories. So far, the only type of gravitational wave detector we've considered is the interferometer. If we want to detect gravitational waves, and learn more about those extreme events, are there any other approaches that could be taken?

In fact, Joseph Weber made the earliest serious attempt to detect gravitational waves using a device known as a resonant bar detector. The idea is that a bar of metal will have some frequency at which it naturally resonates or 'rings'; if a gravitational wave passes through it will deform the bar and thus cause it to change from its characteristic ringing frequency. Weber invented his first resonant bar detector in the 1960s and research into these devices has continued.

Unfortunately, it's no easier to detect gravitational waves using a resonant bar detector than it is to use an interferometer. One problem is that the atoms in the bar will have a random motion simply because the bar is at some temperature, and this random, or thermal, 'jiggling' will generally create a much larger signal than that created by a passing gravitational wave. To counter this, the bar – which should possess a mass of several tons; the larger the better – must be cooled to very low temperatures. Another problem involves the sensitive electronics required to convert the ringing of the bar into electrical signals that can be analyzed. Needless to say, at the time of writing none of the resonant bar detectors in operation have detected a gravitational wave.

Another possibility is to make use of the cosmic clocks that Nature has kindly provided: pulsars. Certain types of pulsar – the millisecond pulsars – spin quickly and emit radiation pulses hundreds of times a second. In many cases the frequency of these pulses varies only a little over long periods of time, and those millisecond pulsars make excellent clocks. The key point is that if a gravitational wave happens to pass between the pulsar and a radio telescope then the time it takes for the pulses to arrive would be seen to vary: such a wave, after all, affects spacetime. If astronomers could measure the deviations in relative timing of lots of millisecond pulsars, spread over the sky, then the presence of a passing gravitational wave could be deduced – as well as its direction of travel. For this technique to work astronomers must study lots of millisecond pulsars and, until recently, this would have been a stumbling block. However, NASA's Fermi spacecraft, which we discussed in the previous chapter, has discovered new pulsars; in early 2010, astronomers announced that Fermi had discovered 17 new millisecond pulsars. That brought the total number of such known objects in the Galaxy to more than 100. Unfortunately, few of them are both bright enough and regular enough to be used in the search for gravitational waves. Further Fermi discoveries, however, should soon bring the number of bright, regular millisecond pulsars to between 20 to 40 – enough for astronomers to have a functioning pulsar timing array. The pulsar technique would work at a frequency that could allow a study of the waves that are generated in the mergers of supermassive black holes at the centers of galaxies. Such a pulsar timing array, then, could soon be joining the hunt for gravitational waves – but at a fraction of the cost of LIGO, ET, LISA, et al.

The race for the first direct observation of a gravitational wave is still on. Sometime within the next decade the race will surely end. And the end will mark the start of a completely new window on the Universe – gravitational wave astronomy will have become a reality.

10

The cosmic-wide web

Most of the Universe is missing. It's not just the form of dark energy and dark matter that's unknown; the present whereabouts of half of the 'normal' matter in the Universe remains something of a mystery. Recently, however, astronomers have discovered hints of the presence of a cosmic web – and it seems that this web is where the missing matter resides. The key to this discovery has been the ultraviolet telescope.

A web that links the cosmos – Forests of the night – UV glasses – The Universe on a WHIM – UV blindness?

S. Webb, *New Eyes on the Universe: Twelve Cosmic Mysteries and the Tools We Need to Solve Them*,
Springer Praxis Books, DOI 10.1007/978-1-4614-2194-8_10, © Springer Science+Business Media, LLC 2012

Our lack of knowledge of the Universe is spectacular. As we saw in the early chapters of this book, about 95% of the Universe is in two unknown forms. Those forms have been given the names 'dark energy' and 'dark matter', but these are simply labels of ignorance. At least we understand the other 5% of the Universe, right? At least astronomers have a handle on the 5% of the Universe that's in the form of 'ordinary' matter?

'Ordinary' matter – the sort of matter that's made up of atoms, which in turn go on to make up you, me and everything we see around us – is often called **baryonic matter**. Baryons are heavy particles such as protons and neutrons, and since they are so much more massive than electrons it's in these particles that nearly all of the mass of atoms resides. So the matter that astronomers observe through telescopes – planets, stars, clouds of interstellar dust and gas – that's all baryonic matter. Well, astronomers have often taken a census of baryonic matter – the 5% of the Universe they think they understand – and every time they do this they observe only about half the amount of material their models tell them should be there. So not only is 95% of the Universe in some unknown form, half of the remainder is missing. Our ignorance about the present-day Universe is almost total.

At least, that was the case until recently. Astronomers are finally learning where that missing matter is located. Of course, knowing the whereabouts of 'ordinary' matter still means that 95% of the Universe remains in a dark form. But that 5% of 'ordinary' matter is important to us humans – and understanding its whereabouts can tell astronomers much about what the visible Universe looks like on the largest scales.

A web that links the cosmos

The Universe today is full of lumps. We live on one of many small lumps of matter that happens to orbit a rather larger lump. Stars group into much larger lumps called galaxies, and even the galaxies convene in larger lumps; our Galaxy, for example, resides with three dozen or so other objects in the **Local Group**. And there are lumps even larger than such local groupings of galaxies. In the 1960s and 1970s it gradually became clear that the Local Group is just one of a hundred or so groups and clusters belonging to the **Virgo Supercluster**. And in the 1980s astronomers began to identify lumps even bigger than superclusters. For example, in 1987 R. Brent Tully discovered that the Virgo Supercluster is part of a gigantic structure he called the **Pisces–Cetus Supercluster Complex**. Two years later, Margaret Geller and

John Huchra discovered a similar-sized structure they called the **CfA2 Great Wall**. Larger than either of these lumps is the **Sloan Great Wall**, whose existence was made public in 2003 after an analysis of data from the Sloan Digital Sky Survey. The Great Wall is about 1.37 billion light years long – which makes it, at the time of writing, the largest known structure in the Universe. In between these structures are immense voids and supervoids, in which galaxies are missing. On large scales the Universe looks rather like a sponge.

Explaining how this lumpiness came into being is not easy because, as we saw in chapter 2, when astronomers look back in time to the recombination epoch they see a cosmic microwave background that was extremely smooth. How did the Universe come to be so lumpy after it started out so featureless? The obvious answer to this question is that gravity pulled on matter and caused it to form clumps, and those clumps attracted other clumps to form more massive chunks, and so on until stars and galaxies and clusters formed. That's a plausible answer and certainly gravity must be involved. However, this can't be the whole story: if the only material in the Universe is the glowing material that astronomers actually see – stars, galaxies, clusters – then there has been insufficient time for gravity to generate the lumps. Some extra factor must be at play.

It's now known what that extra factor must be, and the answer was hinted at in chapter 3: dark matter lets gravity do the job. The advent of supercomputers in the 1980s meant that astronomers could to test their models of large scale structure formation using software. They could build a virtual Universe: they could pick a set of initial conditions that might have been in place just after the Big Bang and then sit back and watch 13 billion years' of their Universe's evolution in just a few days. By comparing the model results with what's actually observed they could begin to understand the initial conditions. At the same time as computer hardware and software advances were taking place, cosmologists were coming to accept that the Universe contains large amounts of cold dark matter. By putting dark matter into the recipe along with baryons it turns out that gravity has had plenty of time to create these structures. Furthermore, in 1996 Dick Bond, Lev Kofman and Dmitri Pogosyan showed how random fluctuations in a very early Universe dominated by cold dark matter would create patches of high density and patches of low density. In turn this pattern of density fluctuations would cause tidal interactions, with nonlinear effects causing the pattern to sharpen. The result would be an early Universe in which there were filaments and voids – a foamy structure that would grow into the walls and voids we see today.

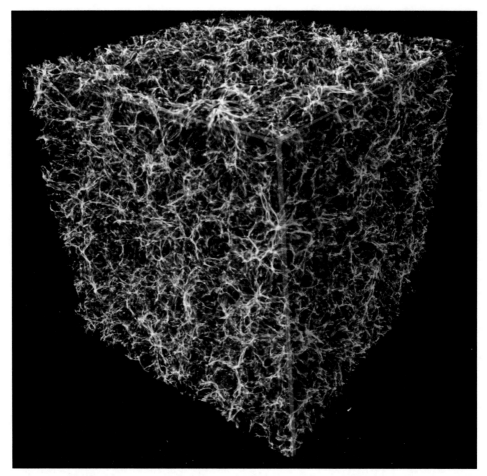

Figure 10.1 *The results of a computer simulation of the spongelike texture of the cosmic web that is our Universe. The filaments along which galaxies cluster contain much more dark matter than the type of matter that shines (so-called 'normal' or baryonic matter). Astronomers have discovered another puzzle involving baryonic matter: the amount of baryonic matter that shines in present-day galaxies is less than the amount of baryonic matter that must have been present in the early Universe. Where did all those baryons go? (NASA, ESA, & E. Hallman, University of Colorado, Boulder)*

Bond, Kofman and Pogosyan called this filamentary structure the **cosmic web**. In the years since they developed the theory, computer simulations of structures in the early Universe have converged in their predictions. The first large structures to form are long threads, or filaments, of dark matter; the threads are connected at their ends by nodes. The first building blocks from which galaxies later develop are formed inside the threads, the filaments of the cosmic web. When these building blocks start to emit light, they trace the path of the otherwise invisible threads – appearing much like bright pearls on the finest nylon string. Over the course of billions of years these early galaxies move along the threads towards the nodes, and it is at the nodes that galaxy clusters form. The Universe thus changes in character as time passes: early on it is the filaments that dominate; later it is the large galaxy clusters that dominate. Traces of those primordial threads are still visible, though, in the great walls – such as CfA2 and Sloan – that connect galaxy clusters.

The cosmic web idea has been extremely successful in modelling the distribution of matter in the early Universe. There remains that puzzle, however: all the baryonic matter that can be seen in galaxies today just doesn't tot up to the amount of baryonic matter that the computer models say must have been around at the start of the Universe. For example, when astronomers take an inventory of baryons in and around the Galaxy they find at best about one quarter of the expected number. The situation is no better in other galaxies. The planets and stars in the present-day Universe seem to contain only a fraction of the protons and neutrons that were created soon after the Big Bang. Most of our baryons are missing. What happened to them as time passed and the Universe evolved? Where could they have got to?

It's a tangled story, but the key to finding the whereabouts of the missing baryons is to study matter at ultraviolet (uv) wavelengths.

Forests of the night

To understand why ultraviolet astronomy is so useful in studying the tiny fraction of the Universe that's so important to all of us, consider the element hydrogen. Hydrogen is the simplest chemical element: a hydrogen atom consists of just a single proton and an orbiting electron. Hydrogen is also by far the most abundant element: not only is it the main constituent of most stars, vast clouds of hydrogen gas are to be found in interstellar space. Those interstellar gas clouds turn out to be of great significance.

If a gas cloud is 'cold' then electrons occupy the lowest possible energy level of the hydrogen atoms. If the clouds are warmed – perhaps by absorbing light energy from nearby stars or even light energy from more distant sources – then the electrons can get kicked up into the next available energy level. Now, as we saw in the introductory chapter, the laws of quantum physics demand that the electron in a hydrogen atom must orbit the proton in particular ways: the electron's wavefunction has to fit a whole number of times into the circumference of the orbit. In the lowest energy level, for example, the electron fits precisely one wavelength in its orbit; the next available energy level sees the electron fit two wavelengths into its orbit; and so on. There is thus a particular energy difference between energy levels in a hydrogen atom. For the two lowest energy levels it turns out that this energy difference corresponds to light with a wavelength of precisely 121.6 nm.

Suppose, then, that you shine some light onto cold hydrogen gas and that the light is energetic enough to contain the wavelength 121.6 nm. The hydrogen will absorb this wavelength, with the energy going into kicking the electron into the higher orbit. So if you then take a spectrum of the light after it has passed through the gas you will find that energy is missing at the wavelength 121.6 nm: there will be an absorption line there. The interaction of light with atoms is a two-way street, so the converse situation exists. If the gas is hot enough then lots of hydrogen atoms will be in the second-lowest energy level; when an atom eventually drops to the lowest energy level it will emit light of wavelength 121.6 nm: there will be an emission line there.

This particular wavelength of 121.6 nm is called the **Lyman-alpha line** (named after the physicist Theodore Lyman, who discovered the emission line in 1906). The Lyman-alpha wavelength is in the ultraviolet part of the spectrum, as are all similar lines that occur when the electron in a hydrogen atom drops from even higher energy levels to the ground level. If the electron falls from the level in which three wavelengths fit into one orbit then radiation with a wavelength of 102.6 nm is emitted – the Lyman-beta line; if it falls from the next highest orbit then radiation with a wavelength of 97.2 nm is emitted – the Lyman-gamma line; and so on. So this is one reason why UV is an important part of the spectrum: amongst other things it enables astronomers to detect clouds of hydrogen gas.

Now, suppose that a distant source is hot enough to emit light at ultraviolet wavelengths. And suppose that the light from the source passes through a cloud of gas on its way to us. If the measured spectrum contains a Lyman-alpha absorption line then that's a clear sign that the gas cloud contained hydrogen in its lowest-energy state. Suppose further that the distant source is a quasar. Since almost all quasars are extremely far away, the expansion of the Universe causes the light we receive from them to be redshifted. Consider, for example, the quasar J0338+0021. This has a redshift of 5, which means that the radiation it emitted is stretched by a factor of 5 by the time

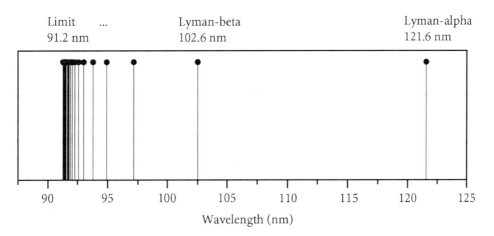

Figure 10.2 *The hydrogen atom contains a series of Lyman emission lines, which occur when an electron falls from an energetic level to the ground level of the atom. All the Lyman lines have an energy (and thus wavelength) that puts them in the ultraviolet part of the spectrum. If a cloud of hydrogen atoms absorbs these particular wavelengths, so that the electron is excited from the ground level to a higher energy level, then a Lyman absorption line is observed.*

we receive it. So although this quasar will have emitted Lyman-alpha photons in the far ultraviolet, with a wavelength of precisely 121.6 nm, by the time the light reaches us the emission line will have a wavelength of 608 nm – an orange color. Since the emission is now in the optical range an ultraviolet telescope isn't required to observe this particular line: a traditional optical telescope will do the job. Thus if a ground-based optical telescope takes a spectrum of J0338+0021 it should see a sharp peak at 608 nm from Lyman-alpha emission, superimposed upon a broad spectrum from redshifted ultraviolet light generated by the quasar in other processes.

The measured spectrum of J0338+0021 is likely to be more complicated than that, however. On its journey to us through intergalactic space, a journey that spanned billions of light years, it's likely that the radiation from the quasar will have passed through *many* clouds of neutral hydrogen. Those hydrogen clouds will not be uniformly distributed throughout space. Rather, they will be found in the threads and filaments of the cosmic web. Since the Universe is expanding, each hydrogen cloud through which the radiation passes will have a different redshift, a different recession velocity as it moves away from Earth. In the case of radiation from J0338+0021 this recession velocity could range from 0.9459 times that of lightspeed at the quasar distance itself down to zero in the immediate cosmic neighborhood. So the broad spectrum of uv radiation that the quasar produces will suffer Lyman-alpha absorption each time it passes through a hydrogen gas cloud – but the location of the absorption line will depend on the particular redshift of the gas cloud. In other words, the Lyman-alpha line will show up in different places of the spectrum. There will be a sequence of Lyman-alpha absorption lines at wavelengths shorter than the quasar's own Lyman-alpha peak.

A quasar spectrum might thus be seen to contain many Lyman-alpha absorption lines below the Lyman-alpha peak, indicating that its light had passed through many cold gas clouds on the line-of-sight between us and the quasar. The closest quasars to us would have relatively few Lyman-alpha absorption lines below the peak, and they'd show up in the blue and ultra-violet part of the spectrum. A really distant quasar, however, might show hundreds of Lyman-alpha absorption lines. This forest of absorption lines is called, appropriately enough, the **Lyman-alpha forest** and it means that a quasar spectrum is quite distinctive. At long wavelengths, above the Lyman-alpha peak, the spectrum is relatively smooth and lacking in structure; below the Lyman-alpha peak lies an irregular tangle of absorption lines that form the forest. Figure 10.3 shows the Lyman-alpha forest in one quasar spectrum.

Observing the Lyman-alpha forest in quasar spectra turns out to be important for several reasons. First, the forest proves that quasar light really has been redshifted. Until relatively recently, some astronomers tried to argue that quasars are nearby objects and that quasar redshifts have some explanation other than cosmic expansion. If those astronomers were correct then much of our modern understanding of cosmology would be wrong. Well, when we see the Lyman-alpha forest we're seeing the intermediate wavelengths that quasar light possessed during its long journey to our telescopes. The alternative – that the complex structure is somehow intrinsic to the quasar – is simply not plausible. When you look at the lines of the Lyman-alpha forest you're looking at the expansion of the Universe.

Second, the forest says something important about dark matter. As we saw in chapter 3, there has been debate over whether the dark matter is 'hot' (in other words, it consists primarily of fast-moving particles) or 'cold' (in other words, it consists primarily slow-moving particles). Computer models that suggest that dark matter causes normal matter to curdle into filaments also suggest that large amounts of the 'hot' variety would have smoothed out the filaments. The existence of the Lyman-alpha forest implies that dark matter must primarily be cold.

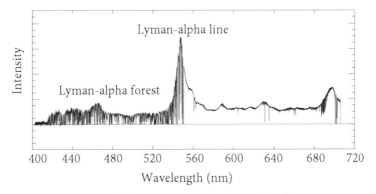

Figure 10.3 *The spectrum of the quasar* Q1159+123. *The characteristic structure of the Lyman alpha forest (namely a smooth spectrum at long wavelengths and a tangle of lines at shorter wavelengths) is clearly evident. Note that the Lyman alpha line is not in its 'laboratory' position because light from the distant quasar is redshifted due to the expansion of the Universe. (A. Songaila)*

A third reason why the forest is so valuable is that it allows astronomers to estimate how much 'heavy' hydrogen was generated in the early Universe. 'Heavy' hydrogen, or deuterium, is a hydrogen atom in which the nucleus contains both a proton and a neutron. It's believed that all the deuterium that now exists was cooked up in the Big Bang, and knowing how much deuterium was produced provides a constraint on models of the early Universe. The simplest way of determining the amount of deuterium, one might guess, is to measure how much of it is around in the present-day Universe. Unfortunately, such an approach is too simple. The problem is that stars burn up deuterium, transforming it into other elements, and at the end of their lives many stars spread those elements over interstellar space. That process has been going on for about 13 billion years. Well, the Lyman alpha forest is a *direct* probe of the ratio of hydrogen to deuterium in the early Universe. The trick here is that the Lyman-alpha line for deuterium has a slightly different wavelength than the line for hydrogen, an effect caused by the extra mass of the deuterium nucleus pushing the electron orbits to slightly higher energies. Thus light passing through a gas cloud containing both hydrogen and deuterium will show an absorption line at the usual wavelength with a small extra dip just away from the main line. The Lyman-alpha forest thus appears even more labyrinthine the closer one looks. Nevertheless, by studying the double-dip of each absorption line astronomers can get an extremely accurate estimate of the amount of deuterium – and from this the density of 'normal' matter – back when the Universe was young.

Finally, and most importantly for this chapter, another application of the forest has to do with the distribution of normal matter throughout the Universe. The tangled structure of the Lyman-alpha forest clearly shows that hydrogen has never been distributed smoothly. It shows the cosmic web – that complicated structure of filaments and voids – was indeed present in the early Universe.

uv glasses

Observations of the Lyman-alpha forest can be made using optical telescopes: what was ultraviolet light when it set off from a distant quasar is stretched by the expansion of the Universe so that it can be blue or green or orange light by the time it reaches Earth. But what about those objects closer to Earth, closer to us in distance and, therefore, closer to us in cosmic time? Such low-redshift objects will still emit and absorb Lyman-alpha radiation – but any uv radiation they emit will still be uv radiation when it hits Earth. And that presents the usual problem for observers: Earth's atmosphere acts as a shield against uv radiation.

Some uv wavelengths can penetrate the atmosphere. Most of the uv radiation that hits Earth originates from the Sun and some of the Sun's long-wavelength uv radiation reaches Earth's surface. One doesn't need a uv telescope to detect this radiation, incidentally: human skin acts as a uv detector, with the severity of sunburn acting as a measure of uv exposure. The amount of uv radiation reaching us has increased dangerously over recent years, due to the depletion of the ozone layer. This radiation is in the near uv region; the Lyman-alpha line, however, lies in the far uv region – to which Earth's atmosphere is completely opaque.

In principle it's not difficult to make a uv telescope: a coating such as silicon carbide can make a mirror sensitive to near uv wavelengths and, for far and extreme uv wavelengths, a Wolter telescope can be used. But the telescope won't work well if it's on the ground. Therefore, as with X-rays and gamma rays, a study of the ultraviolet part of the spectrum really requires space-based telescopes.

Note that a space-based uv telescope would be useful for much more than just observations of the Sun or of the Lyman-alpha line. Matter glows because it has a temperature, and at certain temperatures – between about

30 000 to 500 000 degrees – matter glows brightly at UV wavelengths. A gas cloud possessing such a temperature would be classified as 'warm-to-hot'; UV telescopes would thus be able to study the 'warm–hot' Universe. Such temperatures are exceptional by most standards, of course; they are warm only by comparison with the most extreme processes that take place in the Universe. If a gas cloud is *really* hot, say a million degrees, then it will glow brightly in the X-ray region of the spectrum. X-ray telescopes thus study the 'hot' Universe. If a gas cloud is at a temperature of 10 000 degrees, say, then it will glow brightly in visible light. Optical telescopes could be said to study the 'lukewarm' Universe. The 'cold' Universe is brightest in the infra-red region.

The 'warm–hot' Universe is an interesting place. If your eyes could see the sky in UV light, the stars you are familiar with would fade from view; instead, you would see stars at the beginning and endpoints of their evolution – young, massive stars and certain types of old star. And, where such light was not blocked by dust, you would see clouds of warm interstellar gas. So an ultraviolet telescope would see phenomena such as supernova remnants, the hottest stars and the corona of stars like the Sun.

In addition to seeing the radiation that arises because matter is warm–hot, an ultraviolet telescope would see certain emission lines. We've already discussed the Lyman-alpha line. Well, similar lines exist for other elements when an electron drops from a higher-energy state to a lower-energy state and emits a UV photon. Furthermore, there are situations in which atoms in the gas clouds of the interstellar medium have their electrons stripped away – perhaps by a supernova explosion. When electrons recombine with the positively charged ions, UV photons are emitted.

For example, an ultraviolet telescope might see the emission line from an ion called O VI ('oxygen six') – an oxygen atom that has had five of its eight electrons stripped away in an extremely high-temperature process. As the gas cools, the recombination of the ions and electrons emits ultraviolet light of a particular frequency. Indeed, some of the commonest elements in the Universe – hydrogen, helium, carbon, nitrogen, oxygen, and silicon – all leave a telltale ultraviolet fingerprint in the form of emission lines. Thus ultraviolet telescopes can provide a wealth of information on the chemical composition of the interstellar medium, and how it evolves through time by the action of processes such as supernova explosions. Small wonder, then, that astronomers long wanted access to UV instruments placed above the Earth's protective blanket.

Astronomers took an ultraviolet spectrum of the Sun as long ago as 1946, using an instrument sent up on a brief rocket flight. It was not until the 1960s, however, that orbiting instruments were launched and during this period astronomers drew up plans for more wide-ranging studies of the ultraviolet sky. One of the most influential of the early orbiting UV observatories was the **International Ultraviolet Explorer** (IUE) – a joint British–American–European mission launched in 1978. The original plan was to run the satellite until 1981; in the end the mission ran until 1996. Rather than taking visual images in ultraviolet light, IUE instead recorded ultraviolet spectra – 104 470 of them over its lifetime, taken of 9600 distinct objects ranging from distant quasars to interstellar gas to planets. Spectra contain vital information but they are less visually attractive than images; perhaps that's one reason why IUE's exceptional output received much less publicity than those beautiful images from, say, Hubble. Nevertheless, IUE's success meant that astronomers could be confident that orbiting UV astronomical telescopes would yield useful data.

In 1999, NASA launched the **Far Ultraviolet Spectroscopic Explorer** (FUSE). As with IUE, this was another case of a three-year mission exceeding its design: scientists managed to extend its working life by an extra five years. The telescope design used by FUSE – four separate mirrors with an associated high-resolution spectrograph – meant that it was able to observe the ultraviolet sky with a sensitivity that had never before been achieved. Its detectors covered a band in the range 91–120 nm, just below the Lyman alpha line, and during its lifetime it made more than 6000 observations of nearly 3000 objects – planets and comets in the Solar System, stars in both the Milky Way and nearby galaxies, and the nuclei of more distant galaxies. Crucially, as we shall soon see, it was able to map the warm gas in interstellar and intergalactic space.

In 2003, NASA launched the **Galaxy Evolution Explorer** (GALEX). This orbiting telescope was planned to operate until 2005, although it's still operating at the time of writing. The idea behind GALEX was to view hundreds of thousands of galaxies in both the near and far UV regions in order to gain information on how galaxies form and evolve over cosmic time.

In addition to IUE, FUSE, and GALEX, Hubble has always had the capacity to make UV observations. Indeed, when Hubble was deployed it became the most powerful telescope ever launched for UV astronomy. For example, with Hubble astronomers could for the first time measure the Lyman-alpha line at low redshift and with reasonable accuracy. In 2009, a final servicing mis-

Figure 10.4 *The* M81 *spiral galaxy in ultraviolet light. This image from* GALEX *shows the sort of information that can be gained from studying the Universe in* UV. GALEX *is sensitive to the light from energetic young stars, shown here as wisps of blue–white.* (NASA/JPL-Caltech/J. Huchra, Harvard-Smithsonian CfA)

sion added the **Cosmic Origins Spectrograph** (COS) to Hubble's suite of science instruments. COS performs spectroscopy in the range 115–320 nm, and provides astronomers with their best instrument for observing faint sources of ultraviolet light.

So since the turn of the millennium, astronomers have been able to peer through the ultraviolet window on the Universe using FUSE, GALEX, and COS. The views have been impressive in a variety of ways. But let's focus on one particular vista that can help answer the question posed at the beginning of the chapter: where is all the missing matter?

Figure 10.5 *The Cosmic Origins Spectrograph being tested at the Kennedy Space Center. When astronauts fitted it to Hubble in May 2009, astronomers suddenly had a wonderful new instrument with which to study the warm–hot Universe.* (NASA)

The Universe on a WHIM

To recap: under the standard model of cosmology, about 95% of the mass–energy content of the Universe is 'dark' and the remaining 5% is baryonic matter. However, there remains a problem with our understanding of the whereabouts of baryonic matter. Observations of absorption lines in quasar spectra imply that the early Universe was home to clouds of baryons in the form of ionized, diffuse intergalactic gas; furthermore, such clouds must have formed 75% or more of the baryon mass of the Universe. The baryons were in the primordial Lyman alpha forest. All this is in agreement with the standard model of cosmology. As the Universe expanded and cooled, many of the baryons underwent gravitational collapse (into stars, galaxies or clusters) while some of the baryons remained as a diffuse gas (in interstellar, intergalactic or intercluster clouds). The problem is that when all this present-day baryonic matter is totted up – the stars, galaxies, clouds of gas – astronomers observe only about half of the matter that once existed. Where is all the 'normal' matter today? What happened to the baryons as the Universe evolved? In some ways this is a more pressing problem than that of identifying dark energy or dark matter, because if we don't understand the role of baryons in the Universe then we understand little indeed.

Well, the prediction coming from increasingly refined computer models is that the rest of the 'normal' matter in the Universe resides in the cosmic web – the tracery of extremely low-density filaments that connect galaxy to galaxy. A further prediction is that when clouds of intergalactic gas fall onto the filaments, or when material pours out from star-forming galaxies and strikes the filaments, the shocks will cause the cosmic web to heat. Some material in the cosmic web will thus be 'warm' (a few tens of thousands of degrees) and some of it will be 'hot' (many hundreds of thousands of degrees). So the computer models suggest, then, that the missing matter isn't missing after all. About half of the matter in the Universe is in the form of a sparse **warm–hot intergalactic medium** (WHIM).

So much for prediction. Has the WHIM actually been observed? Do we know for certain that the mystery of the missing baryons has been solved? Well, the warm matter in the WHIM can best be studied by UV telescopes. In recent years, observations of quasars by FUSE and Hubble (using the precursor instrument to the COS) showed the telltale signatures of hydrogen and oxygen superimposed on the quasar spectra. A detailed analysis of the spectra showed that the hydrogen and oxygen formed long trails along the quasar line of sight – filaments of the cosmic web. With the addition of the COS to its armory, Hubble will probe the cosmic web in even more detail. One of the first observations made by the COS was of the quasar PKS 0405-123, an object so distant it takes light more than 7 billion years to reach us. By analyzing that quasar's light, astronomers found evidence for a trail of gas clouds at different distances along the line of sight. See figure 10.6. This observation had been made before, but the COS found up to five times more low-density filaments than other instruments had seen along that line of sight. Furthermore, the COS detected the presence of glowing clouds of oxygen and nitrogen – evidence of the shocks that are supposed to heat the diffuse filaments. These are recent results, but it seems as if at least some of the missing baryons have indeed been found: they form the warm component of the WHIM.

The remaining baryons are predicted to form the hot component of the WHIM. Observing this hot component requires X-ray telescopes rather than UV telescopes, but the observations are difficult because the WHIM is so tenu-

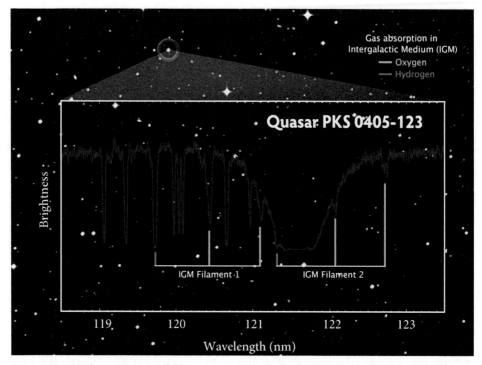

Figure 10.6 *The* COS *on Hubble was used to study the light from the distant quasar* PKS 0405-123. *The spectrum revealed the absorption lines of elements that make up the intervening gas clouds traversed by the quasar's light.* COS *was able to detect many more low-density filaments of hydrogen in the cosmic web than previous instruments had seen.* (NASA, ESA, *Hubble* SM4 ERO *Team, &* DSS)

ous: the density of the WHIM is almost a million times less than the density of the typical interstellar medium so it's quite easy to look straight through it. Nevertheless, Chandra and XMM-Newton have joined forces to look for these hot baryons. Both of these X-ray telescopes studied the same active galactic nucleus, an object that lies about 2 billion light years away and pumps out prodigious amounts of X-rays as it eats matter. About 400 million light years away along the line of sight to this AGN lies the Sculptor Wall. This structure is a smaller version of the Sloan Great Wall, the CfA2 Great Wall and the Pisces–Cetus Supercluster Complex mentioned earlier in the chapter. The Sculptor Wall contains thousands of galaxies stretching across millions of light years: it is a small thread of the cosmic web. The Sculptor Wall should therefore contain some of the WHIM, and the WHIM should absorb some of

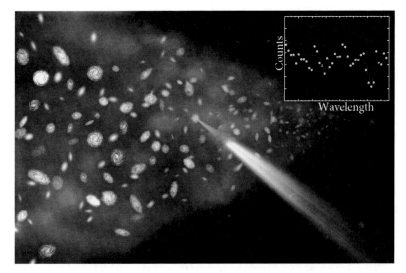

Figure 10.7 An artist's impression of the warm–hot intergalactic medium in the Sculptor Wall, which was studied by Chandra and XMM-Newton. Since X-ray emission from the WHIM was too faint for Chandra and XMM-Newton to detect, the satellites instead studied radiation from an active galactic nucleus in the background. The X-ray spectrum from the AGN (shown in the inset) shows a dip consistent with absorption by oxygen atoms in the WHIM within the Sculptor Wall. (WHIM: XMM-Newton/Chandra; Spectrum: NASA/CXC/Univ. of California Irvine/T. Fang; illustration: CXC/M. Weiss)

the X-rays emitted by the distant active galactic nucleus. In 2010, astronomers announced that both Chandra and XMM-Newton had indeed spotted absorption of X-rays by oxygen atoms and the results were consistent with predictions from computer models of the cosmic web. See figure 10.7 for an illustration of this.

Thus the combined observations of UV and X-ray telescopes seem to confirm the existence and extent of the WHIM. If the results stand then astronomers for the first time will know for sure where the matter lies in the Universe today. Some of the matter shines – in stars and galaxies. Some of it is in the form of cold gas clouds. And the rest of it is in the WHIM. Astronomers will know more about the role of 'normal' matter than they have ever done before. Of course, that 'normal' matter constitutes only about 5% of the energy density of the Universe. But at least it's a start…

uv blindness?

Astronomers and cosmologists are investigating the form of the cosmic web using a variety of methods. For example, gravitational lensing and large-scale surveys of galaxies (which were discussed in chapter 4) can reveal features in the cosmic web. In order to observe the WHIM directly, however, astronomers really need to use uv and X-ray instruments. And at the time of writing, Hubble (with the addition of the COS) is perhaps the observatory best placed to map the structure of the cosmic web. Using the COS, astronomers will observe thousands of quasars in the ultraviolet and, along the line of sight to each quasar, they will detect filaments. Line-of-sight by line-of-sight, the COS will map the cosmic web as no instrument has done before. It will determine not only the *form* of the cosmic web but also its temperature, density and chemical composition. Astronomers will then better be able to understand how the matter in the nearby web – both the WHIM and the shining matter we see in galaxies – is being recycled and how it has evolved over time.

Ultraviolet observations are not undertaken purely to study the cosmic web, of course. The spectra taken by COS will contain information on the chemical evolution of stars and galaxies. By detecting young, hot stars that are currently invisible in the shrouds of dust that surround them, the COS will help reveal how stars are formed. By studying massive stars and the surroundings of supernovae, the COS will clarify how the heavy elements – the chemical elements necessary for life – were created and how they get to be in the dust grains from which planetary systems form. And closer to home, the COS will teach us more about the atmospheres of the giant planets and the comets in our own Solar System.

It's clear, then, that the uv window offers a valuable view of the Universe. Unfortunately, it's a view that astronomers may lose for a while. The COS won't last forever. The plan is for Hubble to continue observing through 2014, but who knows what will happen after that?

Satellite-based astronomy, as ultraviolet astronomy must inevitably be, is 'big' science: it takes many years to design a telescope, agree mission aims, obtain funding, construct the telescope, and then launch it. To illustrate the point, consider the genuine successor to Hubble: the **Advanced Technology Large Aperture Space Telescope** (ATLAST). If ATLAST is launched it will have an ultraviolet and infrared capability in addition to optical – just like Hubble. It will be a tremendous instrument, with the capacity to do marvellous science; investigating the WHIM will be just one of a host of projects that ATLAST will progress. But even if ATLAST is funded, it won't launch until the mid-2020s at the earliest; the problems facing the James Webb Space Telescope (which are discussed in the next chapter) imply that the deployment of an instrument such as ATLAST may be significantly delayed. At the time of writing, there is no planned successor to dedicated UV instruments such as FUSE, GALEX or the COS. There may well be a decade-long hiatus in UV astronomy.

Astronomers can of course investigate the hot component of the WHIM using X-ray telescopes (which were the focus of chapter 5). In addition to those X-ray telescopes already discussed, some mission concepts that are optimized for WHIM studies have been pitched to the various space agencies. For example, the **Diffuse Intergalactic Oxygen Surveyor** (DIOS) is a planned Japanese mission which, if all goes well, will launch in 2016. If approved by its American, European and Japanese collaborating partners, a much larger mission called XENIA (which thankfully is not an acronym, but comes from the Greek word for 'hospitality') would launch some time after DIOS. Both the XENIA and DIOS missions would have the capacity to study the WHIM in soft X-rays.

Other mission concepts have been discussed. For example, **Pharos** would observe the afterglow of gamma-ray bursts at X-ray and UV wavelengths. It would thus be using a gamma-ray burst as a type of 'cosmic lighthouse', illuminating the WHIM. The **Baryonic Structure Probe**, meanwhile, would sit at L2 as it looked for Lyman-alpha and O VI lines, using these as tracers of the cosmic web. By looking at very small redshifts the probe would map the filaments of the cosmic web after it had evolved to its modern structure. The **Telescope for Habitable Exoplanets and Interstellar/Intergalactic Astronomy** (THEIA), which would observe in both the optical and ultraviolet, would have as one of its core science goals an improved understanding of the cosmic web. However, it's not clear whether or when any of these missions will fly.

One project that is much further advanced is the **World Space Observatory – Ultraviolet** (WSO-UV). Figure 10.8 shows an artist's impression of what this orbiting observatory will look like. The Russian-led WSO-UV collaboration hopes to construct a telescope having a mirror 1.7 m in diameter and use it in conjunction with three different spectrographs to form an observatory that would exceed the COS in sensitivity. The WSO-UV would have the same goals as the COS: to learn more about the cosmic web, about galaxy

Figure 10.8 *An artist's impression of* wso-uv. *(*wso-uv *Collaboration)*

and stellar evolution, and about our own Solar System. However, progress with this ambitious mission has been slow because for many years it had to survive on a shoestring budget. If the wso-uv launches as hoped in 2016 then, during its planned ten-year mission, it will provide astronomers with a clear view of the ultraviolet sky – and explore in detail where the 'normal' matter in the Universe resides. If missions such as wso-uv and the others mentioned above don't fly, for whatever reason, there'll be a period during which astronomy is effectively blind at this wavelength.

11

Nurseries in space

Are there other planetary systems out there in space? It's an old question: Aristotle and Epicurus were debating it more than two thousand years ago. We now know the answer: it's 'yes'. But how do stars and planets actually form? Over the past three decades astronomers have made more progress in understanding stellar and planetary system formation than in the rest of history. The key to this improved understanding has been the deployment of infrared telescopes. Although there is still much to learn, the next generation of sensitive infrared instruments will surely fill in the remaining gaps in our knowledge.

Seeing red but keeping cool – A star is born – When telescopes take flight – A new window in the spectrum

S. Webb, *New Eyes on the Universe: Twelve Cosmic Mysteries and the Tools We Need to Solve Them*,   247
Springer Praxis Books, DOI 10.1007/978-1-4614-2194-8_11, © Springer Science+Business Media, LLC 2012

Ａll those stars that we see in the night sky – are they accompanied by planets? Or is the Sun unique in possessing a planetary system? Until recently it seemed impossible to answer such questions by direct observation. After all, how could a telescope possibly image the reflected light from a planet when its parent star is so very much brighter? Theoretical arguments seemed no more capable than observation of settling the question. Two hypotheses of how planets might form – the nebular hypothesis and the encounter hypothesis – generated quite different predictions for the number of planets that might be out there. Under the nebular hypothesis, planetary formation is a natural concomitant of star formation; one expects to see a plurality of worlds. Under the encounter hypothesis, planets form when stars pass close by each other; since such encounters are rare one expects a paucity of worlds. That there were two hypotheses regarding planetary formation was even more troubling when one considered the fact that both of them seemed to be clearly wrong.

The nebular hypothesis began life in 1734, when Emanual Swedenborg proposed that slowly rotating gaseous clouds – nebulae – gradually contract and flatten under their own gravity to form stars and planets. The idea was developed and refined in 1796 by Pierre-Simon Laplace. Unfortunately, Laplace's nebular hypothesis doesn't work. Observations indicate that more than 99% of the Solar System's mass is in the Sun, while about 99% of its angular momentum is in the planets. It's a simple, basic fact. But Laplace's hypothesis couldn't account for it.

The encounter hypothesis was first mooted in 1745, by Georges Buffon, who argued that material was ripped from the Sun as a comet passed close by; that material then condensed into the planets. The idea was refined in the twentieth century by Harold Jeffreys and James Jeans. They proposed that, during a close encounter, a passing star pulled a cigar-shaped gaseous filament from the Sun. The fat section of the cigar condensed into the four giant planets, and the thin end sections condensed into the terrestrial planets. However, further analysis showed that this idea doesn't work either: the gas pulled from the Sun would simply dissipate rather than condense to form planets.

The situation was, then, that astronomers couldn't observe planets outside our Solar System nor had they a realistic model of how planets might form. At least, that's the way things stood until relatively recently. Nowadays the discovery of an **exoplanet** is so commonplace it no longer counts as news. Furthermore, astronomers have settled on a model of planetary system formation; the model still has parts that require refinement, but the overall picture seems settled.

So – how *do* stars and planets form? Infrared telescopes have been in the forefront of answering this question.

Seeing red but keeping cool

Celestial objects emit some infrared radiation simply because they possess a temperature greater than absolute zero. The amount of infrared radiation emitted depends upon temperature. If an object possesses a temperature between about 750–5000 K, which is the typical range of temperature for cool stars such as red dwarfs and red giants, then the peak of the radiation will lie somewhere in the region between about 0.7–5 µm. This is the near infrared part of the infrared spectrum, and it lies just below the range of wavelengths to which the eye is sensitive. If an object possesses a temperature between about 100–750 K, which is the typical range of temperature for planets, comets and asteroids in our Solar System, then the radiation will peak somewhere in the region between about 5–30 µm. This is the mid infrared part of the spectrum and it lies well away from the visible wavelengths. And if the objects are even cooler, possessing a temperature in the range between about 10–100 K, then the radiation will peak somewhere in the region between about 30–300 µm. This is the far infrared part of the spectrum, which involves emission from very cold dust or very cold molecular clouds out in space.

The wavelengths involved in near infrared astronomy are not very different from visible wavelengths, so astronomers don't need to develop completely new types of telescope to work in the near infrared: once suitable adjustments are made the panoply of instruments used in optical astronomy, such as mirrors, gratings, and cameras, can also be used for near infrared astronomy. One could be forgiven, then, for assuming that infrared astronomy must have made rapid advances during the twentieth century and allowed the study of objects ranging from dust clouds through to cool stars. The reality is somewhat more complicated.

The development of increasingly sensitive instruments for measuring infrared radiation did indeed lead to new astronomical knowledge. In the 1920s, for example, Seth Nicholson and Edison Pettitt used new infrared technology to measure the diameter of gas giant stars. To give another example, twenty years before Neil Armstrong set foot on the Moon infrared studies had shown that the surface of our twin planet is covered with dust. Yet another example: in the early 1960s, Harold Johnson studied the radiation from cool stars by using newly developed infrared photometers. Nevertheless, two barriers exist to observing the sky in the infrared spectrum and they hindered the development of the subject.

The first barrier is technological. Astronomers can detect all infrared wavelengths by using a device called a germanium bolometer, which was developed by Frank Low in 1961. The device is simply a thin strip of the metal germanium housed in a shell that contains a small opening. If infrared radiation passes through the opening and strikes the germanium, the metal gets warmer and its resistance changes. By measuring the change in resistance, astronomers have a direct measure of the amount of infrared radiation that entered the shell. The placement of an infrared filter in front of the opening allows the study of particular infrared wavelengths. More recently, for near infrared wavelengths, detectors consisting of $HgCdTe$ and $InSb$ cells are often used. Since large arrays of such cells can be constructed, these instruments have the capacity to produce extremely detailed infrared images. Placing a bolometer or an infrared array detector at the business end of a large optical telescope thus opens up the possibility of studying distant objects in the infrared.

However, there's a problem with a detector that's designed to capture infrared radiation from the cosmos: all the matter that's close to the detector will be pumping out infrared radiation of its own. An infrared telescope will emit infrared radiation simply because it operates at room temperature; indeed, the telescope's own emission would be enough to swamp the infrared detectors. There's a simple solution to the problem: immerse the instruments in liquid helium, so they are working at just a few degrees above absolute zero. This works, but it means infrared telescopes are inherently more high maintenance than optical telescopes.

The second barrier is the same old one that makes it difficult to observe in the ultraviolet, X-ray and gamma-ray parts of the spectrum: Earth's atmosphere.

The atmosphere presents a twofold problem for infrared astronomers. The first is that atmospheric water vapor and carbon dioxide absorb most of the infrared radiation from space. There are a few 'windows' in the infrared spectrum, a small number of wavelengths at which the atmosphere is transparent, so at least at those wavelengths infrared radiation can reach ground-based telescopes. In general, though, trying to observe the infrared sky from the ground is like trying to view stars on a cloudy night. The second problem is that the atmosphere itself emits infrared radiation, with a peak at a wavelength of about 10 μm. So astronomers must take care not to have their observations swamped by atmospheric emission.

From the above discussion it's clear that a ground-based infrared telescope equipped with suitable cooling equipment can observe at a few select wavelengths, particularly if it's situated in a dry location at an altitude above most of the atmospheric water vapor. The several telescopes at the **Mauna Kea Observatory** in Hawaii, for example, are near the summit of a volcano 4205 m above sea level: this is one of the best locations in the world to place an infrared telescope. Antarctica is another ideal location for infrared observatories. Nevertheless, the fact remains that the atmosphere is almost opaque to most infrared wavelengths. Although dedicated, ground-based infrared telescopes can peek through the gaps where the atmosphere is transparent, and can do so with exquisite resolution, a space-based telescope is needed if you want to study the entirety of the infrared spectrum.

The first space-based observatory designed to work at infrared wavelengths was the **Infrared Astronomical Satellite** (IRAS), a joint project between the UK, the US and the Netherlands. The satellite was launched in 1983 and put into orbit 900 km above Earth's surface. On board the satellite was a supply of superfluid helium that kept an array of 62 infrared detectors cooled by evaporation; the telescope operated at just two degrees above absolute zero. The helium ran out after ten months and, once the telescope had heated up, no further observations were possible. Nevertheless, that short lifespan was long enough for IRAS to map the sky four times at four different infrared wavelengths. In doing so IRAS discovered hundreds of thousands of infrared sources, many of which have still not been identified; it took a host of impressive images, including that of the Galaxy's core; and in general it had an impact on most areas of astronomy.

One particularly significant discovery made by IRAS seems at first glance rather prosaic: it spotted the presence of a disk of dust grains surrounding the brilliant star Vega. As with so many discoveries, this one came about serendipitously. Fred Gillett and George Aumann, the mission scientists charged with calibrating the telescope, chose Vega as the primary calibration star against which all the other stars would be compared. It turned out that Vega was unsuitable for this purpose because it was emitting a large excess of far infrared radiation. The two astronomers pointed IRAS at other hot stars and these all turned out to be 'better-behaved' – the infrared excess was missing

Figure 11.1 *The UK Infrared Telescope is the world's largest telescope dedicated purely to infrared astronomy. It observes in the near to mid infrared range of the spectrum. It's based at the Mauna Kea Observatory, which is the home to a dozen other world-class telescopes. (UK Science and Technology Facilities Council)*

– and therefore these stars were superior for calibration purposes. However, the fact that all those other stars lacked an infrared excess suggested to Gillett and Aumann that the excess they spotted for Vega was real – it wasn't just some artefact of the observations. They decided to investigate Vega further and found that there were two different sources of radiation: Vega itself pumped out brilliant white light along with only a little infrared radiation; something else near the star pumped out most of the infrared radiation. Calculations soon showed that the temperature of that 'something' was at roughly the same temperature as Saturn (about 90 K) and that it was about 80 times as far from Vega as Earth is from the Sun. They looked even closer and found that a ring of dust orbits the star. They deduced that Vega's bright light heats the dust grains, which then radiate this heat in the infrared.

The presence of dust grains in space was nothing new for astronomers. Indeed, astronomers had struggled with the problem of interstellar dust for decades: microscopic dust grains absorb blue light from distant stars, and cause those stars to appear fainter and redder. You're probably familiar with the phenomenon: the setting Sun appears redder than normal when its light passes through dust in the atmosphere. What surprised Gillett and Aumann was the size of the grains. The IRAS data showed that the dust grains around Vega were a thousand times larger than interstellar dust grains; if you could hold them in the palm of your hand you'd be able to see them. This implied some sort of accumulation process had taken place: microscopic dust from the primordial material that formed the star must have stuck together to form larger particles. This in turn implied that even larger particles – rocks, asteroids, perhaps even planets – might have formed. The data showed something equally intriguing. IRAS saw lots of dust a long way away from Vega but no dust close to the star. Of course, radiation from Vega might have pushed the dust away – but it was also possible that planets close to the star had swept the region clean. Perhaps, then, this observation hinted at the existence of a planetary system around a star other than the Sun.

Before its helium ran out, IRAS found three other stars with disks of dust around them: Epsilon Eridani, Fomalhaut and Beta Pictoris. Within another year, astronomers knew of more than 40 stars with such disks. In all likelihood, then, such disks were common. IRAS had *not* found other planetary systems but it *had* obtained the first observational, if circumstantial, evidence that other systems might exist. The search for other Earths, once seemingly an impossible task, now seemed feasible. The race to find exoplanets began in earnest, and it was IRAS – an infrared instrument – that fired the starting pistol.

Twelve years after the IRAS mission ended, astronomers finally showed definitively that other planetary systems exist around Sun-like stars. It was the traditional optical telescope, rather than its infrared counterpart, that provided the proof. Using a variety of techniques, which are described in the next chapter, independent groups of astronomers all found evidence for exoplanets. After a search lasting over a century, the issue was finally settled. Other planetary systems exist, and they exist in large numbers.

But we are left with a mystery. If the nebular hypothesis of planetary formation is wrong and the encounter hypothesis is wrong – how, then, do planetary systems form? To solve the mystery, astronomers needed to know how stars themselves form.

A star is born

The success of IRAS prompted the development of ever-more sensitive infrared telescopes. The results of a 1995 follow-up mission, called the **Infrared Space Observatory** (ISO), proved to be as influential as those of IRAS. In 1997, an infrared camera was installed on Hubble and provided a superb complement to its other science instruments. And 2003 saw the launch of the **Spitzer Space Telescope**, the fourth and last of NASA's 'Great Observatories'. Until its 360 liter supply of liquid helium ran out in 2009, Spitzer gave astronomers their most in-depth view of the infrared sky to date.

Infrared observations from telescopes such as ISO, Hubble, and Spitzer impacted on all areas of astronomy. One of their achievements was to identify the birthplace of stars. They found regions of space that contain clouds of cold, molecular hydrogen – clouds that can be more than 60 light years in diameter and that can possess more mass than 300 000 stars like the Sun. These vast gas clouds are the cocoons in which stars form. See figure 11.2.

Over millions of years a gas cloud will inevitably experience disturbances: it might suffer shock waves from nearby supernovae, for example, or undergo a close encounter with another cloud. A large enough shock will cause regions of the cloud to fragment, and those fragments can form small, relatively dense cores called protostellar nebulae. Such nebulae show a range of masses and sizes. The tiniest might possess only a fraction of the Sun's mass, with a diameter as small as 2000 times the Earth–Sun distance. The largest might possess several times the mass of the Sun, with a diameter more than 20 000 times the Earth–Sun distance.

Gravity will cause a protostellar nebula to collapse and it is from this collapse that a star eventually forms. As the nebula contracts it spins ever-faster and begins to flatten, so that incoming material falls in more easily at the poles than at the rapidly spinning equator. The core of the nebula collapses rather quickly and, as it does so, the core gets hotter; the collapse continues until the heat pumping out of the core balances the gravitational self-attraction of the material. As the core continues to attract material from the surrounding disk it gradually gains in mass and increases in temperature until it becomes a hot **protostar**. In short, the gravitational collapse of a protostellar nebula results in a protostar surrounded by a leftover disk of gas and dust. See figure 11.3. The timescale of the collapse depends on the nebula's mass. The protostellar nebula that gave birth to the Sun, for example, would have taken about 100 000 years to collapse.

Figure 11.2 *The infrared telescope on board Spitzer spotted this vast cloud of cool gas and dust, the shape of which has been formed by solar winds blown from a nearby massive star. Astronomers call this cloud the 'Mountains of Creation'. In the far future, stars will form from this cloud.* (NASA/JPL-Caltech/L. Allen & J. Hora)

In the early stages of collapse, a protostar and its disk are enveloped by infalling gas and dust and are quite hidden from view. At least, they are hidden from optical telescopes. But not to infrared telescopes. The longer wavelengths of infrared radiation are less likely to be scattered by tiny dust particles. Spitzer, Hubble, ISO, and other infrared telescopes have enabled astronomers to peer into the heart of these star-forming regions and investigate the details of protostars. Infrared telescopes have watched how a protostar begins to produce energy by fusing first deuterium and then ordinary hydrogen until the core 'turns on' and becomes a star. After a million years or so, the enveloping cloud of gas will have vanished and the star is then fully established; astronomers call such an object a **T Tauri star**. Over the next ten million years or so the disk falls onto the central star and then, over the next hundred million years or so, the T Tauri star evolves into an 'ordinary' star like the Sun.

Figure 11.3 An artist's representation of the spinning disk of dust and gas surrounding the protostar IRS 46. The object is difficult to study at visible wavelengths, but infrared telescopes such as Spitzer have a good view of such objects. One interesting discovery regarding the IRS 46 system is the presence of organic compounds in the inner part of the disk. (T. Pyle, SSC; JPL-Caltech, NASA)

So that's how stars form. How do planets form within the scenario outlined above?

The picture that's painted of planetary formation is not clear in all its details, but the main features are now relatively well understood. That understanding dates back to work by Victor Safronov in the 1960s and 1970s, work that was later developed by George Wetherill. The key to planetary formation lies in the presence of those disks that orbit young stellar objects. Indeed, the surrounding disk is called a **protoplanetary disk**. See figure 11.4.

Let's consider first the formation of rocky planets like Earth. The formation of rocky planets occurs within the so-called 'snow line', where the temperature of the disk is high enough to prevent water and other substances

Figure 11.4 *An artist's depiction of the gas giant planet LkCa 15b inside a protoplanetary disk. In late 2011 the twin Keck telescopes took an image of LkCa 15b; it was the first direct image of a planet still being formed. (Karen L. Teramura, UH IfA)*

condensing into ice; for a star like the Sun the snow line has a radius of perhaps three or four times the present Earth–Sun distance. Outwith the snow line the disk consists primarily of hydrogen and helium; within the snow line there's heavier material – dust grains made of elements such as iron, nickel and silicon. Observations made by infrared telescopes have shown that the dust grains inside the snow line undergo innumerable collisions during which they stick together, growing in size to produce particles the size of a thumbnail. Note that the ability of dust grains to grow to a large size is known to anyone who has laid laminate flooring and subsequently failed to vacuum often enough behind the sofa. Air currents swirl the dust around, and dust bunnies are formed – structures made up of hair, lint, skin as well as dust, which are held together by static electricity and tangled fibres. A dust bunny has a large volume compared to its mass, and this makes it effective at collecting more debris: once one has grown to a certain size it is easier for it to grow even larger.

The creation of dust bunnies is just the first of several stages in planetary formation. The second stage occurs as these thumbnail-sized particles somehow combine to form **planetesimals** – objects that are a kilometer or so across. The details of this process are uncertain; we shall return to this point later.

A typical disk will contain many trillions of planetesimals, which makes the disk sound a crowded place. It isn't. Planetesimals will be separated on average by thousands of kilometers. Nevertheless these 'sticky' collisions continue and the rate at which planetesimals increase in size accelerates; it's a runaway process, with larger bodies growing at the expense of smaller ones. This runaway process doesn't last long on a cosmic scale – somewhere between 10 000 and 100 000 years – and ends when the largest bodies are a thousand kilometers or so in diameter. Planetesimal collisions can also generate dust and debris, and detailed observations of Vega by Spitzer suggest that the infrared excess spotted all those years ago by IRAS was due to precisely such debris. Vega is thus surrounded by a debris disk rather than, as originally thought, a primordial protoplanetary disk. The dust IRAS saw was the remnant of a relatively recent collision between planetesimals.

The largest of the planetesimals at the end of the runaway stage are called **oligarchs**, and a few hundred of these oligarchs continue to attract the remaining smaller planetesimals through their gravitational attraction. As in life, so in planetary formation: the larger an oligarch becomes, the more it wants and the greater its capacity for getting what it wants. In this stage it is only the oligarchs that can grow – and they continue to grow until the supply of planetesimals becomes scarce. When the planetesimals are no longer in plentiful supply oligarchs can only get bigger if they merge. This stage of planetary formation ends when there are a hundred or so **protoplanets**, the smallest being about the size of the Moon and the largest perhaps as large as Mars.

We are almost there. The final stage occurs when the protoplanets are massive enough to introduce perturbations in each other's orbits. Their orbits become chaotic and collisions between protoplanets take place. According to computer simulations of this process, the result is the formation of a small number of planets the size of Earth – on average, between two and five rocky planets – which settle down into stable orbits. None of the bodies in the inner part of the system have sufficient mass to retain hydrogen and helium, which are the lightest gases; only the heavier, slower-moving gases lack the energy to escape from the gravitational grasp of these small, rocky planets.

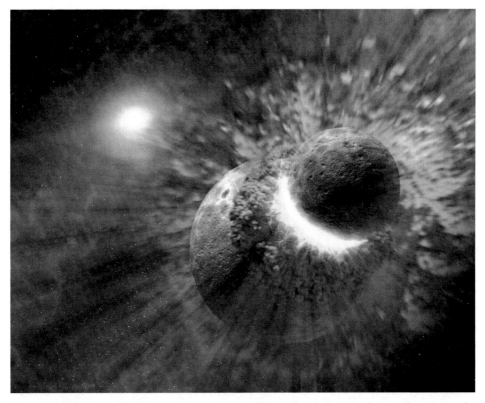

Figure 11.5 *An artist's conception of the collision between two large planetesimals. Planetesimals occur in protoplanetary disks as a stage in the process of planetary formation. They can also occur around older stars with the result that a debris disk forms. The collision pictured here, for example, represents a collision that might have taken place recently around Vega. Subsequent collisions between smaller objects would have created fine dust particles, which Vega's brilliant light would have blown away to form a disk at large distances from the star. It is this debris disk that* IRAS *first spotted and that Spitzer has since observed in detail. (T. Pyle,* SSC; JPL-*Caltech,* NASA)

Our own Solar System, of course, has four such bodies: Mercury, Venus, Earth and Mars. Some of the many trillions of original planetesimals may survive the process of planetary formation: asteroids are the remnants of planetesimals from the inner part of the disk while comets are planetesimals from the outer reaches of the disk.

What about the giant planets? How do planets such as Jupiter, Saturn, Uranus and Neptune form? Astronomers are less confident about their understanding of giant planet formation. In the currently favored scenario, the formation of a giant planet's core takes place beyond the snow line and proceeds along the same lines as described above for rocky planets: planetesimals are somehow formed, they undergo runaway growth to form oligarchs, which then continue to increase in mass but at a slower rate. However, although the scenarios are similar they are not identical. First, beyond the snow line the planetesimals will consist mainly of ice, which allows for more massive cores to form. Second, the oligarchs are less likely to merge because there's more space for them to move unimpeded in the outer regions of the disk. In any case, once a core has reached a threshold mass – about ten Earth masses – its gravitational influence is such that it starts to sweep up gas from the surrounding disk. It stops only when the supply of gas is exhausted. This is a rapid process: in our Solar System, the giants Jupiter and Saturn would have accumulated most of their mass during a period of only 10 000 years, which is an eyeblink in astronomical terms. In this picture, the planets Uranus and Neptune are failed cores: they began sweeping up gas only when almost all the gas had gone. That's why they are so much less massive than Jupiter.

As mentioned above, this picture of planetary formation is not without its puzzles. For example, it's clear that the model developed by Safronov and Wetherill is a variant of the old nebular hypothesis. But Laplace's idea was killed by the angular momentum problem. To put it bluntly: where did the Sun's angular momentum go? Doesn't the angular momentum dilemma also kill this updated version of the nebular hypothesis? Fortunately, it doesn't. In the model developed by Safronov and Wetherill there are various ways in which the Sun could have lost angular momentum – it might have been carried away by ejected planetesimals, for example, or magnetic fields might have transferred the angular momentum to the outer reaches of the system. Although astronomers don't yet know which mechanism represents what actually happened, astronomers need not reject the model on the basis of the angular momentum problem.

A more pressing problem regards the formation of gas giants. The challenge for astronomers is to explain how a protoplanetary core can reach the threshold mass of about ten Earth masses that's necessary for it to start sucking in hydrogen and helium gas from the surrounding disk. The problem is that it would seem to take a *long* time for a core to reach this threshold mass: computer simulations suggest that the protoplanetary disk would dissipate before the core was complete. So there are gaps in astronomers' understanding of this process.

A further puzzle occurs when astronomers look at the structure of other planetary systems. On the basis of the picture of planetary formation painted above you might expect that all planetary systems would follow the same overall plan as our own with large, massive, gas giants inhabiting the outer reaches of the system and small, low-mass, rocky planets orbiting much closer to the star. That's not what happens, however. Many exoplanetary systems don't follow the script. Astronomers have seen planets with a mass far greater than Jupiter moving in near-circular orbits very close to their central star. Since no one knows how these so-called 'hot Jupiters' could *form* so close to a star, the current consensus is that they form beyond the snow line and then migrate inwards. Is there perhaps a final stage in planetary system formation in which catastrophically close encounters cause some planets to change orbits and others even to be hurled into the space between stars? Another problem is that most of the exoplanets seen so far have rather eccentric orbits; the planets in our Solar System, on the other hand, have near-circular orbits. We might expect to see the planets in any system orbiting their star in the same direction and aligned in the same plane, but about a third of exoplanets have orbits that are in some way misaligned. There's more: astronomers expected to see relatively few Neptune-sized planets (in other words, 10–20 Earth masses); it turns out that such planets seem to be common. It isn't necessarily the case that the accepted theory of planetary formation is wrong; but the models that astronomers have based on that theory seem to be lacking.

Perhaps the thorniest puzzle with the model lies in the second stage of planetary formation: how can dust bunnies – objects the size of a sugar cube – combine to form a planetesimal that's a kilometer or more in diameter? The mechanism that enables dust grains to adhere and form larger structures breaks down once those structures get to be about one centimeter in size. Some other mechanism must be in play, but the details are not understood.

So problems remain with the current models of planetary formation. To solve these and other problems, astronomers need a clearer view inside star-forming regions of space – those vast clouds of cold gas and dust that hide their secrets from optical telescopes. They need to observe more planets orbiting young stars, they need to find more cases of the debris that surrounds protostars, and they need to observe many more dusty cocoons that give rise to stars and planets. What's needed are bigger, better, more sensitive infrared telescopes.

When telescopes take flight

Our view of the infrared sky is set to improve markedly in the next decade, thanks to a number of new telescopes.

One of the most interesting of the new telescopes is on board the **Stratospheric Observatory for Infrared Astronomy** (sofia), an American–German facility which, after a long development delay, finally saw first light in 2010. What makes sofia particularly interesting is that it is neither a ground-based nor space-based observatory. Instead, it flies on a modified Boeing airplane. The airplane appears to have a gaping hole just behind its left wing, but the hole is in fact a door that can open and allow a reflecting telescope a good view of the sky. Since the airplane can cruise in the stratosphere at an altitude of about 12 km, the telescope's view is almost unhindered by infrared-absorbing water vapor.

The idea of an airborne observatory such as sofia is not new; nasa has been flying telescopes in airplanes since the 1960s and balloon experiments go back even further. What *is* new is the sheer size of the primary instrument: although it's only a medium-sized telescope compared to ground-based instruments, this is by far the largest telescope to be flown in an aircraft. Edwin Hubble discovered the expansion of the Universe using the 2.5 m Hooker telescope at the Mount Wilson observatory; at the time, it was considered to be a giant telescope. The sofia instrument is the same size as the Hooker telescope.

Observing with an aircraft-borne telescope presents challenges, of course. For example, air turbulence and engine vibration could jog the telescope; for that reason a sophisticated system of 'air springs' had to be developed to isolate the instrument from the bumps, shakes and shudders that would make precision measurements difficult. And, unlike space-based telescopes, the duration of an observing session is limited by the amount of fuel the plane can carry. Nevertheless, there are many benefits to airborne missions. For example, starting from about 2014 the plan is for sofia to fly three or four times per week – and it will do so for the next 20 years. That's a long time for an infrared telescope. Space-based infrared telescopes such as iras or Spitzer inevitably had a limited lifespan: when their supply of helium boiled away the mission was over. There's no such limitation with sofia: when the airplane lands the coolant can be topped up. And it's not just coolant that can be topped up: astronomers can take the chance to fit sofia with the latest technology, so that while she flies she will never age. This will always be

Figure 11.6 *This Boeing 747SP widebody airplane has been modified to contain a 2.5 m reflecting telescope. In observing mode the airplane flies in the stratosphere, at an altitude of about 12 km. When the door is lowered, the telescope enjoys a view of the sky that is above most of the water vapor in Earth's atmosphere. (NASA)*

a state-of-the-art observatory. Compare that with a space-based telescope, which can be difficult or even impossible to maintain; nowadays a space-based observatory might employ old technology even as it's being launched.

The advantages of the SOFIA telescope are many; the observatory will surely produce wonderful science. But that's not to say that astronomers have given up on the idea of a space-based infrared technology. After all, if you want your telescope to be above *all* the infrared-absorbing molecules in the atmosphere then you must go into space.

In 2009 ESA launched the aptly named **Herschel** infrared telescope; its primary mirror has a diameter of 3.5 m, which makes it the largest astronomical telescope ever launched to date. With such a large mirror, Herschel captures about twenty times more light than any previous space-based infrared telescope. The observatory possesses three extremely sensitive infrared detectors that must, of course, be cooled. Herschel stays cool firstly by operating about 1.5 million kilometers away from the heat emitted by the Earth and Moon: as with the Planck observatory, which we discussed in an earlier chapter, Herschel orbits the L2 point of the Earth–Sun system and has shades that protect it from solar radiation. Secondly, Herschel has a cryostat that employs 2300 liters of liquid helium; the cryostat keeps the instruments cooled to almost absolute zero. The mission will last as long as the cryostat contains helium; with luck, Herschel will do science until 2014.

In just its first year of operation Herschel discovered many phenomena that will illuminate the process of stellar and planetary formation. For example, Herschel looked at a bright cloud of gas known as NGC 1999, which shines by reflecting light from a three-star system called V380 Orionis. This gas cloud contains within it a black patch of sky. The assumption has always been that such black patches of sky consist of dense dust clouds that block the transmission of light. When Herschel studied this particular patch, however, it found – quite literally – nothing. In other words, this patch of sky is dark because it really is empty: it's a hole in space. It seems, then, that one of the young protostars that formed V380 Orionis emitted jets that cleared the hole. Further study of this object should give astronomers clues as to how, precisely, newborn stars scatter the clouds that gave them birth. And in 2011, Herschel imaged a hitherto unknown cradle for stars close to the center of the Galaxy. The cradle is a 600-light-year long twisted tube of dust and cold (15 K) gas. How the tube came to be twisted is not yet understood, but further studies of the structure will tell astronomers not only about star formation but also much about the central regions of our home Galaxy. All in all, then, Herschel is an excellent telescope for discovery.

Figure 11.7 Top: the 3.5 m diameter mirror of the Herschel telescope is the largest so far put into space. The mirror and its associated instrumentation are protected by 'sunshades', an example of which can be seen here behind the mirror, that shield them from solar radiation and infrared radiation from the Earth and Moon. Bottom: one of Herschel's discoveries. The bright reflection nebula NGC 1999 contains a dark patch in the middle with a width that's about ten thousand times the distance between the Earth and Sun. In 2010, Herschel showed that this patch of sky is truly empty and contains none of the cold dust and gas that astronomers expected to see. This discovery should eventually help astronomers understand how new stars scatter the material from their birth clouds. (Top: ESA; bottom: NASA)

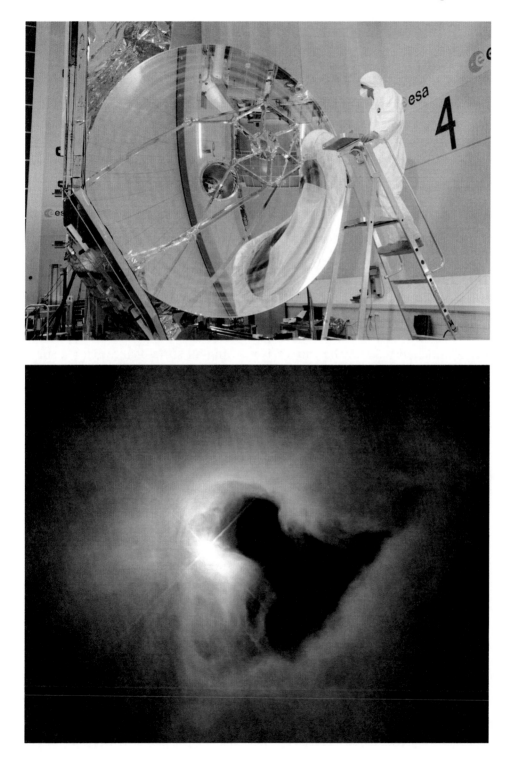

Herschel is not the only major space-based infrared telescope that astronomers hope will unlock the secrets of stellar and planetary formation. If all goes well, the **James Webb Space Telescope** (jwst) should be in space by 2018. It must be said that not everything has gone well to date. The jwst project has suffered from several slips in its schedule and the cost overruns involved have been immense. The extra funding required by jwst has jeopardized some of the other missions discussed in this book, and certain voices in the House of Representatives even called for jwst itself to be scrapped. However, at the time of writing it seems as if the jwst will go ahead – and eventually it will join Herschel, Planck, and wmap to reside at the Earth–Sun L2 point.

In some ways, jwst is designed to be the successor to Hubble. But whereas Hubble observed in the visible and ultraviolet as well as the infrared, jwst will observe primarily in the infrared – its instruments will be sensitive to wavelengths in the range 0.6–28 μm. The jwst mirror is vast. Its 6.5 m primary mirror would be considered quite large even for a ground-based telescope; for a space-based mirror it goes far beyond anything that has ever been launched before. The mirror is pushing all sorts of limits for a space telescope. For example, it follows a new design in which hexagonal segments are folded up to fit inside a launch vehicle and then unfolded once the telescope is in space. Furthermore it is made out of beryllium, which is a new material for telescope construction. The advantages of beryllium are that it is lightweight, which means that the telescope is not so heavy that it can't be launched; it is strong; and it holds its shape even at very cold temperatures.

The mirror and the detectors must, of course, be kept cold – no hotter than about 40 K. In large part this will be achieved through the use of a parasol, a sunshield the size of a tennis court that will shade the instruments from the Sun and allow heat to radiate away. Some instruments will need to be even cooler, but not so cold that helium must be employed. The lifetime of jwst will thus not be limited by the helium supply; the jwst team plan for a lifetime of five years but hope to use the telescope for at least ten years.

Figure 11.8 Top: the first six of 18 gold-plated beryllium segments that will form the jwst primary mirror. The nasa engineer standing before the segments gives an idea of scale: Webb's 6.5 m mirror dwarfs the Hubble 2.4 m mirror. Bottom: an artist's impression of the jwst in space. The sunshield will be the size of a tennis court. (Top: nasa/msfc/David Higginbottom; bottom: esa)

Having a monster such as JWST with which to observe the skies, astronomers expect to make further progress in understanding the history of the cosmos since the reionization epoch – including, of course, the birth of stars and protoplanetary systems. Over the next few years, then, infrared telescopes such as SOFIA, Herschel and (fingers crossed) JWST will transform our knowledge of how stars and planets form.

A new window in the spectrum

As with all the other mysteries discussed in this book, the puzzle of how stars and planets form is one that can be investigated in several different ways; infrared telescopes are not the only tool for this. The more exotic instruments – those that try to detect dark matter, for example, or those that capture cosmic rays or neutrinos – those telescopes have little to offer in this work. But telescopes that detect electromagnetic radiation – at whatever wavelength – have the potential to illuminate the journey from gas cloud to T Tauri star. For example, radio telescopes (which are the focus of chapter 13) are ideal for studying the processes involved in star formation; after all, stars and planets are born in cold environments, where temperatures are such that matter emits mainly at the longest wavelengths. However, astronomers also need to observe at shorter wavelengths if they want a complete picture. Optical telescopes (which are discussed in the next chapter) can be used in these studies once the dust around young stars has gone. And ultraviolet telescopes (discussed in the previous chapter) can be used to investigate starbursts – episodes of massive star formation that take place in some galaxies. Even X-ray telescopes (see chapter 5) can be used to observe star-forming regions: various processes – strong magnetic fields; shocks that heat part of the gas cloud; radiation from very young massive stars – can generate high-energy photons. Thus although X-rays won't play an important role in the initial formation of a star, they may well affect the development of its protoplanetary disk.

There's another window in the electromagnetic spectrum that might illuminate the processes involved in planetary formation. It's a region that wasn't highlighted in the introductory chapter (although of course, to some extent, the boundaries between different regions are artificial). Squeezed between the far infrared and microwave parts of the spectrum lies the **submillimeter band**, where the radiation has a wavelength between about 0.3–1 mm. The submillimeter wavelengths can be used to study gas and dust clouds: not

Figure 11.9 *Three of the antennae in the Atacama Large Millimeter/submillimeter Array. Simply transporting these telescopes to the final site poses a significant technical challenge. Nevertheless, although conditions at* ALMA *are uncomfortable for humans they are ideal for ground-based observations in the submillimeter range. When it is complete,* ALMA *will have 66 antennae – and be one of the most powerful observatories in the world.* (ALMA (ESO/NAOJ/NRAO) *and Carlos Padilla)*

only are some clouds extremely cold – just a few degrees above the cosmic background, which therefore generates submillimeter wavelengths – but far infrared radiation from dust clouds in distant galaxies can be redshifted to submillimeter wavelengths.

The same instruments that are used to observe in the traditional infrared wavelengths can often be used for these longer wavelengths. Herschel, for example, will observe in the submillimeter band, as will SOFIA. However, just as ground-based infrared astronomy is possible, so is ground-based submillimeter astronomy. Three of the telescopes at the Mauna Kea Observatory are dedicated to submillimeter observations, and there are one or two other mountain-top sites that house submillimeter telescopes. The **South Pole Telescope**, which is based at the Amundsen–Scott station in Antarctica, can observe in the submillimeter band. But by far the most impressive instrument for submillimeter observations is the **Atacama Large Millimeter/submillimeter Array** (ALMA).

ALMA is housed at the **Llano de Chajnantor Observatory**. This facility, located at an altitude of 5104 m in the Atacama desert in Chile, is the site for some of the largest and most expensive telescope projects in the world. It's an inhospitable place in which to work. The arid atmosphere parches the skin, chaps the lips, causes the nose to bleed. Visitors must pass a medical check before ascending to such a height, and the symptoms of altitude sickness are an occupational hazard. The scientists and engineers who are constructing ALMA often have to carry an oxygen supply; without the extra oxygen, the thin Atacama air would slow their thinking – with potentially catastrophic results when one of the 115 tonne antennae is being lowered into position. Nevertheless, inhospitable as it may be, this high, dry desert region is one of the best places on the planet to site a telescope.

ALMA is a global collaboration, and one of the largest ground-based astronomy projects ever undertaken. By the summer of 2011, 16 antennae had been transported to the Chajnantor plateau; this was a sufficient number to begin observations, and in late 2011 the ALMA team released the first images from the array. When the array is completed in 2013, ALMA will consist of 66 antennae – 54 of them with a 12 m mirror and 12 of them with a 7 m mirror. The telescopes will be linked together to form an interferometer. Furthermore, the telescopes will be movable: the array can be as small as 150 m in width or as large as 14 km in width. With power such as this, ALMA is certain to make some wonderful discoveries.

Of course, ALMA – along with other infrared instruments such as Herschel, SOFIA, and JWST – will study much more than just star and planet formation. Even so, one of the key science goals of these instruments is to enable astronomers to better understand how stars and planets came into being. ALMA will possess such a high angular resolution that when it looks at a star-forming region it will able to distinguish the different phases in the evolution of a planetary system, from the initial collapse of a cloud to form a protostar through to the condensation of objects that will eventually form planets. It will allow astronomers to study how different physical and chemical conditions affect the formation of stars and planets. The submillimeter window will open up whole new vistas on newly forming planetary systems – and in doing so will help us learn how our own Solar System came into being.

12

Other Earths

We know now that there are planets out there orbiting stars other than the Sun. But surely what we really want to know is whether other Earth-like planets exist – are there planets out there with roughly Earth's mass, containing liquid water, and that might, just might, support life? Perhaps the next generation of optical telescopes will answer this age-old question.

The discovery of exoplanets – The Kepler Mission – Astronomers go large – Other Earths, ATLAST? – Exoplanets galore

S. Webb, *New Eyes on the Universe: Twelve Cosmic Mysteries and the Tools We Need to Solve Them*,
Springer Praxis Books, DOI 10.1007/978-1-4614-2194-8_12, © Springer Science+Business Media, LLC 2012

Most of humankind's knowledge of the cosmos has been gained through observation with optical telescopes. The names of large optical telescopes – Herschel's Leviathan; Hooker, Hale and Hubble – are some of the most evocative in all astronomy. An optical telescope is still what people picture when they hear the word 'telescope'. Despite all the different windows through which astronomers now peer out at the Universe, the visible band remains the key part of the spectrum.

One example that shows the power of the optical telescope is the discovery of exoplanets.

The discovery of exoplanets

As I write this, the *Extrasolar Planets Encyclopaedia* lists 759 extrasolar planets (exoplanets for short). By the time you read this, many more exoplanets will be known. How, though, do astronomers detect a planet outside the Solar System?

Astronomers discovered a majority of those 759 exoplanets using a technique called the **Doppler spectroscopy method** (also known as the **radial velocity method**). The first success of the Doppler spectroscopy method came in 1995, when Michel Mayor and Didier Queloz announced the discovery of a Jupiter-sized planet orbiting 51 Pegasi, a Sun-like star about 60 light years distant from Earth. The planet has the name 51 Pegasi b. The lowercase 'b' indicates that it is was the first planet discovered orbiting its parent star. If another planet is discovered around the star then it will be called 51 Pegasi c; subsequent planets will be designated 'd', 'e' and so on. Mayor and Queloz were not, in fact, the first to find an exoplanet: three years earlier, the radio astronomers Aleksander Wolszczan and Dale Frail discovered a planetary system around a pulsar. But although planets orbiting a pulsar are an interesting quirk, it seems extremely unlikely that any such planet could be the abode of life. What everyone really wants to know about are planets orbiting a 'normal' main sequence star – and 51 Pegasi b was the first such planet to be found.

The Doppler spectroscopy method works because a planet does not in fact orbit around a star; rather, both the star and the planet orbit around a common center of gravity. Stars are much more massive than planets, of course, so the common center of gravity is always closer to the star than to the planet. In the case of Earth and Sun, for example, the center of gravity lies deep inside the star – which is why it makes sense to say that Earth orbits

Sun. But the common center of gravity of the Jupiter–Sun system lies just above the Sun's surface, and the Sun exhibits a definite 'wobble' as it moves around the center of gravity. The same applies to other planetary systems: any star that possesses a planet will 'wobble' around the common center of gravity. The more massive the planet is in relation to its star, and the closer it is to the star, then the greater the wobble will be. Depending on the alignment of the orbit, the star will sometimes be moving away from us and sometimes be moving toward us as it responds to its planet's tugging. The Doppler effect means that movement away from us will cause the star's light to be redshifted while movement towards us will cause the star's light to be blueshifted. This, then, is the basis of the Doppler spectroscopy method: highly sensitive spectrographs, used in conjunction with optical telescopes, study particular lines in a star's spectrum and look for periodic shifts towards the red, then the blue, then back again to red. A shift that repeats itself regularly almost certainly means that the star is being pulled back and forth by an unseen companion. If that companion has a low enough mass, then it's a planet.

That's the method Mayor and Queloz used to discover 51 Pegasi b. They observed the star using a reflecting telescope with a 1.93 m diameter primary mirror to which was attached a high-resolution spectrograph. Since 51 Pegasi b has an orbital period of just 4.2 days, they didn't need to observe for too long to make their discovery.

Just three months later, Geoff Marcy and Paul Butler used a similar technique with the 3 m telescope at Lick Observatory to confirm the original discovery and also to announce the discovery of planets around 47 Ursae Majoris (a Sun-like star that's about 46 light years away) and 70 Virginis (a Sun-like star about 59 light years distant). Marcy went on to discover seventy of the first hundred known exoplanets.

Technology has improved since these original discoveries. Spectrographs are now so sensitive that they can sense whether a star has a radial velocity as slow as one meter per second – that's your walking speed when you're ambling along rather than rushing. Discovering new planets using the Doppler spectroscopy technique is now almost routine. Nevertheless, despite its success, the technique has drawbacks. First, although it's great at finding a high-mass planet orbiting close to its parent star, it's less good at finding a low-mass planet orbiting far from its parent star. If alien astronomers used the technique to detect Earth's motion around the Sun, for example, then they'd have to be using a spectrograph capable of detecting a star moving at a snail's pace. Human astronomers will develop that technology one day, but it's not available yet. Furthermore, since a planet that is far from its star will have a long orbital period, the necessary observations could take many years. Thus the Doppler spectroscopy technique finds lots of 'hot Jupiters' – massive planets orbiting close to their parent star – but it doesn't follow that such planets are representative of the overall planetary population. Astronomers are simply seeing the objects that this technique allows them to see.

Another problem with Doppler spectroscopy is that the size of the signal depends on the inclination of the planet's orbit to our line of sight. If the orbit is precisely edge-on to our line of sight then the signal is as large as it can be. But if the orbit is face-on then the signal is zero. In the latter case the star could be moving like a demented yoyo and the Doppler method wouldn't pick it up because there'd be no component towards us or away from us; there'd be no observable Doppler effect. In most cases, of course, the orbit will be neither fully edge-on nor fully side-on; instead, the orbit will be at some angle to our line of sight and there'll be some component of velocity along that line of sight. But since in most cases astronomers won't know the orbital angle, this means that the Doppler spectroscopy technique can only estimate a planet's minimum mass.

Fortunately, astronomers have another technique in their toolkit that can be used to detect planets around main sequence stars: the **transit photometry method**. A transit occurs when a planet crosses in front of its star as viewed by an observer. Transits thus occur in the Solar System. Observations of the transit of Venus, for example, have historically been important in astronomy, since they led to improved estimates of the Earth–Sun distance. Well, if a planet transits then the star's brightness is going to drop; as long as the planet is in front of the star's disk, the star will be slightly dimmer than usual.

A car's headlamp dims slightly whenever a midge buzzes across it, but it's not necessarily the sort of effect you might think could be measured. Well, astronomers attempt the same thing when they look for exoplanetary transits, and technology is now up to the task. The precise amount of dimming depends on the relative sizes of star and planet: a large planet transiting a small star causes a bigger drop in brightness than a small planet transiting a large star. Even in the best case, though, the brightness changes are small and thus difficult to measure. Difficult – but not impossible. Thus transits provide a clear signal: a transit will cause a small drop in brightness; the dimming will be absolutely periodic; and the dimming will always

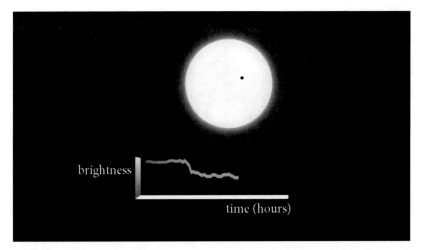

Figure 12.1 *When a planet moves in front of its star's disk it blocks some light from the star, which therefore dims slightly. The amount by which the star dims depends on the relative size of the star and the transiting planet.* (NASA)

have the same duration – typically, a transit will last a few hours. If astronomers observe a particular star to possess this regular behavior then they can be pretty sure they have observed a transit – and thus they can deduce the presence of a planet around the star. The first planet to be discovered using the transit method was OGLE-TR-56 b, whose discovery was announced in 2003 by a team of astronomers led by Dimitar Sasselov. The planet is about 5000 light years away, which is far more distant than any planet found by Doppler spectroscopy.

The transit method has its drawbacks. Perhaps its main disadvantage is that to observe a transit the planet's orbit has to be aligned just right: the larger the orbit, the less likely it is that we'll see a transit. Nevertheless, if astronomers are lucky enough to observe a planetary transit then they have a powerful tool to learn more about that planet. For example, since it's known how long the planet takes to orbit the star and since the mass of main sequence stars of different spectral types is well known, astronomers have enough information to calculate the size of the planet's orbit. This means they can tell whether the planet lies in the star's **habitable zone** – the region where water might exist in the liquid phase. There's more. Since the distance to the planet is likely to be known (it's the same as the distance to the star, which in most cases can be determined fairly accurately), the size of the planet itself can be calculated. If the Doppler spectroscopy method can be applied in tandem to the same star, so that the planet's mass can be estimated, then astronomers can determine the planet's density – and thus tell whether it's a gas giant or a terrestrial planet. Several planets have been subject to this double technique.

The transit method can sometimes deliver even more information. For example, it's sometimes possible to determine how much infrared radiation is being emitted by a planet: it's the difference in measured radiation between when the planet transits and when it disappears behind the star. This in turn allows the planet's temperature to be determined. This has been done for the planet OGLE-TR-56b, for example, and it's believed that this planet is so hot that it could rain iron there. Furthermore, by studying a star's light as it passes through the atmosphere of a transiting planet it should be possible, at least in principle, to determine the chemical composition of the planet's atmosphere.

It's quite a thought: even from a distance of many light years astronomers can sometimes tell whether a planet is rocky or gaseous, whether it's roasting or frigid, and what elements make up its atmosphere!

A third method for detecting exoplanets was hinted at in chapter 4: gravitational microlensing. If Earth, a foreground star and a distant background star happen to align, then the foreground star will act as a gravitational lens and brighten the light from the background star. If the foreground star happens to possess a planet that's appropriately aligned then the planet can also produce a detectable lensing effect. Such lensing events don't last long, and in any event they are rare because of the precise nature of the alignment that's required. Furthermore, they are one-off events: since all three elements in the arrangement are in motion, the alignment won't repeat – you can't use microlensing to study an exoplanet in depth. Nevertheless, if astronomers observe patches of sky that are rich in background stars they do occasionally see microlensing events – and can thus detect the existence of planets at large distances from us. This method is more sensitive than others to small-mass planets such as Earth and, as we saw in chapter 4, the OGLE program has – as a byproduct – discovered 14 exoplanets.

In addition to the transit, Doppler spectroscopy and microlensing methods there's another technique that could be used to detect the presence of planets orbiting Sun-like stars: **astrometry**. This is actually the oldest of all methods of searching for planets. It works by detecting miniscule but regular changes in the position of a star, changes that might indicate the presence of a planet tugging on its parent. Every so often an astronomer claims to have detected an exoplanet using the astrometric method. In 1943, Kaj Strand announced the presence of a planet orbiting the star 61 Cygni. In 1963, Peter Van de Kamp, announced the discovery of a planet orbiting Barnard's Star. In 1996, George Gatewood published details of a planetary system around the nearby star Lalande 21185. And in 2009, NASA astronomers used 12 years' of astrometric observation using the Hale telescope to announce the discovery of a high-mass planet orbiting VB10, which is one of the smallest stars known. However, none of the claims based on the astrometric method have ever been verified using other techniques. The astronomical community thus regards all of these claims with a degree of scepticism.

The problem with the astrometric method is that the observations required are so incredibly difficult to make. Astronomers must measure the position of a star with immense precision over a period of years or even decades. Until recently, the available technology has just not been up to the task.

The Kepler Mission

As I began planning this book the most productive method for finding planets had been Doppler spectroscopy. Second was the transit method. Microlensing and astrometry were well behind. That order is changing as I write: a mission using the transit method is discovering exoplanets by the bucketload. In 2009, NASA launched its **Kepler Mission** – a space-based observatory with the aim of discovering terrestrial exoplanets. The instrument on board Kepler is of a special design: rather than being optimized for taking detailed images, its 0.95 m telescope is designed to measure the light intensity of lots of stars at once. Kepler thus has an extremely wide field of view – about the area of your hand if you hold it at arm's length; this is *much* larger than is typical for an astronomical telescope. The instrument stares unwaveringly at the same patch of sky, a field in the constellations of Cygnus, Draco and Lyra, and as it does so monitors the brightness of about 150 000 main sequence stars. Since this field is in the direction of the Sun's motion around the galactic center, the stars that Kepler observes are all roughly the same distance from the center as the Sun is. This latter fact is important. It's possible, perhaps, that life has developed on Earth because our Solar System occupies a particular and special region in the Galaxy. In other words, it's possible that there is a 'galactic habitable zone'. If such a zone exists then, by studying stars that share the Sun's orbit around the center of the Galaxy, Kepler is presumably looking at stars that are in that habitable zone.

In 2010, the Kepler Mission team released data on almost all of its target stars. About 7500 of those stars turned out to be variable, changing regularly in brightness because of internal changes rather than because of transiting planets, and were thus removed from the list of objects that Kepler keeps its eye on. The intention is that Kepler will monitor the remaining stars in the list until at least August 2013; if things go well its telescope will continue observing for longer. By continuously measuring these stellar brightnesses, Kepler can apply the transit method of planet hunting. The chance of any particular star possessing a transiting planet is not great. Furthermore, the smaller the transiting planet and the larger its orbit, the less chance there is that Kepler will see it transit. But those odds don't matter: Kepler is viewing so many stars at once that it's certain to see transits. Astronomers can't know in advance which particular star will exhibit a transit, but once they've crunched the data they know they'll find *lots* of transits. And once a planet has been observed in this way, follow-up observations using the Doppler spectroscopy method become possible.

The numbers stack up quite impressively in Kepler's favor. Suppose it's common that stars possess an Earth-sized planet orbiting at the same distance as Earth orbits the Sun. In that case simple probability dictates that, by looking at so many stars, Kepler should see about 50 of these other Earths. If it turns out that rocky planets tend to be larger than our world, say with a radius 30% larger than Earth, then Kepler should see about 185 of them. If Earth-like planets tend to be closer to their parent star and orbit with periods less than a year then that would increase the number of detectable planets still further. As for giant planets – well, they should show up in droves. Thanks to Kepler, the transit method is overtaking the Doppler method as the most fruitful technique for finding exoplanets.

In 2011, the Kepler Mission team published its analysis of the first four months of observation. The results were astounding: in that short period of observing time, Kepler identified 1235 planetary candidates orbiting 997 stars.

The list of Kepler candidates contains some interesting objects. The star Kepler-10, for example, has a transiting planet whose radius is just 1.4 times greater than Earth's: Kepler-10b is about five times more massive than Earth, and is thus about as dense as iron. The planet orbits extremely close to its star and thus always shows the same side to the star: its day side will possess oceans of molten rock while its night side will be solid. A star called Kepler-11 has no less than six transiting planets; all of them are more mas-

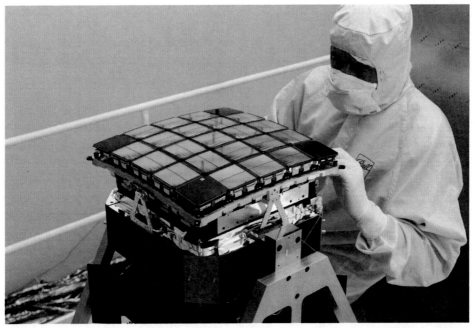

Figure 12.2 *The Kepler photometer has a total of 42 charge coupled devices (CCD), which work together in pairs to monitor continuously the brightness of main sequence stars. Each CCD pair covers one square patch of sky. In February 2011, the Kepler Mission team released data on 156 453 stars that the telescope had been monitoring. (NASA and Ball Aerospace)*

sive than Earth and five of them possess an orbit smaller than Mercury's. That's quite a packed planetary system. And in December 2011, the team announced the discovery of Kepler-20e and Kepler-20f, the first truly Earth-sized planets found orbiting a Sun-like star. Although Earth-sized they are not Earth-like: Kepler-20e has a surface temperature of 760 °C, Kepler-20f a temperature of 427 °C. These are not habitable planets.

In total, the 1235 Kepler candidates announced in 2011 comprise 68 that have a radius less than 1.25 times that of Earth (and thus may be called Earth-like planets); 288 that have a radius between 1.25–2 times that of Earth (so-called super-Earths); 662 that have a radius between 2–6 times that of Earth (and may be categorized as Neptune-like planets); 165 that have a radius between 6–15 times that of Earth (and are thus Jupiter-size planets); and 19 that have a radius between 15–22 times that of Earth (and so are super-Jupiters). On the basis of these figures, the Kepler Mission team estimate

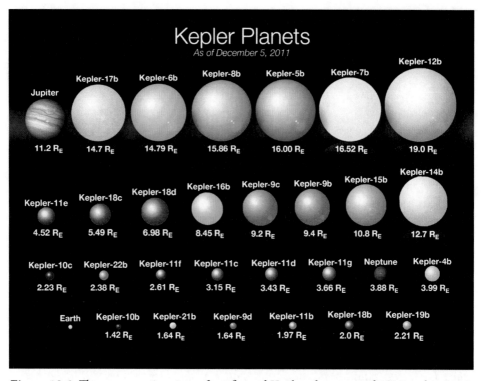

Figure 12.3 *The comparative sizes of confirmed Kepler planets as of 5 December 2011. Jupiter and Earth are also shown for scale. Note that Kepler-20e and Kepler20-f were discovered after this chart was created. (NASA/Kepler Mission/Wendy Stenzel)*

that 6% of stars may possess Earth-like planets. An even more important result is that 54 of the planetary candidates orbit in their star's habitable zone – the orbital region where temperatures are such that liquid water may exist. And five of *those* candidates are less than twice the size of Earth. Kepler tells us that the number of potentially life-bearing planets in the Galaxy is large.

It's worth noting that not all of those 1235 Kepler candidates will turn out to be planets. The trouble, as with many other situations in science, is that it's easy to be fooled. In this case, various phenomena might cause the periodic dimming of a star. The team must weed out the 'false positives' – the stars that appear to change regularly in brightness but do so because of some reason other than a planetary transit. For example, a false positive could occur if a target star happens to lie very close in the sky to a distant binary system. In that case Kepler would record a changing brightness level for the target star, but the fluctuations wouldn't be caused by a planet. Another situation that would generate a false positive is when two eclipsing stars just graze each other's edges. Once the Kepler team has eliminated the imposters they'll be left with fewer than 1235 planets. But they'll be left with loads. And that's just with four months' worth of observation.

How, though, can the Kepler Mission team identify those false positives? How can astronomers confirm the Kepler candidates as being true exoplanets rather than, say, grazing eclipsing binaries? The good old-fashioned optical telescope – the sort of instrument that, until relatively recently in astronomical history, was the mainstay of research – comes into its own here. The Kepler Mission follow-up team has access to several of the world's best optical telescopes and can use them for detailed observations of candidate stars. Furthermore, recent advances in technology mean that these telescopes can perform better now than they did when they were first constructed. These telescopes will reveal the true nature of the Kepler candidates.

For example, one of the instruments available to the Kepler Mission is the venerable 5 m Hale telescope at the Mount Palomar Observatory. When it was completed in 1948 the Hale telescope was the largest in the world, a record it held until 1975, when Russian scientists constructed an instrument with a 6 m mirror. (The Russian instrument never worked as well as had been hoped, however, and so effectively the Hale telescope was the world's largest until 1993.) The Hale telescope is a more effective instrument now, though, than it ever was when it was the 'world's biggest'. In 1999, astronomers installed a package of **adaptive optics** onto Hale and it was as if the telescope had been given spectacles to correct a sight defect. The Hale telescope is now a formidable instrument for planet hunting.

Adaptive optics compensates for one of the main disadvantages of ground-based telescopes: the blurring of an image caused by turbulence in Earth's atmosphere. The technique involves taking light from a suitable guide star and, using a device called a wavefront sensor, determining the distortions introduced by the atmosphere. A computer calculates the optimum shape that a mirror would need to correct for these distortions. A small deformable mirror is then pushed and pulled by devices called actuators and made to take on precisely this shape. A millisecond or so later the process is repeated. This operation takes place so rapidly that the mirror changes shape more quickly than the atmosphere itself can change. The result is a razor-sharp image that's almost as clear as if the atmosphere were absent. It's like having a large space-based telescope without the hassle and expense of actually going into space.

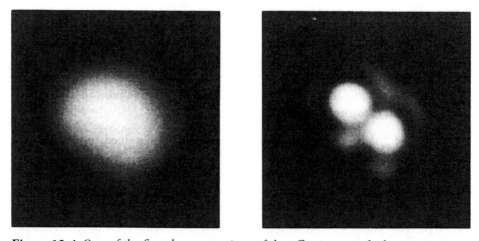

Figure 12.4 *One of the first demonstrations of the effectiveness of adaptive optics: in 1990, astronomers imaged a double star using a 3.6 m telescope first without correction (left-hand image) and then with correction. The corrective system used in that early work had 19 actuators deforming a mirror 100 times per second. More modern systems have thousands of actuators deforming a mirror on even smaller timescales. Such technology means that large ground-based telescopes can outperform the space-based telescopes currently in operation. (ESO)*

In the early days of adaptive optics astronomers ran into a common problem: it's often the case that a suitable guide star is not available near the object of interest. Even that limitation has been circumvented: astronomers simply create their own guide stars. Many observatories now shine lasers into the night sky. The lasers are tuned to a particular frequency such that, at an altitude of about 90 km, sodium gas in the atmosphere starts to glow. This glowing gas acts as a guide star. The light from this artificial guide passes through the same slice of atmosphere as light from the object of interest – and so the adaptive optics package can make its corrections.

Figure 12.4 illustrates how adaptive optics improved a telescope such as Hale; the technique has moved on since the work shown in that image. In 2010, astronomers used a device called a coronograph to block light from the star HR8799; a 1.5 m portion of the Hale mirror, suitably boosted by adaptive optics, was then able to take a direct image of exoplanets around the star. This wasn't the first time a telescope had imaged an exoplanet, but it was the first time an image had been taken without one of the new breed of telescope.

Thus following up the Kepler candidates with Hale and other optical telescopes will confirm whether planets have been discovered. Astronomers

Figure 12.5 *The **William Herschel Telescope** fires a laser beam that excites sodium gas high in the atmosphere. The gas glows and light from this new 'guide star' can then be used by the adaptive optics package that is installed on the telescope. (Javier Méndez, Isaac Newton Group of Telescopes, La Palma)*

are confident that this process will work because it has already succeeded: in 2010 astronomers published details of the first five confirmed planets discovered by Kepler. The planets made their presence obvious in just six weeks' worth of Kepler observations during 2009. Not surprisingly, these planets were *so* obvious to spot and subsequently confirm because they are hot Jupiters – large planets orbiting their parent star in just a few days. Such planets produce the biggest effects.

For the *really* important candidate stars – those that observations suggest may have Earth-like planets orbiting them – the Kepler Mission team has access to the **WM Keck Observatory** on the summit of the dormant volcano

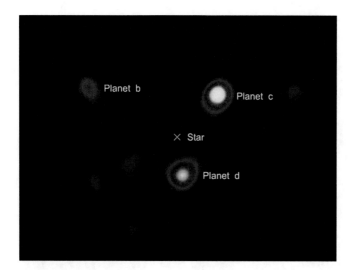

Figure 12.6 *A part of the Hale telescope was used to take an image of the planets b, c and d in orbit around the star* HR 8799. *The planets are all gas giants orbiting at large distances from the star.* (NASA/JPL-*Caltech/ Palomar Observatory*)

Figure 12.7 *The summit of Mauna Kea, Hawaii, which is at an altitude of 4200 m, is home to a dozen or so telescopes. This is one of the best locations in the world for sub-millimeter, infrared and optical observations of the cosmos. The instrument on the far left of this photograph is the 8.3 m optical/infrared Subaru Telescope, which is operated by Japan. The instrument on the far right is the* NASA *Infrared Telescope Facility. The two telescopes in the middle are the twin 8.3 m Keck Telescopes. To give an idea of their scale, each Keck telescope is eight storeys tall. Although each weighs 300 tons, they are capable of being moved with nanometer-levels of precision. (Commons)*

Mauna Kea. Keck has two telescopes, both of them having a whopping 10 m diameter mirror. It was the opening of these twin monsters in 1993 that finally saw the Hale telescope demoted from its position as the telescope with the world's largest effective mirror. The two Keck telescopes are 85 m apart. They can work together as an interferometer, which then creates an instrument with a quite astonishing acuity.

The Keck telescopes have adaptive optics systems installed, of course, along with the ability to create laser guide stars. They also have several instruments attached to them, one of the most important of which is the high resolution echelle spectrometer (HIRES) on Keck 1. The HIRES instrument splits starlight into thousands of component wavelengths and measures the intensity of each with the utmost precision. HIRES can detect the ambulatory laps made by a star being pulled by an orbiting planet; the instrument can tell whether a star moves towards us or away from us by as little as $1\,\mathrm{m\,s^{-1}}$. Any Kepler candidates that may be home to an Earth-sized planet are studied by Keck and HIRES. As already mentioned, with this super-acute vision Earth-sized planets have been found: one day it may find Earth-like planets.

The search for planets that could be home to life is surely one of the most exciting quests in science. However, the difficulty of finding habitable planets should not be underestimated. The story of Gliese 581 g is a reminder of how hard it is to gain definitive knowledge in this area.

The story begins in 2010, when astronomers using Keck 1 to advance the **Lick–Carnegie Exoplanet Survey** (a program that has found hundreds of confirmed exoplanets) announced the discovery of a fifth and sixth planet orbiting the red dwarf star Gliese 581. Keck had been observing this star for a decade, and from a detailed analysis of its motion over that period astronomers on the survey argued that Gliese 581 g has a mass that's between 3.1 to 4.3 times that of Earth and lies just 0.146 AU from its star. The planet would be a world unlike our own. Its year would last just 37 days and, being so close to its star, it would be tidally locked: the length of its day would match the length of its year and so it would always present the same face to its star, just as the Moon always presents the same face to Earth. In other words, one half of the planet would be in perpetual light, the other half in perpetual darkness; the bright side would be hot, the dark side cold. But since the planet is much more massive than Earth, its atmosphere should be much thicker – and that would moderate the effect of these temperature differences. There would likely be zones on the planet where water flowed as a liquid. And if there's water, perhaps there could be life.

Gliese 581 is just 20.5 light years away from Earth. So – we have a potentially habitable planet in the cosmic neighborhood? Well, maybe not. The trouble is that a different exoplanet survey – the **High Accuracy Radial Velocity Planet Searcher** (HARPS) – has also studied the Gliese 581 system and it sees no evidence for the existence of planets f and g. So, as I write this, the planet's discovery is classed as unconfirmed. Gliese 581 g might exist and orbit in the habitable zone – but then again it might not.

The Kepler Mission, combined with follow-up observation by Hale, Keck and others, has already confirmed the existence of one planet in a star's habitable zone: Kepler-22b. This discovery was announced in December 2011. The planet has a radius that's about 2.4 times that of Earth, so it may be more Neptune-like than Earth-like. But finding truly Earth-size planets orbiting in the habitable zone of a Sun-like star – well, even with Kepler helping out that remains a difficult task. Detecting the presence of life on such twin Earths remains a profoundly difficult task. To continue the search bigger telescopes and better technologies are required. Plans are in place, of course, to build those bigger and better instruments.

Astronomers go large

At the time of writing, two telescopes vie for the title 'largest optical telescope in the world'. The telescope with the largest single mirror is the **Gran Telescopio Canarias** (GTC), based on the island of La Palma. Since it's practically impossible to make a usable telescope mirror that's much larger than about 8 m in diameter, the GTC has 36 hexagonal segments that fit together form a mirror with a diameter of 10.4 m, thus making it slightly bigger than the similarly designed Keck telescopes. The **Large Binocular Telescope** (LBT) consists of two mirrors, each with a diameter of 8.4 m, mounted on a

Figure 12.8 Some of the new generation of large telescopes. (a) An artist's depiction of how the Giant Magellan Telescope will look. Its seven mirrors, each of which will have a diameter of 8.4 m, will combine to form an instrument that's equivalent to single telescope with a mirror larger than 20 m in diameter. (b) The two mirrors of the Large Binocular Telescope. (c) The Gran Telescopio Canarias. This is, at the time of writing, the world's largest single-aperture optical telescope. (d) The four 8.2 m instruments of the Very Large Telescope. Superimposed on this image are the paths that likes take when the VLT operates in interferometric mode. In this mode the VLT is the world's biggest optical telescope. ((a) GMTO Corporation; (b) NASA; (c) Commons; (d) ESO)

common base. Unlike the larger Keck mirrors, the LBT mirrors act together as a true binocular telescope; in terms of light collection it's like having a single mirror with a diameter of 11.7 m and in terms of resolving power it's the equivalent of a single mirror with a diameter of 22.8 m. Furthermore, since the LBT has one of the most advanced adaptive optics packages in the world, it's a telescope that at certain wavelengths can see more even sharply than Hubble. Both GTC and LBT are set for long careers at the leading edge of astronomical research. By 2018, however, just as they are hitting their stride as workhorses for observational astronomers, it's possible that they'll be considered rather small.

The **Giant Magellan Telescope** (GMT), whose developers include several large US universities along with partners from Australia and Korea, is to be based in Chile. The design of the GMT is unique. When it is completed it will have seven mirrors, each of which will be 8.4 m in diameter, arranged so that they form the equivalent of a single mirror with an effective diameter of 21.4 m or 24.5 m, depending on whether its light-gathering capacity or its resolving power is being discussed. Once it's in operation it will exceed, at least in terms of effective size, all current optical telescopes. Nevertheless, even as it comes into operation in 2018, assuming everything goes well, a quite different type of optical telescope will surpass it in size.

A collaboration of institutions in America and Canada, with varying levels of involvement from Japan, China and India, has plans to build an instrument called the **Thirty Meter Telescope** (TMT). As its name implies, the telescope will have a primary mirror with a diameter of 30 m – about the length of a blue whale. This Leviathan will have a single giant mirror consisting of 492 smaller hexagonal mirrors whose position will be constantly monitored and then shifted as required. The design of the TMT builds on the design of the Keck telescopes which, if all goes well, it will join in 2018: Mauna Kea has been chosen as the site for the TMT.

Even the TMT may not be the largest telescope in 2018. The primary mirror on the **European Extremely Large Telescope** (E-ELT) will be even larger. The initial E-ELT design was for a primary mirror 42 m in diameter – roughly the width of a football pitch – with a secondary 5.2 m mirror. At the end of 2011, however, the decision was taken to scale down the E-ELT to a 39.3 m primary mirror and a 4.2 m secondary mirror. The scaled-down instrument will cost less to build. Nevertheless, even with the smaller design the E-ELT will still be the world's largest optical telescope. Of course, the E-ELT main mirror cannot be a single object; rather, it will consist of many smaller hex-

Figure 12.9 An artist's impression of how the European Extremely Large Telescope would compare against the current Very Large Telescope at Paranal. In order to accommodate the huge mirror the main building is vast: the Brandenburger Tor is shown to illustrate the scale of this project. (ESO)

agonal mirrors. The 39.3 m design will be made of 798 hexagonal mirrors, each of them 1.45 m wide. It's the mass production of these smaller mirrors that will enable astronomers to keep down the cost of such a gigantic telescope. The E-ELT will be built on Cerro Armazones in Chile's Atacama Desert, a mountain that experiences cloud on only a dozen nights each year; it will be close to the existing **Paranal Observatory**. Note that the Europeans already have an impressive telescope just 20 km away: the **Very Large Telescope** (VLT), which consists of four separate telescopes each possessing a primary mirror that's 8.2 m in diameter. In 2010, one of the VLT telescopes discovered that a small, dim patch of light on a Hubble photograph was the most distant object that had ever been seen: a galaxy that lies at a redshift of 8.55. The light that VLT saw set out from the galaxy just 600 million years after the Big Bang. If an 8.2 m telescope can observe the most distant object known to humankind, just think what a 39.3 m telescope will be capable of observing. Indeed, the E-ELT will gather more light than all of the world's current 8–10 m telescopes put together! The E-ELT will, of course, have an adaptive optics system; there would be little point in building such a huge telescope if it didn't have such a system installed. But the E-ELT will have the most advanced adaptive optics system ever built: its 6400 actuators will be able to deform a correcting mirror by as little as 160 μm and do so one thousand times every second. The images it will take will be astounding.

It's worth noting that European astronomers once had hopes for an instrument that would have been much, much larger even than the E-ELT. They had plans to build an **Overwhelmingly Large Telescope** (OWL), an instrument whose primary mirror would have a diameter of 100 m! The OWL design team decided, however, that this project was simply too ambitious: an Eiffel Tower's worth of steel would have been required to support the mirror, yet astronomers would need to maneuver the structure with absolute precision. The OWL was simply too expensive. Such a telescope may yet be built – but not soon. The OWL design team instead recommended the construction of the E-ELT.

By the end of this century's second decade, then, the intention is that astronomers will be able to observe the sky with three extremely large telescopes (such an instrument being defined as one possessing an aperture greater than 20 m): the GMT, the TMT, and the E-ELT. One could also class some existing instruments as extremely large telescopes when they are used in interferometric mode. The LBT, for example, can have its two mirrors act together to create an instrument with an effective aperture greater than 20 m. These giant eyes will see clearly into many of the mysteries discussed in this book – the nature of dark energy, dark matter, and so on. One of the mysteries these telescopes will be asked to address is whether *truly* Earth-like planets exist around Sun-like stars – small, rocky planets with liquid water, and an atmosphere, and a stable atmosphere. But will even these telescopes be up to that task?

The extremely large telescopes should be able to directly image Jupiter-size planets and perhaps even analyze the chemical composition of their atmospheres. They should also allow astronomers to measure the to-and-fro wobble in a star's motion caused by the minute tug of an Earth-mass planet. By comparing the details of a sufficient number of planetary systems, astronomers will be able to understand the details of how these systems form and subsequently evolve. They will provide a better understanding of how common planets such as Earth are in the cosmos. Even these telescopes, however, will be operating at their limits when astronomers use them to investigate Earth-sized planets. For example, it would be a stretch for them to look routinely for biosignatures – such as the presence of water, methane or molecular oxygen. For that, even these giant instruments might need help from other sources.

Other Earths, ATLAST?

Astrometry – the measurement of the positions and motions of celestial objects – is perhaps the oldest of all sciences. Even the builders of Stonehenge must have realized that the stars appear to be fixed on the celestial sphere while the Sun, Moon and planets seem to wander across the sky. Hipparchus, who was possibly the greatest astronomer of antiquity, developed several instruments to improve astrometrical observations and he used them to make important advances such as the discovery of Earth's precession. In the early nineteenth century, precise astrometrical observations led to an improved understanding of the distance scale to stars. Nowadays, as mentioned earlier, astrometry has the potential of revealing the existence of exoplanets. However, the precision required of astrometrical observations for exoplanetary discover is astonishing. That's one of the reasons why claims of exoplanet discovery by means of astrometry are often met with scepticism.

Astronomers interested in using astrometry to detect planets had pinned their hopes on the **Space Interferometry Mission Lite** (SIM Lite), a NASA project that was slated for a launch in 2017. As we know, an interferometer is capable of making measurements of quite exquisite precision. The plan was for SIM Lite to use an optical interferometer – an instrument that can combine a star's light from two mirrors to generate interference fringes – in order to measure a star's position with unparalleled precision. Imagine being able to look up at the Moon and measuring the length of a flea. That's the sort of precision SIM Lite was designed to achieve. The hope was that SIM Lite would be able to study as many as fifty nearby Sun-like stars and discover whether they were home to Earth-size planets in the habitable zone. If any were found, these would have been subject to detailed follow-up observations by the new generation of extremely large telescopes and it's certain that every effort would have been made to detect biosignatures. However, in 2010 the decadal survey on astronomy and astrophysics, a publication that provides a roadmap for US research in the forthcoming decade, made no mention of SIM Lite. It seems that this mission will not go ahead. The survey did propose that a dark energy mission should be fitted with the capability of detecting exoplanets using gravitational microlensing (as discussed in chapter 4). Unfortunately, any exoplanets that such a mission discovers are likely to be 20 000 light years away, or further. This is too distant for follow-up observations by even the largest telescopes. Astrometry, it seems, will not be making the contribution to exoplanet discovery that was hoped.

The key technology of any mission such as SIM Lite is optical interferometry, since interferometers can provide the necessary precision. There's another way that interferometers could be used to detect exoplanets: they could be employed in a technique called **nulling interferometry**, which would allow telescopes to take direct images of exoplanets. We have already seen how the Hale telescope has been used to image exoplanets, and telescopes such as the VLT also have this capacity. In general, however, it's extremely difficult to take direct images of exoplanets because the reflected starlight from a planet gets drowned out by the much brighter light coming from the star itself.

A nulling interferometer works by using two telescopes to observe the light from a star. However, before the interferometer brings the two light beams together, it delays the light from one of the beams by precisely half of one wavelength. Since the light beams will be in phase when they hit the two telescopes it follows that the pair of beams will be out of phase when they are combined in the interferometer; at least, they will be if the positions of the telescopes can be controlled with the required precision. The result is destructive interference: the peaks from one beam align with the troughs from the other beam. The starlight cancels, giving a null result – hence the name of the technique. Why should this help detect planets? Well, light beams that enter the telescopes from a planet orbiting the star will arrive at a very tiny angle compared to the light beams from the star itself. It's a tiny, tiny difference – but a difference to which an interferometer might be sensitive. The half-wavelength shift will not cause the planet light to cancel. A nulling interferometer thus causes the star to vanish, leaving the planet to shine by itself.

Two space-based missions have been proposed that would use nulling interferometry to search for, and take images of, Earth-sized exoplanets. In Europe, ESA has investigated a mission called **Darwin** while in America NASA has investigated a mission called the **Terrestrial Planet Finder** (TPF). Unfortunately, as with SIM Lite, both the Darwin and TPF programmes have been postponed indefinitely – which is a polite way of saying they are dead. One can hope that these programmes will one day be reanimated but, until they are, the hunt for *truly* Earth-like planets will struggle. Is there any other way of continuing the hunt? And – crucially – if such planets are found is there a way of searching for signs of life on them?

The ATLAST mission, as we learned in chapter 10, is a NASA concept study for the true successor of Hubble. The idea is that ATLAST will observe in the optical region as well as the ultraviolet and the near infrared. This is the range over which Hubble has made such an impact. The JWST is often described as the replacement for Hubble, but JWST will observe primarily in the infrared. The astronomers involved in the study hope that, if it launches, ATLAST will be the prime space-based observatory in the years 2025–2035.

The ATLAST study team have looked at three designs: a telescope with a single, 8 m diameter mirror; a telescope with a 9.2 m segmented mirror; and a telescope with a 16.8 m segmented mirror. A space-based telescope with these sorts of mirrors would outperform Hubble and the JWST in all respects. With such an instrument, it becomes possible to search for biosignatures on Earth-like planets.

Let's briefly recap on the difficulties inherent in looking for other Earths. Suppose for a moment that there exist alien astronomers on a planet that's roughly 33 light years away from us, and that they want to study us. The first difficulty they'd have to overcome is the faintness of Earth: when Earth is at its maximum elongation from the Sun it would be so faint that an alien Hubble telescope would be unable to observe it. If those alien astronomers wanted to take a spectrum, looking for biosignatures such as the presence of molecular oxygen in Earth's atmosphere, then that would add to the challenges they'd face. The second difficulty is that Earth would appear very close to the Sun; the aliens' telescope would have to resolve extremely tiny angles. The third difficulty is that light from the Sun would drown out the reflected light from Earth; our alien friends would need to block sunlight without blocking Earthlight. Finally, all of the above assumes that the aliens knew about the existence of Earth in the first place; in practice, if Earth-like planets are relatively scarce, they'd be in the business of having to search the habitable zones of lots of stars. Human astronomers, of course, face exactly the same problems if they hope to be able to study an Earth-twin that's a few tens of light years away.

If a 16 m ATLAST mission were deployed, and if suitable methods were used to block starlight while allowing the light from planets to pass, then a detailed study of an Earth-twin would be possible. From the details of a planet's spectra astronomers could look for biosignatures. They could look for the presence of oxygen and ozone; they could check for the presence of vegetation on continents; they could look for the amount of cloud cover and determine the rotational period of the planet; and, crucially, they could check for the presence of water. The ATLAST mission might, at last, answer the question of whether life exists on other planets.

Exoplanets galore

We have seen how optical telescopes, both ground- and space-based, are finding exoplanets galore. Can telescopes working at other wavelengths join in the hunt? It turns out that they can.

For example, infrared telescopes (discussed in the previous chapter) can be used to directly image a planet. Spitzer, indeed, was the first instrument to observe light coming directly from an exoplanet. The difficulty for optical telescopes is that a planet only emits reflected visible light from its much brighter star; it emits its own infrared light, however, by virtue of the fact that it possesses some temperature. Of course the star is millions of times brighter than a planet in the infrared, but at least the planet emits its own active signal – and the signal can be detected: as we saw earlier, the total amount of infrared radiation coming from a stellar system is greater when a planet is transiting its star than when it is eclipsed by the star. It's a small variation, but space-based infrared telescopes can detect it.

Even radio telescopes (which are the focus of the concluding chapter) can be used to investigate exoplanets. Jupiter emits radio waves from an aurora; so does Saturn. It seems likely, then, that giant exoplanets would also emit radio waves from aurorae – and some of the new generation of radio telescopes might have the sensitivity to detect them. Indeed, the detection of radio emission from exoplanetary aurorae is a detection technique that could work even when the Doppler and transit methods fail.

We already know that planets are common. In the next few years the new generation of giant optical telescopes – and those other instruments we've mentioned – will find yet more of them. It surely won't be long before we find out whether other Earths – planets *truly* like our own – exist.

13

Listening out for life

It seems almost certain that truly Earth-like planets exist somewhere out there. It's possible that such Earth-like planets are home to life, and future observatories should be able to search for biosignatures. But we all know what the big question is, the most tantalising mystery of all: do alien intelligences exist? For fifty years now astronomers have searched for signs of extraterrestrial intelligence using radio telescopes. The new generation of radio telescopes will search with more sensitivity than ever before. Might they solve the riddle?

The strangest puzzle – Radio, radio – Shouting at the Universe – New ears on the Universe – The silent Universe

S. Webb, *New Eyes on the Universe: Twelve Cosmic Mysteries and the Tools We Need to Solve Them*,   295
Springer Praxis Books, DOI 10.1007/978-1-4614-2194-8_13, © Springer Science+Business Media, LLC 2012

If you've read the previous chapters you'll know that telescopes have collected staggering amounts of data about the Universe. Nothing in those myriad observations contradicts the simple ΛCDM model of cosmology, a construct that is surely a pinnacle of human ingenuity. Everything astronomers have seen can in principle be explained by the cold equations of physics. Of course, the ΛCDM model of the Universe contains within it lots of puzzles (which, after all, have formed the basis for this book) and astronomers will certainly have to grapple with further conundrums before they can claim to understand fully the processes that take place in stars, galaxies and the spaces in between them. Nevertheless, no signal has ever been observed that could not be explained in terms of natural phenomena. We see no signs that intelligent life has disturbed the Universe.

Isn't that rather odd?

The strangest puzzle

Consider the recent technological history of humankind. The first sustained, controlled, heavier-than-air powered flight took place in 1903; less than 40 years later a V-2 rocket became the first object to achieve sub-orbital spaceflight; a few decades after that and Voyager 1 was approaching the edge of the Solar System. In roughly one century, then, humankind developed from being an essentially Earthbound species to one that has sent out physical messengers heading into interstellar space. A century, though, is but an eyeblink in cosmic terms. If our technological civilization continues to learn and progress over the next few centuries or millennia – well, who knows what it could achieve? Of course, there's an unfortunately large 'if' in the previous sentence, but let's not be too pessimistic about our future. Surely human civilization in AD 40 000, say, would possess a sufficiently advanced technology to make its presence known throughout the Galaxy? But even 40 millennia is just an instant when we consider astronomical timescales.

Figure 13.1 Top: the first powered manned flight, here with Orville Wright at the controls, took place in 1903. Bottom left: a German V-2 rocket, here being fired by the British from a launchpad in Germany in 1945. The Germans had already been using the rockets for years. Bottom right: the launch of Voyager 1 atop a Titan rocket in 1977. Voyager is now approaching the edge of interstellar space. Where will we be in a thousand years? (Top: USAF; bottom left: Crown Copyright 1946; bottom right: NASA)

As we have discussed in earlier chapters, it now seems likely that there are hundreds of billions of planets in the Galaxy. If just a small fraction of those planets are home to extraterrestrial civilizations then presumably at least some of those civilizations will be older than ours. Some of them won't be just a few centuries or a few millennia older than us; they'll likely be many *millions* of years older than us. Of what fantastic feats will our species be capable in a million years from now? Then ask the same question of those extraterrestrial civilizations: what pinnacles would an alien species reach if it were a million years in advance of our own?

The Russian astrophysicist Nikolai Kardashev classified advanced civilizations in terms of their technological development: Type I (able to access the energy resources of an entire planet), Type II (able to access the energy resources of an entire star) and Type III (able to access the energy resources of an entire galaxy). For what it's worth, humankind is not yet a Type I civilization; currently we rate about 0.7 on the Kardashev scale. If we continue to progress, we might reach the status of a Type I civilization in a couple of centuries. But truly ancient civilizations – those that are millions of years in advance of us – will, if they continued to develop technologically, be Type II or Type III civilizations. They will have access to energy resources that we can scarcely imagine.

So – where are all these ancient, incredibly advanced, extraterrestrial civilizations? Why aren't they here, right now? Even if they chose not to colonize the Galaxy, why don't we see some sign of their astroengineering projects, such as infrared emission from Dyson spheres that surround stars; emission from vehicles burning antimatter; traces in stellar light curves of the transiting of artificial objects or traces in stellar spectra of artificial lines; or, perhaps more likely, some signal whose production is clearly artificial but through a means that is beyond our immediate comprehension? To date, astronomers have seen no traces of activity from advanced extraterrestrial civilizations. Isn't that puzzling?

Well, perhaps you consider the lack of evidence for astroengineering projects to be not too puzzling. After all, with very few exceptions, astronomers have been using telescopes to do astronomy rather than search for the byproducts of extraterrestrial civilizations. But there's another aspect to all this, an aspect that surely *is* odd: where are the deliberately aimed *signals* from those advanced civilizations? Even if we don't see incidental signs of their presence we should be able to hear them calling – shouldn't we?

Radio, radio

If an extraterrestrial intelligence wanted to communicate its presence to the Universe, perhaps even try to initiate a conversation over interstellar distances, how would it set about the task? The honest answer must be that we can't possibly know how an alien intelligence would reason. Nevertheless, if such intelligences are indeed out there then presumably they are bound by the same laws of physics as we are. If they want to communicate with other intelligences then the universality of physics narrows the range of options.

Our putative extraterrestrials would presumably decide to use a signal carrier that travelled at the fastest possible speed; that went where it was pointed; and that was as cheap as possible to generate. Furthermore, they would presumably choose a signal carrier that they believed could be detected by as wide a range of listeners as possible. Well, the need for speed implies that large physical objects won't be used to carry the signal; instead they'll have to consider electromagnetic or gravitational waves, which travel at the fastest possible speed, or perhaps neutrinos or the highest energy cosmic rays, which travel slightly slower. The presence of random magnetic fields in space means that it's difficult to know where charged particles are going to end up; so that rules out cosmic rays. The combination of requiring that signals be cheap to generate and easily detectable by all civilizations rules out communication by gravitational waves and neutrino beam modulation. That leaves us with electromagnetic waves as the method of choice for communicating over interstellar or intergalactic distances.

Perhaps it's possible to narrow the possibilities even further. An advanced civilization will presumably argue that its audience, or at least the less technologically advanced among the audience, will be housed on a planet underneath the protective blanket of an atmosphere. (Or will they argue that way? Are we guilty here, as perhaps elsewhere, of rampant anthropocentric thinking?) As we know, an atmosphere such as our own blocks out the shorter electromagnetic wavelengths. In essence, it's just the visible wavelengths and certain radio wavelengths that are sure to reach ground-based telescopes. Thus we can reasonably argue that the **search for extraterrestrial intelligence** (seti) should concentrate its efforts in the radio and visible parts of the electromagnetic spectrum.

Historically, the major thrust of seti has been in the radio region. Information-carrying radio waves are cheap and easy to produce: one simply needs a source of electrical energy to produce an alternating current with a desired frequency, a modulator with which to impress a signal, and an antenna to convert the alternating current into an electromagnetic wave. This is all so simple that physicists have been playing with such systems for well over a century. Perhaps this familiarity is one reason why radio wavelengths, rather than optical, remain the favored place to search for signs of intelligent life. And since most searches have taken place in the radio spectrum, in the rest of this chapter we'll concentrate manly on radio SETI.

Before discussing the use of radio telescopes in SETI it makes sense to look briefly at their more conventional scientific applications, because the use of radio waves for astronomy is well established. The development of radio astronomy has been extremely rapid. Karl Jansky discovered the first astronomical radio source in the early 1930s; by the early 1950s astronomers were engaged in mapping the entire radio sky; nowadays, radio telescopes are a key element in the astronomer's toolkit. Studying the sky in the radio part of the spectrum has led to many advances in astronomical knowledge (quasars and pulsars, to give just two examples, were first spotted by radio telescopes) and radio beautifully complements the view of the sky that is afforded by visible light. Many celestial objects emit radio waves, and different objects emit radio waves through different processes. For example, planets emit radio waves through thermal radiation; fast-moving electrons in weak magnetic fields emit synchrotron radiation; molecular clouds and the envelopes of old stars are places where masers can sometimes be found. By analyzing all the data contained within these radio signals, astronomers can get clues about many important physical processes taking place in the Universe – clues that simply are not forthcoming from visible light.

How, though, do radio telescopes work? How can astronomers get at those clues hiding in radio signals from the cosmos?

Radio telescopes vary widely in form but, in the simplest set-ups, they are much like optical telescopes. A radio telescope consists firstly of an antenna, often a parabolic-shaped dish, which collects the radio waves from a celestial object. The antenna focuses the waves onto a receiver, which amplifies the signal by perhaps a millionfold. The receiver then sends the signal to a recorder, which stores the data for processing by computer. The set-up, then, is simply: antenna, receiver, recorder.

When talking about optical telescopes, bigger is usually better. It's the same with radio antennae. A bigger dish can collect more waves and therefore detect fainter and more distant objects. But there's another reason why size is important: for a telescope to be effective it needs to have good resolv-

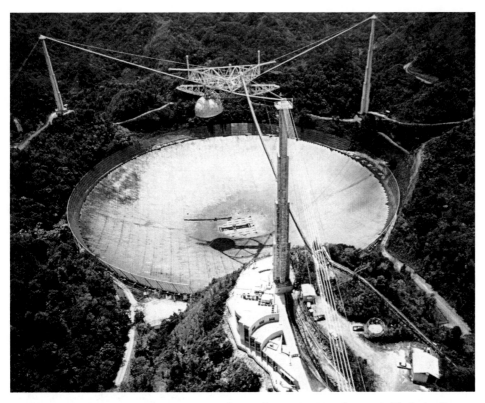

Figure 13.2 *Since its construction in the early 1960s, in a karst sinkhole in Puerto Rico, the Arecibo Observatory has been home to the world's largest single-dish telescope. A new Chinese telescope will overtake Arecibo in size, but Arecibo remains a formidable instrument. (H. Schweiker/WIYN & NOAO/AURA/NSF)*

ing power – in other words, it must be able to distinguish two distant, proximate objects as being distinct. Resolving power depends upon the ratio of the radiation wavelength to the diameter of the telescope (the mirror diameter in the case of an optical telescope, the antenna diameter in the case of a radio telescope). Well, if an optical telescope uses a 60 cm diameter mirror then that's about a million times larger than the wavelength of light. That means the telescope will have good resolving power: if it were on the south coast of England, for example, then in principle it could quite easily distinguish two people standing a meter apart on the northern tip of mainland Scotland. Radio waves, however, have a much longer wavelength than visible light: they can be a few centimeters long or hundreds of kilometers long. If we were observing at a 1 cm wavelength then, in order to have the same

Figure 13.3 The 27 dishes of the VLA are shown here in a tight Y-shaped arrangement. The dishes can move into different configurations. Studying the same radio source using different configurations allows one to obtain a lot of information about the source. (NRAO/ AUI; Dave Finley)

Figure 13.4 The ten telescopes of the Very Long Baseline Array are separated by large distances. By combining signals from the different dishes, the array has an effective baseline measured in thousands of kilometers. Such a large baseline can produce images with incredibly sensitive resolution. If your eyes were that sensitive then you could be standing in Cairo and be able to make out a London underground sign. (NRAO/AUI)

theoretical resolving power as a 60 cm optical telescope, a radio telescope dish would need to have a 10 km diameter. Such radio dishes simply cannot be constructed with current technology.

At present the world's largest single radio dish is at the **Arecibo Observatory** in Puerto Rico (see figure 13.2); its antenna has a diameter of 305 m. The **Five-hundred-meter Aperture Spherical Telescope** (FAST) is a Chinese radio telescope, due to be completed in 2013. FAST will, as its name implies, have a 500 m diameter dish. Arecibo and FAST are exceptions. A typical radio dish will have a diameter of 30 m and, if it observes at a wavelength of 1 cm, it will just about be able to distinguish two people standing one meter apart so long as they are no farther than the length of a football pitch away.

The poor resolution of single-dish radio telescopes was their main limitation in the early days of radio astronomy. After all, it's of little use capturing cosmic radio emission if you can't tell what object is emitting the radiation. There is a way around this limitation, however. A single radio dish will inevitably have poor angular resolution compared to an optical telescope; but by combining the signals from several widely separated dishes it's possible to achieve extremely high resolving power, since the resolving power of an interferometer depends upon the distance between the dishes rather than the size of the individual dishes themselves. For example, the **Very Large Array** (VLA) in Socorro, New Mexico, is an interferometer consisting of 27 antennae. Each antenna has a dish diameter of 25 m. The dishes are arranged along the three arms of a Y, with each arm consisting of a railway line that's 21 km long; the dishes can be moved to different points along the line. When the dishes are at their largest separation, and their signals are combined, the result in terms of resolving power is equivalent to having a single dish that's 36 km wide. The VLA was inaugurated in 1980. The **Expanded Very Large Array** (EVLA) is a project to upgrade the old VLA with new electronics, producing an instrument that possesses even more sensitivity. The **Very Long Baseline Array** (VLBA), which was inaugurated in 1993, consists of ten 25 m dishes situated all over the US and its territories, ranging from Mauna Kea in Hawaii to St Croix in the Virgin Islands. Sometimes Arecibo, the VLA and a 100 m dish in Germany join the observations. The maximum baseline of the VLBA is 8000 km; when the array is observing at its shortest wavelength, 3.5 mm, it can achieve a sensitivity of resolution that far surpasses even that of Hubble. Of all branches of astronomy it turns out, perhaps surprisingly, that radio astronomy provides the highest resolution images.

Thus radio telescopes, particularly when they are linked together in arrays, can be superb tools for astronomical observation. They have observed the center of the Galaxy, a place that is totally obscured at optical wavelengths. They are used to study the jets that emanate from pulsars and from active galactic nuclei. They can be used to study how galaxy clusters merge. They can be used to study supernova remnants. Closer to home, they can observe objects in the Solar System. All of these objects emit radio waves at different frequencies and with quite different properties. The Universe is *full* of radio waves. That's great for the radio astronomer – but what about the SETI scientist? What about the extraterrestrial intelligence that wants to make its voice heard? How would an alien intelligence go about transmitting a radio signal, and how should we go about trying to hear a signal above all the din?

Shouting at the Universe

Let's assume there are extraterrestrial intelligences out there using radio to shout at the Universe. We haven't yet heard from them. Why? Well, the problem of searching for radio signals is often compared to looking for a minuscule, straw-colored needle in a giant haystack. But it's much more difficult than that. For example, we have no idea where the signal will come from. Which of the several hundreds of billions of stars in the Galaxy should we focus on? Indeed, *should* we focus on particular target stars (in the hope that a civilization there was once pointing its transmitter to us) or should we look simultaneously at as many stars as possible (in the hope that the larger number of targets outweighs the much-reduced sensitivity)? There's a related question: is it more likely we'll hear from a weak transmitter that's relatively nearby, or from a very powerful transmitter that's very far away? The answer depends on how common extraterrestrial civilizations are. If it turns out they are widespread then a targeted approach may work. If they are rare then it makes more sense to survey as many stars and galaxies as possible, and hope that we'll detect an extremely powerful beacon from an extremely advanced civilization in the same way that radio astronomers were able to find rare quasars more easily than common sources such as stellar coronas: the power of quasars more than compensates for their scarcity. Project Phoenix, which focused on about 800 stars within a 200 light year radius, was an example of a deep, targeted search. Project BETA, which analyzed lots of sources from across the heavens, was an example of a shallow, wide-sky survey. But perhaps a hybrid approach makes more sense?

A second difficulty is that the signal might be faint. Whether it's an intrinsically bright signal whose source is extremely distant, or an intrinsically dim signal that's relatively nearby, our receivers may need to be extremely sensitive (and of course they must be pointing in the right direction at the right time).

A third difficulty is that we don't know which channel they'll be using. We *do* know that radio waves from natural phenomena are generally spread out over a wide range of frequencies. Nature, as far as we know, does not produce narrowband radio signals – in other words, those less than about 300 Hz in width. On the other hand, it is straightforward for a technically advanced civilization to produce narrowband radio signals: humankind has been doing this routinely for several decades. This is the logic behind looking for a narrowband signal: it's not something that Nature generates and it is something that a civilization can generate. Furthermore, if you squeeze a relatively small amount of power into a small enough channel the signal becomes bright. You can easily buy equipment that can pump a 1 kW signal into a 1 Hz channel and, in that particular channel, the signal would outshine the Sun. The Sun pours out huge amounts of radio energy, of course, but that radiation is spread out over a huge range. The point is that at any particular radio frequency it's possible to surpass the Sun in brightness. Much of the SETI program, then, has been predicated upon listening for a narrowband signal, say 1 Hz or so in width.

But what frequency would extraterrestrial civilizations choose for their communications? What part of the radio dial should we be tuning into? The problem here is that there are, as the astronomer Carl Sagan used to say, 'billions and billions' of radio frequencies to choose from. The radio spectrum is *vast*.

We can, perhaps, employ some guides to help us narrow the spectrum of search possibilities. As we have seen, radio noise from various natural phenomena fills the sky. However, there's a 'window' in the radio spectrum, a place where the background murmur is particularly low. Furthermore, within this window there are two frequencies – 1.42 GHz (caused by excited hydrogen atoms) and 1.65 GHz (caused by hydroxyl ions) – that are of particular interest. Hydrogen (H) and hydroxyl ions (HO) can combine to create water – H_2O. The reasoning goes that any extraterrestrial civilization will know about both this quiet region and the importance of water – and will further know that other civilizations will know that too. Therefore a good place to shout at the Universe, and a good place at which to listen, would be the region 1.42–1.65 GHz, the so-called **waterhole**. Even in the waterhole there are a huge number of channels in which one could miss a narrowband signal. Thus a search program such as Project BETA decided to analyze billions of channels. Fortunately, developments in computing and electronics make such an approach feasible.

There are several other difficulties faced by SETI, but the point should have been made: to pick up a radio signal from ET requires, at the very least, a radio telescope to be pointing in the right place at the right time and be listening at the right frequency. The scale of the challenge is immense.

Frank Drake initiated the first SETI project in 1960, when he used a 26 m antenna tuned to the 1.42 GHz frequency to listen for signals from a couple of nearby Sun-like stars. Subsequent searches possessed improved sensitivity, or looked at more than one channel, or both. For example, Project BETA and Project Phoenix were both launched in 1995. Project BETA had a multi-million channel receiver, and for four years surveyed the northern sky through the whole of the waterhole. Project Phoenix looked for narrowband signals from 800 nearby stars in the bandwidth between 1–3 GHz; after nine years of searching without success, the project team concluded that we perhaps live in a quiet neighborhood. In 1999, astronomers launched Southern SERENDIP, a southern hemisphere counterpart to a long-standing northern hemisphere project; in these experiments, the SETI program 'piggybacks' on traditional observations of the sky being carried out by large radio telescopes. In total, about one hundred SETI projects have taken place since 1960.

The equipment being used by modern SETI projects allows for observations that are up to one hundred trillion times more sensitive than Drake's first attempt. And it is not just the sensitivity of the search that has increased; the computing power now available for the search would have been unimaginable when Drake started out. In 1999, for example, the popular SETI@home project was launched. Members of the public were able to download a free screensaver that enabled their internet-connected computers to analyze data from the Arecibo Observatory. Each personal computer possesses relatively little power, but when millions of them run together in parallel the result is one of the most powerful supercomputers on Earth – and it's dedicated to looking for any narrowband signals that might be lurking in the Arecibo observations.

So: if both the sensitivity of SETI observations and the computing power that's available to the SETI effort have increased by such large factors – why haven't we heard anything from anyone? Usually, when you improve equipment by a factor of a trillion, you see something different. Why has no one spotted a signal?

Perhaps we simply need to listen longer, perhaps for many more decades. It could be that the scale of the challenge is so great that we can't expect success until many more observations have been made. Perhaps, though, we need other instruments to join in the search.

New ears on the Universe

We saw in chapter 6 how the Low Frequency Radio Array (LOFAR) is being employed to study ultra-high energy cosmic rays. Unlike an array of traditional parabolic dish antennae, LOFAR consists of large numbers of cheap dipole antennae grouped in stations spread out over Europe. As the name implies, LOFAR has been designed to study the Universe at low radio frequencies – from about 10 MHz up to about 240 MHz. In addition to providing clues about the origin of ultra-high energy cosmic rays, LOFAR will be used to learn more about what happened at reionization, about magnetic fields in the Universe, about solar weather and about much else besides. But it won't only be used for such traditional astronomical purposes. It will also be used in SETI.

Television and radio broadcasters on Earth often transmit in the band covered by LOFAR; so do certain military communications set-ups. Well, a telescope such as LOFAR may possess sufficient sensitivity to detect leaked radiation from the television, radio or military transmitters of nearby extraterrestrial civilizations. Furthermore, if those extraterrestrial signals are rare or intermittent in nature then LOFAR is well placed to detect them since it surveys large parts of the sky simultaneously. Of course, one can argue that it is exceedingly unlikely that an extraterrestrial civilization will be at the same level of technological development as humankind; it scarcely seems possible that they'll be watching broadcast television, listening to broadcast radio or employing radio for military purposes. Increasingly even we get our telecommunications in other ways, such as through cable, dish or fiber. Earth itself may soon become 'quiet' in the radio region. Even if it's unlikely that we could pick up stray television programmes, there may be reasons why extraterrestrials transmit their signals at low frequencies; we don't know what those reasons are, but then we aren't aliens. And the one certainty is that if we don't listen we can't possibly hear. To date, SETI has not ventured into the range of low frequencies. With LOFAR it becomes possible to continue the search for extraterrestrial intelligence in an unexplored part of the spectrum.

It was hoped that a more familiar search program would be provided by a joint project of the SETI Institute and the University of California Berkeley: the **Allen Telescope Array** (ATA) consists of 42 relatively small antennae (a number that, it was planned, would eventually grow to 350 antennae) spread out over a large area at the **Hat Creek Radio Observatory** in the northern

Figure 13.5 *The antennae of the Allen Telescope Array are here being prepared for operation. In the initial phase of operations the array contained 42 antennae; it was hoped that the number of antennae would eventually rise to 350. In 2011, however, funding difficulties forced the* SETI *Institute to put the system into hibernation. Although the telescope has subsequently been brought out of hibernation, its long-term future is unclear. (Colby Gutierrez-Kraybill)*

part of California. Each antenna has a diameter of 6.1 m. If the observations of 350 antennae were combined, the instrument would be the equivalent of having a single dish with a collecting area of one hectare. It turns out that it's much cheaper, as well as being more flexible and efficient, to combine the observations of a large number of small antennae than it is to build a single huge dish. Indeed, the cost is almost within the reach of individuals – albeit extremely wealthy individuals. The SETI Institute had plans to construct a telescope with an area of one hectare. Paul Allen, who along with Bill Gates founded Microsoft, agreed to fund the first phase of the project; thus the proposed instrument, instead of being called the one hectare telescope, became known as the ATA.

Unfortunately, even though the construction of the ATA is relatively cheap, it seems that no program is immune from review in these times of economic turbulence. In early 2011 the SETI Institute announced that funding difficulties meant the ATA was to be placed into hibernation. It wasn't a huge amount of money that was needed to keep the ATA running – it was about the size of an investment banker's bonus – but the funding shortfall was enough to cause the instrument to be mothballed. The SETI Institute initiated a campaign to raise cash for the program; contributions from thousands of individuals, including people such as the actress Jodie Foster, the former astronaut Bill Anders, and the author Larry Niven, raised enough to restart observations just four months after the ATA was closed. But this is clearly not a sustainable funding model, and the future of the ATA remains unclear. All this came at a bad time since, as we learned in the previous chapter, the Kepler Mission is currently finding lots of candidate systems to target.

The ATA, even in its initial phase, had several advantages over previous instruments. One advantage was that it could observe large areas of sky at the same time; it had a field of view 17 times larger than that of the VLA, for example. Another feature was its frequency coverage: it could observe all the way from 500 MHz up to 11.2 GHz. The array, with its 42 antennae, was also being used for conventional radio astronomy – studying active galactic nuclei, black hole accretion, gravitational lenses and much else besides. Thus radio astronomers decided where to point the array, which objects to target, and then SETI enthusiasts were able to study those stars in that large field of view that might be of interest. By splitting the signal before it was processed both traditional radio astronomers and SETI scientists got data to study: radio astronomy and the search for extraterrestrial intelligence could take place simultaneously with the ATA.

The intention was for the full ATA to survey a million stars in the region 1–10 GHz with a sensitivity great enough that it could detect signals from an Arecibo-sized transmitter as far away as 1000 light years. And in the region of 40 billion stars would be surveyed in the waterhole as the array looked for high-power transmitters there. It would have been the most comprehensive SETI project yet carried out. Those interested in the search for extraterrestrial intelligence can only hope that new funding channels will allow ATA to fulfil its potential.

The ATA, while being an invaluable observatory in its own right, was also important because it prototyped some technologies that can be employed in a proposed giant radio telescope: the **Square Kilometer Array** (SKA). The SKA is a worldwide collaborative effort, with funding probably coming in equal measures from Europe, America and the rest of the world. As with the ATA, the SKA will consist of a large number of small antennae arranged over a large area. That area will be centred primarily in a desert region, either in Australia or South Africa; it seems that South Africa is the favored venue, but at the time of writing the final decision about location has yet to be taken. While the ATA in its final form was planned to have 350 antennae, the SKA will have *thousands* of antennae; this will make it the equivalent of a single dish with a collecting area, as the name implies, of one square kilometer. A radio telescope with a collecting area more than twice the land area of Vatican City will possess a sensitivity ten times better than Arecibo and fifty times better than EVLA. The plan is for the SKA to cover an even wider range of frequencies than the ATA – from 70 MHz up to 25 GHz and beyond, eventually. And the SKA will have an even larger field of view than the ATA, enabling it to survey the sky much more quickly than any existing radio telescope.

By any standards, when the SKA becomes operational some time in the 2020s it will be a tremendously powerful telescope. Such an instrument will be brought to bear primarily on the other scientific questions discussed in this book. However, the scientists involved in pitching the potential of the SKA have advertised the instrument's ability to answer questions about the existence of life in the Universe. Such a giant telescope will be able to study protoplanetary disks and improve astronomers' understanding of how Earth-like planets form. It will be able to detect whether molecules thought to be necessary for life are present in the clouds that collapse to form proto-planetary disks. And of course it will be the most sensitive telescope for SETI activities; where LOFAR would struggle to watch alien television, a telescope such as the SKA could certainly detect signals from broadcast transmitter on a planet a few light years distant. Furthermore, the SKA will have the capacity to search for general 'leakage' signals from an extraterrestrial civilization over much larger volumes of space than current telescopes can manage.

Scientists in China are currently planning their own version of SKA, so the world may eventually have two giant radio telescopes. As already mentioned, engineers in southwest China are constructing FAST – the world's largest single dish telescope. The telescope is being built in a karst, a distinctive type of bowl-shaped sinkhole. (Arecibo was likewise constructed in a karst sinkhole.) The plan, though, is for FAST to be joined by a further thirty dishes, each of a diameter of about 200 m, built in karst valleys. Taken together these dishes will form the **Kilometer-square Area Radio Synthesis Telescope** (KARST).

With these mega-telescopes – SKA and KARST – astronomers will have instruments that can continue the search for extraterrestrial intelligence into hitherto unexplored regions.

The silent Universe

Are there any alternatives to radio SETI? Well, as already mentioned, optical wavelengths provide another region in which it makes sense to undertake the search. Until relatively recently, for reasons that aren't entirely clear to me, the astronomical community eschewed optical SETI. This is strange because as long ago as 1961 Charles Townes (one of the developers of the laser) and Robert Schwartz pointed out that optical lasers would allow the transmission of large amounts of information over interstellar distances. A type II civilization, say, would have sufficient energy resources at its disposal to transmit *Encyclopaedia Galactica* amounts of information in a laser beam and the technical capacity to point a beam with sufficient accuracy and precision to ensure that it swept across the habitable zones of target stars. Perhaps one reason why the SETI community has tended to concentrate on radio is that scientists have had decades more time to understand the possibilities of the longer wavelengths. When looked at in cosmic terms, however, decades are as milliseconds: we shouldn't view cosmic possibilities through the prism of human technological development. Whatever the reason, nowadays several SETI programs are taking place at optical wavelengths – they are looking for laser flashes from a distant civilization. While optical searches are likely to remain the junior partner when it comes to SETI, they will increasingly complement traditional radio techniques.

Other wavelengths can be used to carry out the search if one looks not for targeted broadcasts but for the byproducts of advanced engineering. For example, it has been suggested that advanced civilizations would be likely to construct artificial habitats for themselves – objects such as Dyson spheres, planetary rings or O'Neill colonies. Others have suggested that advanced civilizations might build immense computational structures, called Matrioshka brains. These extraterrestrial megastructures would inevitably emit infrared radiation as a byproduct: infrared telescopes (which were discussed in chapter 11) could thus be used for SETI purposes. Even gamma-ray telescopes (discussed in chapter 8) can be used to rule out the existence of certain extraterrestrial technologies. It has been suggested that the burning of antimatter would be an efficient means of interstellar propulsion for a sufficiently advanced civilization. Well, antimatter burning would produce copious amounts of gamma-rays with a signature that marks it as artificial. Already data from Compton has been analyzed with this in mind and, needless to say, no signs were seen of antimatter propulsion systems.

The various instruments described in this book – from radio telescopes all the way up to gamma-ray telescopes in the electromagnetic spectrum; particle 'telescopes' that are used to detect cosmic rays and neutrinos; more tentative 'telescopes' that are attempting to detect gravitational waves and dark matter – these instruments have given astronomers and cosmologists a wealth of information. Data from these telescopes have enabled scientists to understand our Universe with a level of detail that was undreamed of in the last century. And things are going to get even better. Over the next two decades, as the next generation of telescopes provide yet more information, astronomers and cosmologists and physicists will report back to the taxpayer with an ever-improving understanding of the Universe.

But what about an improved understanding of our *place* in the Universe?

It may be that this question is beyond the reach of even the most powerful telescopes. If radio and optical SETI fail to find signs of an extraterrestrial civilization in the next few decades, if no signs of macroengineering are detected, if the Universe remains silent, would we conclude that high intelligence is confined to this small planet? Or would we look for some other reason why we haven't heard from them? Surely our scientific appreciation of the cosmos will be incomplete until we understand the role played by consciousness in the Universe. Have the intricate patterns of evolution produced only one sentient species?

Are we alone?

Glossary of terms

absorption line A dark line at a particular wavelength superimposed upon a bright, continuous spectrum. Such a spectral line can be formed when electromagnetic radiation, while travelling on its way to an observer, meets a substance; if that substance can absorb energy at that particular wavelength then the observer sees an absorption line. Compare with emission line.

accretion disk A disk of gas or dust orbiting a massive object such as a star, a stellar-mass black hole or an active galactic nucleus. An accretion disk plays an important role in the formation of a planetary system around a young star. An accretion disk around a supermassive black hole is thought to be the key mechanism powering an active galactic nucleus.

active galactic nucleus (AGN) A compact region at the center of a galaxy that emits vast amounts of electromagnetic radiation and fast-moving jets of particles; an AGN can outshine the rest of the galaxy despite being hardly larger in volume than the Solar System. Various classes of AGN exist, including quasars and Seyfert galaxies, but in each case the energy is believed to be generated as matter accretes onto a supermassive black hole.

adaptive optics A technique used by large ground-based optical telescopes to remove the blurring affects caused by Earth's atmosphere. Light from a guide star is used as a calibration source; a complicated system of software and hardware then deforms a small mirror to correct for atmospheric distortions. The mirror shape changes more quickly than the atmosphere itself fluctuates.

astrometry The branch of astronomy dealing with the precise measurement of the position and motion of celestial bodies. Historically, astrometry has played an important role in developing our understanding of cosmic distance scales, and it continues to play a key role in modern research; for example, many dark energy surveys require knowledge of the precise position and motion of galaxies.

axion A hypothetical elementary particle, first postulated by Roberto Peccei and Helen Quinn in 1977 as a means of solving a problem in the theory of quantum chromodynamics. If axions exist then they have no electric charge and will react scarcely at all with matter: they are more elusive even than the neutrino. In some models, however, the Big Bang would have created vast numbers of low-mass axions that would still be around. This makes them a possible dark matter candidate.

baryon acoustic oscillation (BAO) The competing forces of gravitational attraction and gas pressure caused acoustic oscillations – essentially sound waves – in the normal matter (baryons) of the primordial plasma. Peaks in the oscillations

S. Webb, *New Eyes on the Universe: Twelve Cosmic Mysteries and the Tools We Need to Solve Them*,
Springer Praxis Books, DOI 10.1007/978-1-4614-2194-8, © Springer Science+Business Media, LLC 2012

correspond to regions of the Universe in which matter is denser than average. This clustering occurs on a well-understood length scale, and thus BAO clustering generates a standard length for use by astronomers. This standard ruler has a length of 480 million light years.

baryonic matter A baryon is a subatomic particle composed of three quarks. The proton and the neutron are both types of baryon. (Particles such as the electron and the neutrino are not baryons; they are a type of fundamental particle called a lepton.) An atom of normal matter consists of some combination of protons, neutrons and electrons. Since protons and neutrons are much more massive than electrons, nearly all of the mass of an atom is contained in the baryons. In astronomy and cosmology, then, 'baryonic matter' refers to matter that is composed mostly of baryons by mass; since that includes every type of atom, baryonic matter includes essentially all the matter that we observe. Most of the mass of the Universe appears to be in the form of non-baryonic matter and, in particular, some form of dark matter.

Big Bang An event that took place 13.7 billion years ago in which the Universe was created. Initially the Universe was in an incredibly hot, dense state. As the Universe expanded, it cooled; the existence of the cosmic microwave background radiation, which indicates that the current temperature of the Universe is 2.7 K, is good evidence for the Big Bang theory.

blackbody An idealized object that absorbs all the radiation that hits it and thus appears perfectly black. A blackbody is a perfect emitter, radiating the maximum amount of energy at any particular temperature. A blackbody spectrum has a distinctive shape and maximum that depends only upon the blackbody temperature.

black hole Any massive body possesses an escape velocity, the speed that something must have in order to leave the gravitational pull of the body. For Earth, the escape velocity is just over 11 kilometers per second. If the mass of the body is crushed into a smaller volume the escape velocity increases. Eventually, when the escape velocity reaches 300 000 kilometers per second, not even light itself can travel fast enough to escape the pull of gravity. The result is a black hole. Although by definition it is impossible to see a black hole, it is possible to deduce its presence by its effects on surrounding matter. Astronomers have deduced the presence of several stellar-mass black holes and it is believed that a supermassive black hole lies at the heart of every galaxy.

Blandford–Znajek mechanism An idea, proposed by Roger Blandford and Roman Znajek, for how energy can be extracted from a rotating black hole and be used to power the relativistic jets that shoot out from active galactic nuclei. The rough idea is that magnetic field lines will be threaded through the matter in the accretion disk surrounding a black hole. A rotating black hole drags space around it, and as matter falls into the black hole the field will become tightened, coiled and twisted. In these circumstances large amounts of energy, in the form of jets, will be emitted from the poles of the black hole.

blazar   A type of active galactic nucleus in which the relativistic jet from a central supermassive black hole is pointing more or less straight towards Earth. The name 'blazar', which is short for blazing quasi-stellar object, was coined in 1978 by Edward Spiegel.

B mode polarization   A distinctive polarization pattern in the cosmic microwave background radiation that would be caused by the interaction of primordial gravitational waves with the background radiation. If this polarization could be observed it would demonstrate that inflation did indeed occur in the very early history of the Universe. The signals are extremely weak, however, and so exquisitely sensitive observations of the microwave background are needed to search for them.

bolometer   An instrument, in essence a very sensitive thermometer, to measure infrared radiation. The first bolometer was invented in 1878 by Samuel Langley, who noted that when a thin, blackened strip of platinum absorbs radiation its electrical resistance changes in a way that can easily be measured. The sensitivity of Langley's original bolometer was sufficient for him to discover absorption lines in the infrared part of the Sun's spectrum. Modern bolometers, of course, are fantastically more sensitive; they can pick up temperature differences on the sky of ten-millionths of a degree.

Cerenkov radiation   When a charged particle moves through certain types of material at a speed greater than light itself is able to move through the material, the charged particle can polarize the molecules in that material. When the molecules return to their ground state, they emit electromagnetic radiation – Cerenkov radiation. It has a continuous spectrum that is more intense at shorter wavelengths: most Cerenkov radiation is in the ultraviolet spectrum, while the visible radiation is seen to be a deep blue.

CfA2 Great Wall   A filament of galaxies, discovered in 1989 by Margaret Geller and John Huchra, that is more than 500 million light years long. Large, filamentary structures such as this are thought by astronomers to form part of the cosmic web.

chirp   The gravitational wave signal expected to come from two inspiraling neutron stars. As the stars spiral inwards, the frequency of the gravitational waves increases in a particular way; the crescendo, if it were converted into sound waves of the same frequency, would sound like the chirp of a large bird.

coded aperture imaging   A technique for taking images when the photons involved have high energies. The photons pass through an aperture mask containing holes arranged in some particular pattern. When the photons hit the detector, a computer applies algorithms that allow the image to be reconstructed.

cold dark matter Several lines of evidence lead to the notion that most of the matter in the Universe cannot be seen by its electromagnetic effects (in other words, it's dark) and that its constituent particles move relatively slowly (in other words, it's cold). At the time of writing it is not known what cold dark matter consists of, but one favored idea is that it consists of some type of weakly interacting massive particle.

collapsar Originally this referred to a collapsed star, the end result of a high-mass star that undergoes gravitational collapse to form a stellar-mass black hole. It is now often used to refer to a model in which a fast-rotating, high-mass Wolf–Rayet star undergoes gravitational collapse to form a black hole. Calculations suggest that relativistic jets shoot out along the poles of the rotating black hole. This model may help explain long gamma-ray bursts.

Compton scattering If an X-ray or gamma-ray photon moves through matter, it is likely to scatter off an electron. The photon loses energy (and thus increases its wavelength) and transfers it to the scattered electron, which escapes and leaves behind an ionized atom. This effect was first observed by Arthur Compton in 1923, who was awarded the 1927 Nobel prize for the discovery.

cosmic microwave background (CMB) Blackbody radiation, with a form corresponding to a temperature of 2.725 K, that fills the present-day Universe. It is relic radiation from an earlier stage in the evolution of the Universe, a time about 380 000 years after the Big Bang, when the dense, ionized, opaque Universe suddenly became transparent.

cosmic ray An energetic, charged particle that hits Earth's atmosphere from outer space. Many low-energy cosmic rays originate from the Sun. The source of ultra-high-energy cosmic rays remains unknown.

cosmic web Structures exist on a variety of scales, from galaxies, to clusters of galaxies, to superclusters of galaxies. The 'walls' of superclusters are separated by huge voids such that the overall structure of the visible matter in the Universe resembles a foam. This foam-like structure is the cosmic web.

cosmological constant A term added by Einstein to his equations of general relativity. The effect of a positive cosmological constant is to drive an acceleration of empty space. It might thus explain the observed acceleration of the expansion of the Universe; in other words, dark energy might be explained by a positive cosmological constant. Understanding the value of the cosmological constant is one of the key unsolved problems in physics.

cross-polarized gravitational wave Gravitational waves possess many of the properties of more familiar types of wave; for example, they can be polarized. There are two fundamental polarization modes. Cross-polarization takes its name because the waves simultaneously oscillate in two directions, 45° diagonals, rather like a cross (×) sign. The passing of a cross-polarized gravitational wave could in principle be detected by its characteristic effect on a ring of test particles: the ring will be alternately stretched in one direction (so that the ring becomes an oval with a north-east/south-west aspect) and then another (with the oval having a north-west/south-east aspect). See also plus-polarized gravitational wave.

dark ages The period in the history of the Universe between the time of recombination (about 380 000 years after the Big Bang) and the formation of the first shining stars (a few hundred million years after the Big Bang). During this time the Universe was 'dark' because there were no sources of light.

dark energy A hypothetical form of energy that is associated with empty space. Dark energy exerts a negative pressure so it would tend to accelerate the expansion of the Universe. Astronomers have measured an accelerating expansion through observations of distant supernovae, but the precise nature of dark energy remains unknown.

dark flow The standard model of cosmology suggests that the motion of galaxy clusters should be randomly distributed with respect to the cosmic microwave background. However, a team of astronomers using WMAP data claim to have found evidence that some galaxy clusters are flowing towards a patch of sky between the constellations of Centaurus and Vela. One proposed explanation for such a coherent flow of galaxy clusters is that the clusters are being pulled by the gravitational attraction of something outside our visible Universe.

dark matter The mass of the matter that can be seen (stars, gas, dust and so on) is insufficient to account for various observed gravitational effects (such as the motion of galaxies in clusters). Astronomers and cosmologists thus infer the existence of matter that cannot be seen. Such dark matter would possess mass (and therefore exert a gravitational influence) but would not interact electromagnetically (and therefore would not be visible).

diffuse supernova neutrino background Core collapse supernovae generate vast amounts of neutrinos. The Universe is thus thought to contain a diffuse background of neutrinos from all those past supernovae.

Doppler effect If a wave detector moves relative to a wave source then the detector observes a wave frequency that is different to the emitted frequency. If the source and the detector approach each other then the observed frequency is higher than the emitted frequency (a blueshift); if they recede then the frequency is lower (a redshift). The effect is often heard in everyday life, when an emergency vehicle passes with its siren sounding: the pitch of the siren is higher when the vehicle approaches, lower when it recedes.

Doppler spectroscopy method Also known as the radial velocity method, this is a technique with which astronomers can discover exoplanets. As a planet orbits a star, the star will 'wobble' due to the gravitational pull of the planet. If the planet orbits in a plane that is edge-on to our line of sight then the star will have a radial component of velocity – it will move periodically towards us and away from us. Lines in the star's spectrum will exhibit a periodic Doppler shift to the blue then to the red – and if that Doppler shift is large enough then it can be observed by using a spectroscope.

drag-free operation A method of precisely controlling a spacecraft's position using special microthrusters. The method is likely to be required by any space-based gravitational wave observatory, which will work by watching the relative movement of widely separated test masses. The test masses will be protected by spacecraft, but in order for the observatory to work the spacecraft will have to move along with the test masses with an accuracy of 10 nm.

Einstein ring An effect, first described in detail by Einstein in 1936, whereby the light from a bright object is bent by the gravitational pull of an intervening massive object so that we observe a ring. The intervening object, which could be a black hole, a galaxy or a group of galaxies, thus acts as a gravitational lens. A perfect Einstein ring has yet to be seen, since the perfect alignment of observer, lens and source is inevitably going to be a rare phenomenon. Nevertheless, less-than-perfect examples have been observed: the first partial Einstein ring was discovered in 1979; the first complete ring in 1998.

electromagnetic wave A wave consisting of oscillating electric and magnetic fields. Radio waves, microwaves, infrared, light, ultraviolet, X-rays and gamma-rays are all types of electromagnetic wave. An electromagnetic wave can be generated by accelerating an electric charge. In a vacuum, electromagnetic waves travel at the speed of light.

electron An elementary particle that carries negative electric charge. Electrons and atomic nuclei constitute the building blocks of atoms; electrons thus play a fundamental role in many important physical phenomena. The electron, like the muon and the tau, is a lepton. The electron, muon and tau have similar properties, and all have an associated neutrino. The leptons differ in mass: the muon is about 200 times more massive than an electron; the tau is about 3500 times more massive.

electron degeneracy pressure A basic principle of quantum mechanics states that two electrons in the same quantum state cannot occupy the same location. When matter is compressed to the point where quantum effects become important, the quantum mechanical resistance of electrons to further compression generates a pressure. This electron degeneracy pressure supports white dwarf stars against further gravitational collapse.

electronVolt A unit of energy defined by the amount of kinetic energy gained by an electron when it accelerates through an electric potential difference of one volt. The electronVolt is a tiny amount of energy on everyday scales, but is a useful unit in particle and atomic physics.

emission line A bright line at a particular wavelength in the spectrum arising from radiating material. It corresponds to the emission of electromagnetic radiation at a certain energy. Compare with absorption line.

E mode polarization A polarization pattern in the cosmic microwave background. The level of polarization is small (one has to measure the background to a precision of a few millionths of a degree in order to see it). It has a pattern similar to that seen with static electric charges, and has been seen in several experiments. B mode polarization of the background, on the other hand, has yet to be detected.

event horizon A boundary around a black hole beyond which no event that takes place can affect an outside observer.

exoplanet An extrasolar planet – a planet outside our Solar System.

extensive air shower A cascade of ionized particles and photons generated when a primary cosmic ray strikes Earth's atmosphere. The shower is 'extensive' because it can be many kilometers wide.

fluorescence The emission of electromagnetic radiation at some frequency by a substance that has absorbed radiation of a different frequency. Usually, the emitted radiation is at a lower frequency (and thus lower energy) than the frequency of the absorbed radiation.

frequency The number of occurrences of something (such as the number of wave peaks passing a given point) in a given time period. The SI unit of frequency is the Hertz (Hz), which is the number of cycles per second of a periodic phenomenon.

galaxy rotation curve A graph on which the orbital velocity of objects in a galaxy (for example, objects such as stars or gas clouds) is plotted against the distance of those objects from the center of the galaxy. Such galaxy rotation curves do not follow the Keplerian decline that is predicted by Newtonian dynamics, and are considered to be evidence of the presence of dark matter.

gamma-ray Electromagnetic radiation of extremely high frequency. Gamma-ray photons, the most energetic form of light, are emitted during radioactive decay.

gamma-ray burst (GRB) A short-lived burst of gamma-ray photons. Bursts are typically detected at the rate of roughly one every few days, and they come from random directions on the sky. For a brief time, a GRB is the brightest source of gamma-rays in the Universe. There seem to be two classes of GRB, 'short' bursts and 'long' bursts, which have different progenitors.

gravitational lens A situation in which the gravitational effects of a large mass (such as a galaxy or cluster of galaxies) bends the path of light from a more distant source as it travels towards us. The effect can cause us to see multiple examples of the same distant object (so-called strong lensing). In weak lensing, the deformation of a single background object cannot be detected but the presence of an intervening mass can be inferred from a statistical analysis of many background objects.

gravitational wave A disturbance in the curvature of spacetime, caused by accelerating masses (such as those in binary neutron star systems), which propagates as a wave from the source. The amplitude of gravitational waves is typically minuscule, which makes their detection extremely challenging.

grazing incidence X-ray photons carry so much energy that they would simply smash into the mirror of a typical reflecting telescope. In order to deflect them and bring them to a focus, the mirror must be at a shallow angle to the incoming X-ray photons. This is the basis of the grazing incidence technique.

GZK cutoff Calculations made in 1966 by Greisen and, independently, Kuzmin and Zetsepin, suggest that a cosmic ray with an energy greater than 5×10^{19} eV loses energy by interacting with photons in the cosmic microwave background. The GKZ cutoff suggests that cosmic rays possessing more than this energy are unlikely to reach us from a distance further than about 160 million light years.

habitable zone The distance from a star at which a planet can possess liquid water on its surface and thus, perhaps, sustain life. (There may also be a 'galactic habitable zone': a planet's star needs to be close enough to the galactic center that there is an abundance of heavy elements but not so close that it is exposed to large amounts of high-energy radiation.)

inflation It is supposed that the very early Universe underwent a brief period of exponential expansion; during this time the volume of the Universe increased by a factor of perhaps 10^{78}. This concept of inflation solves several problems with the original Big Bang idea, and is an integral part of the standard model of cosmology.

infrared A form of electromagnetic radiation with a wavelength between about 700 nm (the red edge of the visible spectrum) to about 0.3 mm (the far infrared region). An object at room temperature will emit radiation with a peak in the infrared region, chiefly in the wavelength range 8–25 μm.

inspiral As binary stars orbit around one another they lose gravitational potential energy by emitting gravitational waves. They thus gradually inspiral – moving closer together and orbiting each other ever faster. The process is not noticeable in standard binary systems, but is measurable when the two stars involved are neutron stars. Gravitational wave observatories hope to detect the waves emitted during the final seconds of an inspiraling binary neutron star system.

interferometer Interferometry refers to several techniques that are widely used in astronomy. Interferometry works by bringing together electromagnetic waves so that they can undergo interference. For example, arrays of optical or radio telescopes can be connected so that waves from an astronomical source interfere. If the individual instruments are separated by a known distance, the position of the source may be determined with high accuracy. Many gravitational wave observatories are built around an interferometer: by splitting a light beam into two, and making those beams traverse paths at right angles before being recombined and undergoing interference, the path lengths taken by the two beams can be determined accurately. Changing path lengths may be an indication that a gravitational wave has passed through the system.

intermediate-mass black hole A black hole with a mass in the region of a hundred to a few thousand times that of the Sun. This compares to stellar-mass black holes (which are formed by the gravitational collapse of stars) and supermassive black holes (which are at the heart of galaxies). It is not yet entirely clear how these mid-mass black holes form.

Lagrange point One of five positions in an orbital configuration at which an object with little mass (such as an artificial satellite) can be stationary relative to two much more massive objects (such as Earth and Moon, or Earth and Sun). A Lagrange point is thus a good place to 'park' a spacecraft. The L2 point of the Earth–Sun system, for example, is home to the Wilkinson Microwave Anisotropy Probe as well as the Herschel and Planck missions.

light curve A graph of the changing brightness of a celestial object over time. The light curve of a Cepheid star, for example, varies periodically: it is a periodic variable star. The light curve of a supernova, on the other hand, shows no periodicity.

lightest supersymmetric particle In many models of supersymmetry, the least massive (or lightest) superparticle is stable. The lightest supersymmetric particle, whatever it might be, is an excellent WIMP dark matter candidate.

Local Group A group of three dozen or so galaxies, gravitationally bound together, which includes our Milky Way galaxy. The two stand-out members of the Group are the Andromeda galaxy and the Milky Way galaxy. The Group is part of the much larger Virgo Supercluster.

Lyman-alpha forest Light that is emitted from a distant object can be absorbed en route to us by gas clouds. Clouds of intergalactic hydrogen gas will absorb radiation at the Lyman-alpha wavelength. However, due to the expansion of the Universe, objects at different distances exhibit different redshifts. Since hydrogen clouds absorb whatever light has been redshifted to the Lyman-alpha wavelength by the time it reaches them, astronomers detect a tangle of Lyman-alpha absorption lines – a so-called forest – rather than a single spectral line.

Lyman-alpha line The spectral line of hydrogen that occurs when an electron falls from the first excited state to the ground state. It corresponds to a wavelength of 121.6 nm, which is in the ultraviolet region. Photons with a wavelength of 121.6 nm are absorbed by cold neutral hydrogen, with the electron in the hydrogen atom being excited from the ground state to the next energy level. When astronomers look at the light from very distant objects, this absorption process can clearly be seen; it gives rise to the Lyman-alpha forest.

magnetar A type of neutron star, with a magnetic field that can be as large as ten billion tesla. (Magnetars thus possess the largest known magnetic fields in the Universe. For comparison, the magnet in a medical MRI device typically operates at 2 T.) The huge magnetic field of a magnetar causes the emission of vast amounts of X-rays and gamma-rays.

massive astrophysical compact halo object (MACHO) A generic name given to any body composed of 'normal' baryonic matter that might explain the flat rotation curves of galaxies. Thus MACHO candidates such as black holes, neutron stars, brown dwarfs or unattached planets have all been put forward as possible explanations for dark matter. The term was coined in opposition to another dark matter hypothesis: the WIMP. While it is clear that MACHOs exist, it seems they cannot exist in the numbers required to account for dark matter.

metallicity If an astronomical object has a high proportion of elements other than hydrogen and helium, it is said to have a high metallicity. Older objects in the Universe tend to have lower metallicity than younger objects, because the early Universe possessed only small amounts of elements heavier than helium.

microlensing When a low-mass object (a planet or star) moves directly in front of a distant object then it forms a gravitational lens. However, unlike in strong or weak lensing, the deformation of the distant object is far too small to be detected. Nevertheless, the gravitational lens formed by a low-mass object can be strong enough to cause a transient brightening of the background object. The brightening will take place on a timescale of seconds to years, rather than megayears. Thus a microlensing event becomes apparent by looking at the light curve of an object.

microquasar An X-ray binary system in which relativistic jets are ejected either side of an accretion disc. It gets its name because it behaves like a quasar, but on a much smaller scale. In both cases, the ultimate power source is a central black hole. In the case of a microquasar the central object is a few solar masses in size, whereas the mass of central object of a quasar can be many million times that of the Sun. Microquasars are a useful 'laboratory' for studying their more massive and more distant cousins.

microwave A form of electromagnetic radiation with a wavelength somewhere in the region of 1 mm (frequency 300 GHz) to 1 m (frequency 300 MHz).

modified gravity theory (MOG) The physicist John Moffatt has proposed a modification of Einstein's theory of general relativity that serves to change the strength of gravity at large distances. Such a modification would explain galaxy rotation curves without the need to invoke dark matter. However, the theory is not widely accepted.

modified Newtonian dynamics (MOND) The physicist Mordehai Milgrom has proposed a modification of Newtonian dynamics in order to explain galaxy rotation curves without the need to invoke dark matter. However, the theory is not widely accepted.

muon An elementary particle that has similar properties to the electron and the tau particle. All three are leptons, and have an associated neutrino. The leptons differ in mass: the muon is about 200 times more massive than the electron, and about 17 times less massive than the tau.

negative pressure Imagine a cylinder that contains a vacuum; a piston caps the cylinder. Suppose further that there is a constant energy density associated with the vacuum. If you pull the piston out, you create more vacuum within the cylinder and thus there's more energy associated with it; you supplied this energy through the force required to pull the piston. The vacuum must have negative pressure; after all, if it had positive pressure it would push the piston out. Now, according to Einstein's theory of general relativity, gravitational effects are caused not only by mass and energy but also by pressure; and a negative pressure generates an effective repulsion. Since the cosmological constant has negative pressure equal to its energy density, it is a favored explanation for the observed acceleration of the Universe.

neutralino A quantum mechanical mix of the photino, the higgsino and the zino (these three particles are respectively the supersymmetric partners of the photon, the Higgs boson and the Z boson). It might be the lightest supersymmetric particle.

neutrino A type of elementary particle. It possesses very little mass and carries zero electric charge. This combination of characteristics makes the neutrino extremely elusive; since it interacts only through gravity and the weak force it is difficult to detect . There are three different types of neutrino, associated with the electron, the muon and the tau; they are all leptons. Neutrinos are created in nuclear reactions such as take place in the centers of stars.

neutron A subatomic particle that occurs alongside protons in the nuclei of atoms. A neutron has slightly more mass than a proton but contains no net electric charge. It consists of one up quark and two down quarks.

neutron degeneracy pressure The Pauli exclusion principle states that two fermions (particles such as electrons or neutrons, possessing a half-integer unit of spin) cannot occupy the same location if they possess the same quantum state. When identical fermions are brought close together, the exclusion principle gives rise to a pressure. In a white dwarf, further gravitational collapse is prevented by electron degeneracy pressure. In a neutron star it is the close proximity of neutrons that give rise to degeneracy pressure.

neutron star When a star runs out of nuclear fuel it undergoes gravitational collapse. Depending upon the mass of the star, it may end up as a white dwarf, a neutron star, or a black hole. In a neutron star, the object is composed primarily of neutrons. Further gravitational collapse is prevented in large part by neutron degeneracy pressure. Beyond a certain mass, however, such pressure cannot prevent further collapse and a black hole forms. Neutron stars are extremely dense and can rotate rapidly. A pulsar is a rapidly rotating neutron star with a large magnetic field that beams electromagnetic radiation in our direction.

nulling interferometry A type of interferometry that combines signals from different telescopes in such a way that the light from a central star is cancelled out, leaving the much fainter light from any orbiting planets much easier to detect. The technique works because the light from a planet hits the telescopes at a slightly different angle to light from the star.

objective A telescope lens or mirror (or indeed a combination of elements) that gathers light from an object and brings it to a focus to produce an image. The larger a telescope's objective, the dimmer the object it can image and the more detail it can resolve.

oligarch In the currently accepted model of planetary formation, dust particles increase in size by gradually accumulating matter until they reach the size of asteroids; in turn these asteroid-sized objects continue to capture material until they reach the size of planets. These planet-sized objects are called oligarchs because their gravitational potential dominates their immediate vicinity. The inner regions of the early Solar System might have contained as many as 30 oligarchs. Some oligarchs joined to become the inner planets that we know today; others would have been ejected from the Solar System.

pair production The simultaneous formation of a particle and its antiparticle. If a photon possesses sufficient energy then it can give up its energy to form such a particle–antiparticle pair when it interacts with matter. For example, an electron has a rest mass of $0.511\,\mathrm{MeV}$; if a photon possesses an energy of more than $1.02\,\mathrm{MeV}$ (making it an X-ray or gamma-ray photon) then it can create an electron–positron pair.

petabyte A unit of information corresponding to 1000 terabytes (or a million gigabytes). For those of us who can remember a time before personal computers became ubiquitous, the petabyte seems too large to be usable or relevant. However, new astronomical observatories such as the Large Synoptic Survey Telescope will be collecting petabytes of data every night.

phantom energy Several possibilities exist for dark energy. A scenario in which dark energy is a cosmological constant fits observations well, but phantom energy – a form of energy that increases in density as the Universe expands – is not yet ruled out by the data. If phantom energy is causing the accelerating expansion then the Universe will end in a 'Big Rip' – a singularity in which the distance between particles increases to infinity in a finite time.

photoabsorption A process in which a photon transfers all its energy to an atom or its nucleus.

photon The quantum particle associated with the electromagnetic interaction. As with other quantum particles, the photon exhibits wave–particle duality. In the radio part of the spectrum the wavelike aspects of the photon are dominant; in the gamma-ray part of the spectrum the particle-like aspects come to the fore.

Pisces–Cetus Supercluster Complex A filament of galaxies, discovered in 1987 by R. Brent Tully, that includes the local supercluster of galaxies (the Virgo Supercluster). It is about one billion light years long. Such large, filamentary structures are thought to form part of the cosmic web.

planetesimal A type of object that is thought to form from dust and rocks in protoplanetary disks. As dust grains stick to them, planetesimals increase in size; when they become kilometer-sized objects they can gravitationally attract one another and stick together to become a protoplanet.

plus-polarized gravitational wave Gravitational waves exhibit many of the properties of more familiar types of wave. For example, they have a wavelength, a frequency, and an amplitude; and, as with other types of wave, they can be polarized. There are two fundamental polarization modes. Plus polarization takes its name because the waves simultaneously oscillate in two directions: up/down and left/right, rather like a plus (+) sign. The passing of a plus-polarized gravitational wave could in principle be detected by its characteristic effect on a ring of test particles. When such a wave passes the ring will be alternately stretched (so that the ring becomes an oval with an up/down aspect) and then squashed (the oval has a left-right aspect). See also cross-polarized gravitational wave.

polarized light Light is a transverse wave that has both electric and magnetic components. In unpolarized light, which comes from sources such as a naked flame or the Sun, the electric and magnetic vibrations take place in many different planes. However, certain processes can cause polarization: in polarized light, the vibrations occur in a single plane.

proper distance The distance between two cosmological objects at a specific moment in time, as measured by a long chain of rulers placed end-to-end between the objects. The proper distance increases as the Universe expands.

proton   A subatomic particle that occurs alongside neutrons in atomic nuclei. A proton has slightly less mass than a neutron; unlike the neutron it possesses a net positive electric charge. It consists of one d quark and two u quarks.

protoplanet   A Moon-sized object thought to form from the collisions of planetesimals in protoplanetary disks. Protoplanets are massive enough to gravitationally perturb each other's orbits, which cause protoplanetary collisions that lead to the formation of planets. The asteroids Ceres, Pallas and Vesta are the remnants of protoplanets in our Solar System.

protoplanetary disk   A rotating disk of gas and dust grains surrounding a young star. A sequence of stages involving planetesimal and protoplanet formation may eventually lead to the formation of a fully fledged planetary system.

protostar   An early stage in the formation of a star, in which a gas cloud in the interstellar medium undergoes gravitational collapse. The end result of the collapse is the formation of a T Tauri star, which in turn evolves into a main sequence star.

PRS 1913+16   A binary system discovered in 1974 by Russell Hulse and Joseph Taylor, in which a rapidly spinning pulsar is orbiting a similar-mass compact companion. As the two objects orbit, the system loses energy through the emission of gravitational waves. The orbit thus decays and the objects gradually inspiral. Hulse and Taylor made careful measurements of the orbital decay and found that it agrees precisely with predictions coming from general relativity.

pulsar   A rapidly rotating neutron star with an extremely large magnetic field that beams electromagnetic radiation. If a beam is pointing in our direction we see it rather like the beam from a lighthouse. Pulsar periods range from a few milliseconds to a few seconds.

quark   An elementary particle that combines in triplets to form objects such as protons and neutrons. There are six types: up, down, charm, strange, top, and bottom.

quasar   The term quasar was coined in 1964 as a shortened form of 'quasi-stellar radio source'. Astronomers now know that quasars are active galactic nuclei as seen from a large distance. These exceptionally luminous objects are powered by an accretion disk around a central supermassive black hole.

quintessence   A hypothetical form of dark energy which, unlike the cosmological constant, has a density that can vary in time and over space.

radial velocity method   A means of finding exoplanets by detecting regular changes in the radial (in other words, line-of-sight) velocity of a star. See Doppler spectroscopy method for more information.

radio   The form of electromagnetic radiation with the longest wavelengths (typically longer than about 10 cm).

recombination The time, about 380 000 years after the Big Bang, when electrons and protons in the Universe first combined to form neutral hydrogen atoms. Before recombination photons could not travel far before scattering off charged particles. After recombination photons were free to travel. We observe these photons today in the form of the cosmic microwave background.

relativistic jet A jet of particles, moving at close to light speed, that can emerge from the poles of an accretion disk around a compact massive object such as a black hole. Jets appear in pairs, moving in opposite directions. Such jets are extremely powerful, and have been seen emanating from active galactic nuclei and micro-quasars. It is thought that spinning black holes form the power sources for relativistic jets, with energy being extracted through, for example, the Blandford–Znajek mechanism.

Sagittarius A* A bright, compact radio source at the center of our Milky Way galaxy. Astronomers believe this to be the site of a supermassive black hole at the heart of the Milky Way.

scintillation A flash of light that occurs when an energetic particle strikes a certain type of material; if the particle ionizes atoms in the material, and those ionized atoms subsequently re-emit photons of visible light, then scintillation has occurred. If physicists connect a light sensor to a scintillator, they have the basis for a scintillation detector. Such detectors can be used to investigate the properties of the original, incoming particle.

search for extraterrestrial intelligence (SETI) An umbrella term for various activities that attempt to detect signs of extraterrestrial intelligence. Frank Drake designed the first 'modern' SETI experiment in 1960; many subsequent experiments followed Drake's lead in employing radio telescopes to search for artificial radio signals. Since then the search has broadened. For example, some scientists have investigated optical SETI while others have looked at the possibilities of detecting byproducts of advanced astroengineering projects. Nevertheless, radio telescopes remain the prime tool for the SETI program.

Seyfert galaxy A particular type of active galactic nuclei named after Carl Seyfert, an American astronomer who first identified such a galaxy in 1943.

singularity A point in spacetime where the gravitational field becomes infinite. General relativity predicts that the gravitational collapse of a sufficiently massive star results in a singularity surrounded by an event horizon – a black hole.

Sloan Great Wall A filament of galaxies, whose existence was discovered in 2003 by the Sloan Digital Sky Survey. The Sloan Great Wall is 1.37 billion light years in length which makes it, as of the time of writing, the largest known structure in the Universe. Large filamentary structures are thought to form part of the cosmic web.

sound horizon   Pressure or 'sound' waves known as baryon acoustic oscillations would have moved through the early Universe. About 380 000 years after the Big Bang, when electrons and protons combined to form hydrogen atoms, these sound waves would have 'frozen' in place. The distance these waves could have travelled, which is given by the speed of sound in the medium that comprised the early Universe multiplied by the age of the Universe when they froze, is the sound horizon. The sound horizon distance today is about 450 million light years, and galaxies are more likely to have formed in these high-density regions. Thus the sound horizon provides a standard length.

standard candle   A standard candle is any class of object that possesses a known luminosity. By comparing the apparent magnitude of the object with its intrinsic magnitude, astronomers can determine the distance to the object (the more distant the candle, the fainter it appears). The identification of standard candles has played an important role in determining cosmic distance scales. Standard candles include Cepheid variable stars and type Ia supernovae.

standard length   A standard length (also called a standard ruler) is any class of object that possesses a known size. By comparing the apparent size of the object with its true size, astronomers can determine the distance to the object (the more distant the standard length, the smaller it appears). In astronomy it has proven to be more difficult to identify standard lengths than standard candles. Recently, however, baryon acoustic oscillations have been used as a standard length; cosmologists hope to use this as a probe of dark energy.

standard model of particle physics   Currently the best theory for answering questions about what the world is made of and how it behaves on the smallest scales. In the standard model, material ultimately consists of 12 fundamental 'matter' particles (six types of quark; and the electron, muon and tau, each with their associated neutrinos). The fundamental particles can interact via the exchange of fundamental 'field' particles (the photon; the W and Z bosons; eight gluons), which are associated with the electroweak and strong interactions.

starburst galaxy   A galaxy in which the rate of star formation is exceptionally high. A starburst is likely to be a temporary stage in a galaxy's existence (it must be temporary because ongoing creation of the new stars would use up all the gas in a galaxy on a timescale much less than the age of the galaxy). A starburst may be created when one galaxy interacts with another in some way, for example through collision.

stellar-mass black hole   At the end of its life a star undergoes gravitational collapse. Depending on its mass it will become a white dwarf, a neutron star, or a stellar-mass black hole. Several objects in our Milky Way galaxy are strongly suspected of being the black-hole remnants of stellar collapse; they range in mass from about 4–14 times that of the Sun.

sterile neutrino A hypothetical particle that arises in certain models of grand unification in particle physics. Unlike a normal neutrino, which can interact via the weak force, a sterile neutrino would interact only via gravity (although it could decay into 'ordinary' neutrinos). The sterile neutrino has been proposed as a possible dark matter particle.

string theory A theory that aims to unite general relativity and quantum mechanics by positing the existence of fundamental objects that are one-dimensional strings rather than point particles. Mathematical consistency demands that the strings must oscillate in 10 or 11 dimensions (the unseen dimensions being extremely tiny, since they have not yet been observed) and that the theory exhibits supersymmetry.

strong lensing When a gravitational lens is extremely massive, and a source is close to it, the light can take different paths to an observer. Thus in strong lensing an observer can see multiple images of the same source object, or the object can appear as a ring or smeared out into an arc.

submillimeter band A form of electromagnetic radiation with a wavelength somewhere in the region of perhaps 0.3 mm (frequency 1000 GHz) to 1 mm (frequency 300 GHz).

supermassive black hole Most (and possible all) galaxies are thought to possess a black hole at their center. The central object in a galaxy is much more massive than mid-mass or stellar-mass black holes: a supermassive black hole can contain a billion times the mass of the Sun. Relativistic jets from supermassive black holes are believed to be the power source of active galactic nuclei. It is still not entirely clear how such supermassive objects form.

supernova An extremely luminous stellar conflagration that can, briefly, outshine an entire galaxy. Supernovae come in two broad types. A type I supernova occurs when a white dwarf accretes matter from a companion star; if the white dwarf becomes so massive that electron degeneracy pressure can no longer support it against gravitational collapse, an uncontrolled nuclear fusion explosion occurs and blows the dwarf to bits. A type II supernova occurs when a massive star reaches the end of its life; the core collapses to form a neutron star or black hole, while the surrounding material is blasted into space.

supersymmetry A proposed symmetry of nature, so far unobserved, that relates elementary particles with different spins: for every particle with an integer spin there is a superpartner that differs by a half unit of spin; and vice versa.

synchrotron radiation Electromagnetic radiation that is generated when fast-moving charged particles are accelerated by magnetic fields. Many astronomical objects generate synchrotron radiation. For example, synchrotron radiation is generated when fast-moving electrons spiral through the magnetic fields that occur in relativistic jets. Synchrotron radiation can be identified through its characteristic spectrum and polarization.

tau An elementary particle that has similar properties to the electron and the muon. All three are leptons, and have an associated neutrino. The leptons differ in mass: the tau is about 17 times more massive than the muon, and about 3500 times more massive than the electron.

thermocouple If two strips of different metals are placed together to form a junction, a voltage is generated with a magnitude that's proportional to the temperature difference between either end of the two strips. A thermocouple made in this way can be used to measure temperature.

transit photometry method A technique through which astronomers can discover exoplanets. If a planet orbits in a plane that is edge-on to our line of sight then it will occasionally transit, or move in front of, the star. During such a transit the amount of light that reaches us from the star drops slightly. Thus by detecting periodic dimming that lasts a fixed length of time astronomers can deduce the presence of an orbiting planet. The method also allows an estimate of the planet's size (though not its mass).

T Tauri star A type of variable star that has yet to begin its life as a main sequence star. Such an object has a relatively low central temperature, and is powered not by nuclear fusion but by the gravitational potential energy released as it contracts towards the main sequence. A T Tauri star thus does not last long: after about 100 million years it becomes a 'true' star and shines through nuclear fusion.

ultraviolet A form of electromagnetic radiation with a wavelength between that of visible light and X-rays (with wavelengths in the range 10–400 nm).

Virgo Supercluster Also known as the Local Supercluster, this is the supercluster to which our own Milky Way galaxy belongs. (The Milky Way is part of the Local Group, which in turn is part of the Virgo Cluster, which in turn is part of the Virgo Supercluster.) It is about 110 million light years in extent, and contains about a hundred galaxy groups and clusters.

warm–hot intergalactic medium (WHIM) A reservoir of extremely dilute gas that exists in the cosmic web. The whim is believed to contain a large fraction of the 'normal' baryonic matter in the present-day Universe.

waterhole A part of the electromagnetic spectrum corresponding to radio wavelengths between about 18–21 cm (1666–1420 MHz). The HO compound (called hydroxyl) radiates strongly at 18 cm; H (neutral hydrogen) radiates strongly at 21 cm. Combine HO and H and you get water, which as far as we know is necessary for the existence of life. Furthermore, the region of the spectrum between 18–21 cm is particularly quiet. Putting all these factors together, the astronomer

Bernard Oliver suggested that this waterhole region would be an obvious band in which extraterrestrial civilizations could broadcast; it is thus relevant in the search for extraterrestrial intelligence.

wavelength The distance between two successive points in a wave that have the same phase (thus the distance between successive peaks of a wave, for example, or between successive troughs).

weak interaction A fundamental force of nature, which is responsible for radioactive decay. According to the standard model of particle physics, the weak interaction is mediated by the exchange of W and Z particles. The same model explains how electromagnetism and the weak interaction are different aspects of an underlying electroweak interaction. Neutrinos interact only through the weak interaction (and, as with all particles, through gravity); this makes neutrinos difficult to spot.

weak lensing When a gravitational lens has relatively little mass and/or a source is far away from it (which is the usual situation in astronomy) then the multiple images, rings and arcs that one sees in strong lensing are absent. Nevertheless, even in the weak lensing case, the presence of the lensing object can be inferred from a statistical analysis of the properties of the background objects surrounding the lens. Weak lensing is a powerful probe of dark energy.

weakly interacting massive particle (WIMP) A massive particle that interacts solely through the weak interaction (and gravity). Thus a WIMP behaves in a similar way to a neutrino, but it has much more mass. Cosmologists introduced the notion of a WIMP as a possible solution to the dark matter problem. There are several plausible WIMP candidates, and some observatories have found signals consistent with the presence of WIMPs, but at the time of writing such particles remain hypothetical.

weakly interacting sub-eV particle (WISP) A particle possessing extremely little mass that interacts solely through the weak interaction (and gravity). Cosmologists introduced the notion of a WISP as a possible solution to the dark matter problem. The axion is a plausible candidate WISP, but so far the axion remains a hypothetical particle.

white dwarf When a star runs out of nuclear fuel it undergoes gravitational collapse. Depending upon the mass of the star, it may end up either as a white dwarf, a neutron star, or a black hole. In a white dwarf, the object is composed primarily of so-called electron degenerate matter. Further gravitational collapse is prevented by the Pauli exclusion principle, which states that two electrons in the same quantum state cannot occupy the same location. Beyond a certain mass, however, this electron degeneracy pressure cannot prevent further gravitational collapse; electrons combine with protons to create neutrons and thus a neutron star is formed. Any star with a mass that is not large enough to generate a supernova explosion will end up as a white dwarf; thus a white dwarf is the final state of almost all stars.

Wolf–Rayet star A type of very massive, very hot star first identified in 1867 by Charles Wolf and Georges Rayet. It is believed that when a massive star (one with perhaps twenty times the mass of the Sun) comes towards the end of its life, the heavier elements that it cooked up in its central regions reach the surface. This causes powerful stellar winds to occur, which are the final stage in the star's life before it explodes in a supernova.

Wolter telescope A telescope that uses grazing incidence in order to bring X-rays to a focus.

X-ray A form of electromagnetic radiation with a wavelength somewhere in the region of 0.01 nm (frequency 3×10^{19} Hz corresponding to an energy of 120 keV) to 10 nm (frequency 3×10^{16} Hz corresponding to an energy of 120 eV).

Glossary of facilities
and experiments

Advanced Gamma-ray Imaging System (AGIS) A US collaboration that hoped to build a next-generation ground-based gamma-ray observatory. In 2011, the AGIS project merged with the European Cerenkov Telescope Array project.

Advanced Satellite for Cosmology and Astrophysics (ASCA) A Japanese X-ray astronomy mission, whose Wolter telescope studied the cosmos between 1993 and 2001. In 2014 the Japan Aerospace Exploration Agency hopes to launch Astro-H; the intention is that it should build on the ASCA observations.

Advanced Technology Large Aperture Space Telescope (ATLAST) A proposed successor to Hubble: its large (possibly 16.8 m) mirror would observe in the infrared, optical and ultraviolet wavelengths. The scientists involved hope that ATLAST will be the flagship astronomy mission in the period 2025–2035.

Advanced Telescope for High Energy Astrophysics (ATHENA) A preliminary study by ESA to investigate the extent to which a European-led mission could deliver the science outcomes of the International X-ray Observatory (IXO), which was cancelled in 2011.

Akeno Giant Air Shower Array (AGASA) A surface array of 111 particle detectors, spread over a hundred square kilometers near the village of Akeno, Japan, used to detect cosmic ray air showers. The array was completed in 1991 and, until the construction of the Pierre Auger Observatory, it was the largest such array in the world.

Allen Telescope Array (ATA) A radio interferometer that, in addition to making astronomical observations, was heavily employed in SETI activities. The first phase of development saw the successful deployment of 42 radio antennae; the intention was to create an array of 350 antennae. However, funding difficulties mean that the future for the ATA is unclear.

Alpha Magnetic Spectrometer (AMS-02) An experiment on the International Space Station. By detecting cosmic rays it looks for signs of antimatter in the Universe and evidence of dark matter and other exotic particles. The experiment was taken into space in 2011 on the last flight of Space Shuttle *Endeavour*.

Arecibo Observatory A radio telescope, constructed in 1963 inside a karst sinkhole in Puerto Rico, that at the time of writing remains the largest single-aperture telescope ever built (the main collecting dish is 305 m in diameter). When completed, the Chinese FAST instrument will have a 500 m dish.

S. Webb, *New Eyes on the Universe: Twelve Cosmic Mysteries and the Tools We Need to Solve Them*, Springer Praxis Books, DOI 10.1007/978-1-4614-2194-8, © Springer Science+Business Media, LLC 2012

Argon Dark Matter Experiment (ArDM) An experiment using a ton of liquid argon to try and detect dark matter particles. Detectors surrounding the argon will try to observe the results of a WIMP scattering off an argon nuclei: namely, electrons from ionization and photons from scintillation. The experiment is situated at the Canfranc underground laboratory, which is in a road tunnel under the El Tobazo mountain in the Spanish Pyrenees.

Astro-H A Japanese satellite, due for launch in 2013/14, which will carry several instruments to detect both soft and hard X-rays from the cosmos.

Astronomy with a Neutrino Telescope and Abyss Environmental Research (ANTARES) A Cerenkov detector, optimized for detecting the muons produced when high-energy neutrinos collide with atoms in water molecules, which is housed on the Mediterranean seabed. The initial detector has a surface area of 0.1 square kilometers; the hope is that the detector will eventually have a surface area of 1 square kilometers.

Atacama B-mode Search (ABS) An experiment based in the Atacama desert in Chile that hopes to detect B mode polarization patterns in the cosmic microwave background. Such patterns would be left behind by gravitational waves from the inflationary epoch.

Atacama Large Millimeter/submillimeter Array (ALMA) An array of movable 12 m antennae, spread out over the Atacama desert in Chile, in such a way that the instrument will work as an interferometer with baselines up to 16 km. An additional compact array of 7 m and 12 m antennae will complement the main instrument. The facility is due to be completed in 2013. ALMA will observe at a wavelength of about a millimeter or just below, and will study phenomena such as protostars and protoplanetary disks.

Axion Dark Matter Experiment (ADMA) An experiment, running since 1995, that searches for axions in the dark matter halo of our Milky Way galaxy.

Balloon Observations of Millimetric Extragalactic Radiation and Geophysics (BOOMERANG) A balloon-borne microwave telescope that flew at an altitude of 42 km over the South Pole in 1998 and 2003. It made several important observations of a part of the cosmic microwave background; BOOMERANG data showed that the Universe is flat and supported the notion of dark energy.

Baryon Oscillation Spectroscopic Survey (BOSS) A program, which is part of the Sloan Digital Sky Survey, to map the distribution in space of about 160 000 quasars and 1.5 million luminous red galaxies. The program began in 2009 and will run until 2014. The data will enable astronomers to study the distance scale that baryon acoustic oscillations imprinted on fluctuations in the cosmic microwave background. These fluctuations subsequently evolved into the present-day tangle of galactic walls and voids. The study of baryon acoustic oscillations allows cosmologists to investigate theories of dark energy.

BeppoSAX An Italian–Dutch satellite for X-ray astronomy, which launched in April 1996 and operated until April 2002. The satellite provided key evidence that gamma-ray bursts lie at cosmological distances and thus must be extremely energetic events. The satellite received the first part of its name from the Nobel-prize winning physicist Beppo Occhialini; the acronym SAX is Italian for 'satellite for X-ray astronomy'.

Baryonic Structure Probe A proposed UV spectroscopy mission to learn more about the cosmic web of intergalactic matter, and how it interacts with galaxies. It is unclear when such a mission might be launched.

Big Bang Observer (BBO) A tentative proposal for a space-based gravitational wave detector. The BBO would be a successor to the Laser Interferometer Space Antenna (LISA). However, since NASA announced in 2011 that it would not continue to investigate LISA, the launch of an instrument like the BBO is likely to be several decades in the future.

BigBOSS A proposal to build on the Baryon Oscillation Spectroscopic Survey (BOSS) and, using a new spectrograph in conjunction with over 500 nights of observing time on the Mayall Telescope at Kitt Peak, map the distribution in space of about 20 million galaxies and quasars. As with BOSS, the results will allow cosmologists to investigate theories of dark energy.

Canada–France–Hawaii Telescope Legacy Survey (CFHTLS) A large project involving the equivalent of 450 nights' observation over five years between 2003 to 2009. The project involved three different surveys: a very wide survey of the Solar System, a wide survey that allowed astronomers to study the large-scale distribution of matter in the Universe using weak gravitational lensing, and a deep survey to identify type Ia supernovae.

Cerenkov Telescope Array (CTA) A proposal for the next generation of ground-based telescopes aimed at observing very-high-energy gamma-rays. The CTA would thus build on the work of telescopes such as HESS, MAGIC and VERITAS. The plan is that the CTA would observe gamma-rays with energies between about 100 GeV to 100 TeV, and with much more sensitivity than previous instruments. The three-year preparatory phase, which involves investigating the possible science as well as the technical aspects of building and operating the array, began in 2010.

Chandra X-ray Observatory A satellite, launched by NASA in 1999 and named after the Nobel-prize winning physicist Subrahmanyan Chandrasekhar. Chandra was the third of four NASA 'Great Observatories' (Hubble, Compton and Spitzer being the others).

Coherent Germanium Neutrino Technology (COGENT) A dark matter experiment, located about 700 m underground in the Soudan Mine in Minnesota, which looks for possible signs of interactions between WIMPs and atoms in its germanium crystals. In 2011 the COGENT collaboration announced signs of a seasonal variation in signal that is consistent with the WIMP hypothesis and that matches a similar signal from the Dark Matter Project (DAMA).

Collaboration of Australia and Nippon for a Gamma Ray Observatory in the Outback (CANGAROO) A gamma-ray telescope, based near the town of Woomera, Australia, that became operational in 1999. The instrument detects Cerenkov radiation from very-high-energy gamma-rays entering Earths' atmosphere (observations that must be made against the much larger background of Cerenkov radiation produced by cosmic rays).

Compton Gamma Ray Observatory A satellite, launched by NASA in 1991 and named after the Nobel-prize winning physicist Arthur Compton. Compton was the second of four NASA 'Great Observatories' (Hubble, Chandra and Spitzer being the others). Until the end of the mission in 2000, the satellite observed the sky over a large range of gamma-ray energies and its results helped astronomers better understand the nature of gamma-ray bursts.

Constellation-X A proposal for a NASA X-ray satellite to build on the legacy of Chandra and XMM-Newton. In 2008, the proposal merged with a similar proposal of ESA and JAXA (a mission called the X-ray Evolving Universe Spectroscopy mission); the result was the proposed International X-ray Observatory.

Cosmic Background Explorer (COBE) A satellite, launched by NASA in 1989, that studied the cosmic microwave background radiation. COBE's discovery that the CMB is an almost perfect blackbody cemented Big Bang cosmology as the preferred theory of the origin of the Universe.

Cosmic Origins Spectrograph (COS) An instrument installed on Hubble in 2009 that is designed to study the sky in ultraviolet light.

Cryogenic Dark Matter Search (CDMS) A dark matter experiment, located about 714 m underground in the Soudan Mine in Minnesota, which looks for WIMPs using cryogenic germanium and silicon detectors.

Cryogenic Rare Event Search with Superconducting Thermometers (CRESST) A dark matter experiment, located about 1400 m underground in the Gran Sasso National Laboratory in Italy, which hopes to detect the small energy of a recoiling nucleus that has been hit by a WIMP. It employs $CaWO_4$ crystals and sensitive detectors that work at temperatures of just a few thousandths of a degree above absolute zero. In the period 2009–11 CRESST detected 67 events consistent with dark matter events. However, this falls short of the number required to claim a discovery.

Dark Energy Survey (DES) A collaboration of more than 100 scientists in 23 institutions from the US, UK, Brazil, Spain and Germany that will use a 570 megapixel camera mounted on a 4 m telescope at the Cerro Tololo Inter-American Observatory to study the accelerating Universe. The five-year survey starts taking data in September 2012.

Dark Matter Experiment using Argon Pulse-shape Discrimination (DEAP) A dark matter experiment, located about 2000 m underground at the SNOLAB facility near Sudbury, Canada, which uses liquid argon as a target for detecting WIMPs. The current DEAP-1 experiment employs 7 kg of argon as the target mass; the successor experiment, DEAP3600, which is under construction, will use 3600 kg of liquid argon.

Dark Matter Project (DAMA) A dark matter experiment, located about 1400 m underground in the Gran Sasso National Laboratory in Italy, which uses thallium-doped sodium iodide (NaI) crystals to try and detect the recoil of interacting WIMP particles. The original DAMA experiment collected data in the period 1996–2002; its successor experiment, DAMA/LIBRA, uses similar technology but employs a larger target mass. The DAMA experiments claim to have seen a dark matter signal that changes annually as the Earth orbits the Sun.

Darwin A proposed ESA mission that would have used four or five space telescopes, arranged in such a way as to form an interferometer, in order to search for Earth-like planets around the thousand or so closest stars and for signs of life on those planets. The Darwin concept was first mooted in 1993, but work on studying the mission design ended in 2007 with no further studies planned. Darwin will thus never launch, but ideas from the study may feed into subsequent missions having a similar scope.

Deci-hertz Interferometer Gravitational Wave Observatory (DECIGO) A design for a space-based gravitational wave observatory that is under active consideration by Japanese scientists. The current plan is for DECIGO to consist of four clusters, each consisting of three drag-free spacecraft at a distance of 1000 km from each other, in heliocentric orbits. The relative displacements would be measured by three pairs of interferometers.

Degree Angular Scale Interferometer (DASI) An interferometer based at the Amundsen–Scott South Pole station that operated in the frequency range 26–36 GHz. The instrument was used to measure the temperature and polarization anisotropy of the cosmic microwave background.

Diffuse Intergalactic Oxygen Surveyor (DIOS) A proposed Japanese satellite that aims to investigate the warm–hot intergalactic medium by observing redshifted spectral lines of oxygen. The scientists involved in the project hope for a launch some time around 2016.

Directional Recoil Identification from Tracks (DRIFT) A dark matter detector, located about 1100 m underground in the Boulby potash mine in North Yorkshire, which aims to identify any WIMPs that might collide with atoms in a test volume of a cubic meter of carbon disulfide gas. The drift detectors have the potential to determine not only the type of particle that collided with the gas atoms, but also the direction from whence the particle came.

Einstein The second in a series of three High Energy Astrophysical Observatories (HEAO-2) launched in 1978 by NASA. The satellite was soon renamed Einstein. Einstein was the first fully-imaging X-ray telescope and operated successfully until 1981. Its predecessor HEAO-1 launched a year earlier and surveyed the X-ray sky almost three times before ceasing to work in 1979.

Einstein@home A computing project, developed by LIGO, in which members of the public volunteer the spare computing power of their internet-connected devices in order to analyze LIGO data and search for gravitational waves. Although the individual computers concerned may not be particularly powerful, in combination they perform as well as a large supercomputer.

Einstein Telescope (ET) A proposed third-generation gravitational wave detector, under active design consideration by a collaboration of several European research groups. The detector would be about a hundred times more sensitive than instruments such as LIGO and VIRGO, and ten times more sensitive than Advanced LIGO. The telescope would differ from previous instruments by being built underground. It would be sensitive to low-frequency gravitational waves, down to about 2 Hz. It is hoped that ET will be completed by 2025.

Euclid An ESA space telescope with a planned launch date in 2017. Euclid will use observations of both weak gravitational lensing and baryon acoustic oscillations in order to study dark energy.

European Extremely Large Telescope (E-ELT) An optical telescope, which is planned to see first light around the year 2018. It will be the largest of the new generation of extremely large telescopes. The E-ELT mirror will have a diameter of 39.3 m, which will allow astronomers to investigate the Universe in unprecedented detail. The telescope will be located at an altitude of 3064 m on the Cerro Armazones mountain in Chile; this has cloudless nights for about 350 days in a year. The E-ELT will be an astonishing instrument.

European Underground Rare Event Calorimeter Array (EURECA) A proposed dark matter experiment, to be located in the Modane Underground Laboratory about 1700 m underneath the Fréjus summit. EURECA will combine the research groups who are currently working on the CRESST and EDELWEISS dark matter experiments, and will thus be a cryogenic detector. The hope is that EURECA will start operations some time soon after 2013.

Expanded Very Large Array (EVLA) A radio observatory, located in New Mexico, which is an upgrade of the Very Large Array (VLA). EVLA consists of 27 antennae, each with a diameter of 25 m, and state-of-the-art receivers and electronics. The antennae can be moved along railroad tracks, and are usually arranged into one of four different configurations. In its largest configuration, EVLA acts as a single antenna with an effective diameter of 36 km.

Expérience pour Détecter les WIMPs en Site Souterrain (EDELWEISS) A dark matter experiment, located in the Modane Underground Laboratory about 1700 m underneath the Frejus summit, which searches for WIMPs by using germanium detectors operating at about 10 mK. In 2011, the EDELWEISS team released data showing some events that may have been dark matter collisions; however, the team could not exclude the possibility that these were background events so no discovery could be claimed. Both EDELWEISS and CRESST will be combined into the forthcoming EURECA experiment.

Explorer 11 An American satellite, launched in 1961, that carried the first gamma-ray telescope into space.

Extended Roentgen Survey with an Imaging Telescope Array (EROSITA) A German X-ray telescope, which will be launched from Baikonur on the Russian Spectrum–Roentgen–Gamma satellite in 2012 and then placed in an L2 orbit. The intention is that EROSITA will perform the first imaging all-sky survey in the medium-energy X-ray spectrum. It will also concentrate on observing the hot intergalactic medium; finding previously hidden accreting black holes in nearby galaxies and millions of new active galactic nuclei; and studying X-ray sources in our own Milky Way galaxy.

Extreme Universe Space Observatory (EUSO) A European space-based mission that planned to investigate cosmic rays and high-energy neutrinos by looking down on the extensive air showers generated when such particles interact with Earth's atmosphere. The concept was not taken beyond the design stage, but then the Japanese space agency became actively involved and renamed the mission JEM-EUSO. The intention is for JEM-EUSO to launch in 2015.

Far Ultraviolet Spectroscopic Explorer (FUSE) A US space-based telescope that operated in low earth orbit between 1999 and 2007. FUSE observed in the far ultraviolet, and in particular studied conditions in interstellar and intergalactic space.

Fermi Gamma-Ray Space Telescope A space-based gamma-ray observatory, which is a collaborative venture between NASA and several other agencies, that launched in 2008. The observatory contains two separate instruments, with which it studies gamma-ray bursts and active galactic nuclei (amongst many other phenomena). The observatory is named in honor of the physicist Enrico Fermi, who won the Nobel Prize in 1938.

Five-hundred-meter Aperture Spherical Telescope (FAST) A Chinese radio telescope, currently under construction and with a planned completion date of 2016. FAST will be the largest single-dish radio telescope in the world.

Fly's Eye A cosmic-ray detector, located in the western desert of Utah, which operated between 1981 and 1993. In 1991, Fly's Eye saw the highest-energy particle ever detected: this cosmic ray carried about 50 J of energy. The Fly's Eye was superseded by the HiRes Fly's Eye detector.

Galaxy Evolution Explorer (GALEX) A US space telescope, launched in 2003, that is surveying thousands of galaxies in ultraviolet light.

GEO600 A British–German gravitational wave detector, located in Ruthe near Hannover in Germany. The detector is an L-shaped interferometer, with each arm being 600 m in length.

Giant Magellan Telescope (GMT) A ground-based optical telescope, to be located at the Las Campanas Observatory in Chile and planned for completion in 2018. It will have seven 8.4 m diameter primary mirror segments (equivalent in terms of resolving power to a single 21.4 m primary mirror).

Ginga A Japanese space-based X-ray telescope that was launched in 1987 and ended its mission in 1991. Ginga is a Japanese word meaning 'galaxy'.

Gran Telescopio Canarias (GTC) A telescope with a 10.4 m segmented primary mirror located in the Canary Islands. The GTC is a Spanish-led initiative. When it saw first light in 2007 it became the world's largest single-aperture optical telescope.

Gravity Probe B A satellite, launched by NASA in 2004, that was designed to measure the amount by which Earth warps the spacetime around it and the amount by which Earth's rotation drags spacetime with it. The existence of these effects is predicted by Einstein's general theory of relativity.

Hard X-ray Modulation Telescope (HXMT) A Chinese space-based instrument that aims to study cosmic X-rays in the energy region 1–250 keV (and particularly in the hard X-ray region above 15 keV). By studying hard X-rays, HXMT should be able to probe the regions around supermassive black holes, where magnetic fields are extremely intense. The telescope will launch in 2015 and has an intended mission lifetime of four years.

Hat Creek Radio Observatory A radio observatory, founded in 1959 and located in northern California at an altitude of 1280 m, that is home to several telescopes including the Allen Telescope Array.

Herschel A telescope, launched by ESA in 2009 into an L2 orbit, which observes in the infrared and sub-millimeter regions. When it was launched, Herschel's 3.5 m diameter mirror was the largest single mirror yet built for a space telescope. Herschel aims to study, amongst other things, the formation of galaxies in the early Universe and the creation of stars and their interaction with the interstellar medium.

High Accuracy Radial Velocity Planet Searcher (HARPS) A 3.6 m telescope, with an attached high-accuracy spectrograph, dedicated to discovering exoplanets. The HARPS facility is located at La Silla in Chile. It can measure radial velocities to an accuracy of better than one meter per second. In the six years following its opening in 2003, HARPS discovered 75 exoplanets.

High Altitude Water Cerenkov Experiment (HAWC) A proposed American–Mexican observatory, to be located in Sierra Negra, Mexico, for detecting very high-energy gamma-rays (from 100 GeV to 100 TeV). The plan is to re-use techniques and equipment from the Milagro experiment; however, HAWC would possess much more sensitivity than Milagro.

High Energy Stereoscopic System (HESS) An imaging atmospheric Cerenkov telescope array, located in Namibia, which detects gamma-rays with energies in the range 100–1000 GeV. It became fully operational in 2003. The telescope is named after the Nobel prize-winning physicist Victor Hess.

High-z Supernova Search An international group of astronomers who used type Ia supernovae to study the expansion history of the Universe. In 1998 this team, along with the Supernova Cosmology Project, found the first evidence for the accelerating expansion of the Universe.

Hi-Res Fly's Eye An observatory for observing ultra-high-energy cosmic rays, located in the western desert of Utah. Hi-Res was a successor to the original Fly's Eye experiment, and operated between 1997 to 2006. The observatory made the first observation of the GZK cutoff, although this result contradicted observations made by AGASA. In order to resolve the differences, teams from both experiments merged to create the Telescope Array Project.

Hubble Space Telescope The first of the NASA Great Observatories (the others being Compton, Chandra and Spitzer). The telescope is named after the famous astronomer Edwin Hubble. Hubble was put into a near-circular low Earth orbit in 1990 and, although it had a troubled start, it proved itself to be a powerful and versatile observatory. Hubble is planned to continue until the launch of the James Webb Space Telescope, so it may function until 2018 or beyond. It has made many important discoveries, and is perhaps the most famous telescope of all time.

Hyabusa A spacecraft, developed by JAXA, that was launched in 2003 and landed on the asteroid Itokawa in 2005. Five years later it returned samples of the asteroid to Earth. Hyabusa is Japanese for 'Peregrine Falcon'.

IceCube Neutrino Observatory A neutrino telescope, located at the Amundsen–Scott South Pole station, whose construction was completed at the end of 2010. The telescope consists of 86 'strings' of optical sensors buried deep in a large volume of Antarctic ice. The sensors search for any high-energy neutrinos that interact with water molecules in the ice. IceCube is the biggest neutrino telescope in the world.

Infrared Astronomical Satellite (IRAS)   An infrared space-based telescope that operated for ten months in 1983. Although its mission lifetime was short, IRAS made many important discoveries and it was the first infrared telescope to complete an all-sky survey. The satellite was a joint project of the US, UK and the Netherlands.

Infrared Space Observatory (ISO)   A European space-based infrared telescope that operated between 1995 and 1998. It made several discoveries relating to stellar and planet formation.

Innovative Interstellar Explorer (IIE)   A NASA-funded study, focusing on how to build a spacecraft that can travel beyond the influence of our Sun and reach the interstellar medium, developed with the IIE concept. A travel distance of at least 200 AU would have to be reached in order for relevant scientific work to be carried out. It is not clear when or if a mission such as IIE will launch.

International Space Station   A research facility in low Earth orbit, the construction of which began in 1998. The station is home to several instruments, including the Alpha Magnetic Spectrometer.

International Ultraviolet Explorer (IUE)   A space-based UV telescope, which launched in 1978 and operated until budgetary constraints caused it to be shut down in 1996. It was a collaboration between US, UK and European agencies. During its mission lifetime the telescope made a variety of discoveries.

International X-ray Observatory (IXO)   A proposed space-based X-ray telescope that combined pre-existing proposals (Constellation-X and X-ray Evolving Universe Spectroscopy) into an international collaboration. In 2011, however, NASA indicated that it would be unable to continue with the mission as planned. In response, ESA announced that it would look at European options for delivering the science outcomes planned for IXO.

James Webb Space Telescope (JWST)   A planned space telescope named after the man who led NASA in the period 1961–1968. NASA hopes to launch the telescope in 2018. (This date, though, is tentative; the JWST mission so far has suffered from several delays and cost overruns.) Webb is optimized for infrared observations. It will have a 6.5 m diameter mirror, which will be shaded from the Sun by a shield the size of a tennis court.

Kamioka Observatory   A neutrino physics laboratory, located about 1000 m underground in the Mozumi mine in Hida, Japan. Several important neutrino experiments have taken place at Kamioka, include the Kamiokande-II experiment which detected 11 neutrinos from SN1987A.

Kepler Mission   A NASA spacecraft, launched in 2009, which is continuously observing over 145 000 main-sequence stars in a particular part of the Milky Way galaxy. Kepler uses the transit photometry method in order to detect the presence of exoplanets around these stars. In 2011, the Kepler Mission team announced the discovery of 1235 candidate exoplanets found in just four months' observing time after launch.

Kilometer-square Area Radio Synthesis Telescope (KARST)   A Chinese radio telescope project, which is planned to consist of 30 dishes each of 200 m diameter. FAST is a forerunner of the technology that will be used in KARST.

Kitt Peak National Observatory   A US observatory, located at an altitude of over 2000 m on Kitt Peak in Arizona. It is home to 26 independent telescopes, including one of the radio telescopes that forms the Very Long Baseline Array.

km³ Neutrino Telescope (KM3NET)   A proposed neutrino telescope that will employ at least one cubic kilometer of Mediterranean seawater to form a Cerenkov detector. If built, KM3NET will be the northern hemisphere cousin of the IceCube Neutrino Observatory.

Large Apparatus Studying Grand Unification and Neutrino Astrophysics (LAGUNA)   A design study for a next-generation underground neutrino detector, to be based somewhere in Europe. Scientists are investigating three possible technologies for the LAGUNA (based on water, liquid argon, or scintillators) and seven possible sites.

Large Binocular Telescope (LBT)   An optical telescope, located at an altitude of about 3300 m on Mount Graham in Arizona. The LBT consists of two 8.4 m diameter mirrors, which are the equivalent in light-gathering power to an 11.8 m diameter single mirror; when used in an interferometric mode the LBT is equivalent to a 22.8 m diameter mirror. The LBT became fully operational in 2008.

Large Hadron Collider (LHC)   A particle accelerator, built by CERN and located in a 27 km-long circular tunnel underneath the France–Switzerland border near Geneva. The construction of the LHC was an example of 'big science': the project involved more than 10 000 scientists and engineers from more than 100 countries. The LHC is the world's highest-energy particle accelerator, and is addressing several fundamental questions in physics.

Large Scale Cryogenic Gravitational Wave Telescope (LCGT)   A proposed gravitational wave detector, which Japanese scientists intend to build at the Kamioka Observatory. The arm length of the L-shaped interferometer will be 3 km.

Large Synoptic Survey Telescope (LSST)   A planned optical telescope, to be located at an altitude of over 2600 m on Cero Pachón in Chile, which will undertake wide-field surveys of the sky. The LSST's extremely large field of view will allow it to photograph all the sky available to it every three nights. A 3.2 gigapixel camera located at the focus of the telescope will take an image every 20 seconds; this will lead to vast amounts of data being collected. This LSST data will be useful in many studies, including weak gravitational lensing studies. First light for LSST is expected in 2015.

Large Underground Xenon Experiment (LUX) A dark matter experiment, to be located about 1500 m underground in the Davis Cavern in the Homestake Mine in South Dakota. The experiment will use 350 kg of liquid xenon as a target mass to search for WIMPs.

Laser Interferometer Gravitational-wave Observatory (LIGO) A gravitational wave observatory, which employs two detectors: one at Hanford in Washington and one at Livingston in Louisiana. Both are L-shaped interferometers with an arm length of 4 km. The observatory became fully operational in 2002. The setup is being upgraded: Advanced LIGO, which scientists hope will be operating in 2014, will be an even more sensitive observatory than the initial LIGO.

Laser Interferometer Space Antenna (LISA) A proposed space-based gravitational wave detector. LISA was to be a joint NASA/ESA project, but in 2011 NASA announced that it was unlikely to be able to participate fully in the project. In 2012, ESA will begin reviewing whether a revised LISA mission is feasible. If a LISA-like mission ever does fly, it is unlikely to be before 2030.

Lick–Carnegie Exoplanet Survey A program that uses the Keck I telescope (part of the WM Keck Observatory) to make sensitive Doppler spectroscopy measurements of nearby stars in order to search for exoplanets. Until the Kepler mission, the Lick–Carnegie Exoplanet Survey had found the majority of known exoplanets.

Llano de Chajnantor Observatory An observatory located at an altitude of over 5100 m in the Atacama desert in Chile. It is home to several telescopes, including the Atacama Large Millimeter Array.

LOFAR Prototype Station (LOPES) A prototype for the LOFAR observatory, located in Karlsruhe in Germany. Lopes demonstrated the feasibility of using radio telescopes to observe cosmic rays.

Low Frequency Radio Array (LOFAR) A Dutch initiative, which began in 2006, with the aim of constructing a linked array of low-cost radio antennae across Europe. To date there are 40 stations in the Netherlands, with other stations in England, France, Germany, and Sweden. The LOFAR stations are able to study the Universe at low frequencies (between about 10–240 MHz).

Major Atmospheric Gamma-ray Imaging Cerenkov Telescope (MAGIC) A system of two Cerenkov telescopes, separated by 85 m and located at altitude of about 2200 m on La Palma, that became fully operational in 2009. MAGIC observes gamma-rays in the energy region 50 GeV to 30 TeV.

Mauna Kea Observatory   A collection of 13 independent telescopes (nine optical/infrared telescopes; three sub-millimeter wavelength telescopes; and one radio telescope) located at an altitude of about 4200 m on the summit of Mauna Kea in Hawaii. The telescopes are some of the most important in the world, including the two telescopes that form the WM Keck Observatory; the westernmost Very Long Baseline Array telescope; and the Canada–France–Hawaii telescope.

Milagro   A gamma-ray telescope, located at an altitude of about 2530 m near Los Alamos, New Mexico, that used a large water tank to observe Cerenkov radiation. Milagro, which is the Spanish word for 'miracle', could detect photons with energies in the TeV range, and also functioned as a cosmic ray observatory. The experiment ran between 2001–2008.

Mount Palomar Observatory   A facility owned and operated by Caltech, which is located at an altitude of over 1700 m in the Palomar Mountain Range in California. The main instrument at the observatory is the famous 5.1 m Hale telescope which, when it saw first light in 1949, was the largest telescope in the world. Edwin Hubble was the first to use the Hale instrument.

Mount Wilson Observatory   An observatory, located at an altitude of over 1700 m in the San Gabriel Mountains range in California, which is home to two instruments that played a key role in twentieth century astronomy. The 1.5 m Hale telescope, completed in 1908, was used in pioneering measurements of stellar parallax and spectroscopic analysis. The 2.5 m Hooker telescope, completed in 1917, was used by Edwin Hubble to discover the expansion of the Universe.

Nearby Supernova Factory (NSF)   A project to find type Ia supernovae with redshifts in the range 0.03–0.08 (and thus at a distance of between about 400 million to 1.1 billion light years); its predecessor, the Supernova Cosmology Project, by contrast, focused on much more distant supernovae, with redshifts typically greater than 1. The NSF has found more than 600 supernovae.

New Generation WIMP-search with an Advanced Gaseous Tracking Device Experiment (NEWAGE)   A Japanese experiment that hopes to be sensitive not only to the presence of WIMPs, but also to the direction from which they arrive. Prototype NEWAGE detectors have been tested in the Kamioka Observatory.

Nuclear Spectroscopic Telescopic Array (NUSTAR)   A NASA mission, scheduled for launch in 2012, that will use a Wolter telescope to focus high-energy X-rays. NUSTAR will study supermassive black holes, AGNs, and supernova remnants.

Optical Gravitational Lensing Experiment (OGLE)   A Polish experiment, which has been running since 1992, that uses gravitational microlensing in order to try and discover dark matter. As a byproduct of this research, OGLE has discovered several exoplanets.

Orbiting Solar Observatory (OSO)   An American program consisting of nine satellites, launched between 1962 and 1975, with the aim of studying the Sun. The OSO missions also made contributions to the study of gamma-ray bursts.

Origins Spectral Interpretations Resource Identification Security Regolith Explorer (OSIRIS-REX)   An American mission, planned for launch in 2016, that aims to rendezvous with a carbonaceous asteroid and return to Earth in about 2023 with a sample of material for analysis.

Overwhelmingly Large Telescope (OWL)   A design for an optical telescope with a 100 m diameter mirror investigated by the European Southern Observatory (ESO). Such an instrument would have possessed quite astonishing resolving power and light-gathering capacity. Unfortunately, OWL would have cost too much to construct; ESO are instead developing the European Extremely Large Telescope.

Panoramic Survey Telescope and Rapid Response System (PANSTARRS)   A design for a wide-field observation of the sky, which uses relatively small mirrors hooked up to large digital cameras. The first PANSTARRS 1.8 m telescope, located on Maui, Hawaii, became operational in 2010; it will be joined by three other similar telescopes. The main goal of PANSTARRS is to search for near-Earth objects that might pose the threat of impact, but the same observations will be used to investigate many other areas of interest, including supernova studies of dark energy.

Paranal Observatory   A facility operated by the European Southern Observatory, which is located at an altitude of over 2350 m on Cerro Paranal in Chile. Paranal is home to the Very Large Telescope, as well as the VLT Survey Telescope and the VISTA Survey Telescope.

Payload for Antimatter Matter Exploration and Light-nuclei Astrophysics (PAMELA)   When it was launched in 2006 PAMELA became the first space-based cosmic ray detector. A mission priority is to study antimatter (particularly positrons and antiprotons) in cosmic rays. The mission claims to have detected an excess of positrons in the energy range 10–60 GeV; this signal is consistent with the annihilation of WIMP dark matter particles.

Pharos   A suggested space-based mission to explore the warm–hot intergalactic medium using high-resolution X-ray and UV spectroscopy.

Pierre Auger Observatory (PAO)   An international observatory for studying ultra-high-energy cosmic rays, located in the Mendoza province of Argentina. The observatory, which is named after the French physicist who discovered air showers, was completed in 2008. It consists of 1600 Cerenkov detectors (spread over 3000 square kilometers) and four fluorescence detectors.

Planck Observatory   An ESA space-based observatory, launched in 2009, which is studying the cosmic microwave background radiation. Cosmology was revolutionized by the findings of the Wilkinson Microwave Anisotropy Probe (WMAP); Planck, which is much more sensitive than WMAP, is expected to shed light on several key questions in cosmology. The observatory is named after Max Planck, who was awarded the Nobel prize in physics in 1918.

Polarization of the Background Radiation (POLARBEAR) An international collaborative experiment, based at the Llano de Chajnantor Observatory in Chile, that from the end of 2011 will use a dedicated telescope in order to make sensitive measurements of the polarization of the cosmic microwave background. The aim is to search for B mode patterns in the polarization.

Q/U Imaging Experiment (QUIET) An international collaborative experiment, based at the Llano de Chajnantor Observatory in Chile, that made sensitive measurements of the polarization of the cosmic microwave background. Observations began in 2008 and ceased at the end of 2010.

Röntgensatellit (ROSAT) A German–American–British space-based X-ray observatory, which was designed for an 18-month mission but in fact operated between June 1990 to February 1999. The satellite was named after Wilhelm Röntgen, who was awarded the first Nobel prize in physics for his discovery of X-rays. The ROSAT all-sky survey catalog contained more than 150 000 objects.

Rossi X-ray Timing Explorer (RXTE) A space-based X-ray observatory, launched by NASA in 1995, which studied black holes and neutron stars until its decommissioning in January 2012. Rossi made several discoveries throughout its long mission. For example, in 1997 it found the first direct physical evidence for 'frame dragging', an effect predicted by general relativity; in 2006 it found one of the first intermediate-mass black holes to be discovered; and in 2008 it discovered the smallest known black hole, an object with just 3.8 times the mass of the Sun.

SETI@home A computing project, developed by the University of California, Berkeley, in which members of the public volunteer the spare computing power of their internet-connected devices in order to analyze data from the Arecibo telescope and search for signals from extraterrestrial civilizations. Although the individual computers concerned may not be particularly powerful, in combination they perform as well as a large supercomputer. The project has been so successful (in demonstrating that volunteer computing works, if not in detecting alien signals) that it has spawned several similar scientific projects.

Sloan Digital Sky Survey (SDSS) An extremely influential survey of astronomical objects over about one quarter of the sky. Sloan employs a dedicated 2.5 m telescope, located at Apache Point Observatory, New Mexico, and since the survey began in 2000 it has generated a three-dimensional map of the cosmos containing almost a million galaxies and more than 120 000 quasars. The survey will continue until at least 2014.

South Pole Telescope A 10 m telescope, optimized for CMB studies, based at the Amundsen–Scott station in Antarctica. It observes in the frequency range 70–300 GHz (microwave to submillimeter region) and began observations in 2007.

Space Interferometry Mission Lite (SIM Lite) A proposed space-based interferometer which would, it was hoped, use astrometric measurements to identify Earth-like planets in the habitable zones of their stars. In 2010, NASA withdrew sponsorship of the SIM Lite project; it is unclear when such a mission might launch.

Spitzer Space Telescope The final instrument in the NASA Great Observatories program (the first three being Hubble, Compton and Chandra). Spitzer observes in the infrared. It was launched in 2003 and was fully operational until 2009, when its supply of liquid helium coolant ran out. One of Spitzer's instruments, the infrared array camera, is capable of operating at certain wavelengths even at relatively high temperatures, and so Spitzer continues to observe (in the so-called 'warm mission').

Square Kilometer Array (SKA) A planned radio telescope with a collecting area of one square kilometer. The project is a collaborative venture of 20 countries and will be the largest and most sensitive radio telescope ever built. In 2011, the SKA collaboration announced that the project headquarters would be the Jodrell Bank Observatory in England. The telescope itself will be located either in Australia or South Africa; the decision as to location will be made in 2012.

Stratospheric Observatory for Infrared Astronomy (SOFIA) A 2.5 m telescope, fitted into a modified Boeing 747SP airplane, which observes the sky in the infrared. The first SOFIA science flights began in 2011.

Sudbury Neutrino Observatory (SNO) A neutrino observatory, located about 2000 m underground in a nickel mine in Sudbury, Ontario. The original SNO experiment ran from 1999 to 2006. It was superseded by SNOLAB, an underground facility that hosts several experiments including the Dark Matter Experiment using Argon Pulse-shape Discrimination.

Supernova Cosmology Project (SCP) An international group of astronomers who used type Ia supernovae to study the expansion history of the Universe. In 1998 this team, along with the High-z Supernova Search, presented evidence for the accelerating expansion of the Universe.

Swift A space-based observatory, launched by NASA in 2004, that employs three different instruments to identify and study gamma-ray bursts and their afterglows in the optical, ultraviolet, X-ray and gamma-ray parts of the spectrum. Since it started observing Swift has detected on average two bursts per week, and it has added significantly to astronomers' understanding of bursts.

Telescope Array (TA) An observatory, located in Millard County, Utah, which (like the Pierre Auger Observatory) combines scintillator ground array and air-fluorescence techniques to study ultra-high-energy cosmic rays. The TA project is a collaboration between institutes in Japan, Korea, Russia, and Belgium as well as America; it follows on from the HiRes Fly's Eye and AGASA experiments. The TA covers an area of about 750 square kilometers.

Telescope for Habitable Exoplanets and Interstellar/Intergalactic Astronomy (THEIA) A concept study, proposed to NASA, for the direct detection of Earth-sized planets in Earth-like orbits around their stars. THEIA would consist of a 4 m telescope, observing in the optical and UV regions, with an associated 40 m diameter 'shield' or occulter to block out unwanted starlight.

Terrestrial Planet Finder (TPF) A proposed NASA mission to detect Earth-like exoplanets. However, funding for the project never materialized and it is effectively cancelled.

Thirty Meter Telescope (TMT) A planned telescope, to be located an altitude of about 4000 m on Mauna Kea in Hawaii, whose primary mirror will be 30 m in diameter and capable of observing from the near ultraviolet through optical to the mid-infrared wavelengths. The TMT is scheduled to be the first of a new generation of extremely large telescopes.

Uhuru An X-ray telescope launched in 1970. This was NASA's first such satellite, and it was the first orbiting mission dedicated entirely to celestial X-ray astronomy. Uhuru (the Swahili for 'freedom') was also known as the Small Astronomical Satellite 1 (SAS-1). It operated until 1973.

VELA A group of twelve satellites, launched throughout the 1960s, that were developed by the US to monitor the effectiveness of the Nuclear Test Ban Treaty. The VELA satellites were the first to detect gamma-ray bursts.

Very Energetic Radiation Imaging Telescope Array System (VERITAS) A ground-based gamma-ray observatory, located in Amado, Arizona, which employs four 12 m Cerenkov telescopes to observe high-energy gamma-rays with energies in the region 50 GeV to 50 TeV. The observatory was completed in 2007. VERITAS complements the Fermi Gamma-Ray Space Telescope.

Very Large Array (VLA) A radio observatory, located in New Mexico, which consisted of 27 antennae, each with a diameter of 25 m. The VLA was formally opened in 1980. It has now been upgraded with better receivers and electronics, a system known as the Expanded Very Large Array.

Very Large Telescope (VLT) A European Southern Observatory facility, located at an altitude of about 2600 m at the Paranal Observatory in Chile, which consists of an array of four 8.2 m optical telescopes. The four telescopes are named Antu, Kueyen, Melipal, and Yepun (respectively Sun, Moon, Southern Cross, and Venus in the local Mapuche language).

Very Long Baseline Array (VLBA) An array of ten 25 m diameter radio telescopes, located at various points across the US, which are linked using the very long baseline interferometry technique. The VLBA began observations in 1993. The longest separation between radio dishes in the array is 8611 km; such large baselines mean that the VLBA can provide the sharpest images of any astronomical telescope.

VIRGO A gravitational wave observatory, located in Cascina, Italy. VIRGO is an L-shaped interferometer with an arm length of 3 km. The observatory became fully operational for science in 2007. It is sensitive to gravitational waves with a frequency in the range 10 Hz to 10 kHz.

Voyager 1 A spacecraft designed to fly past and photograph the planets Jupiter and Saturn, launched by NASA in 1977. At the time of writing it remains operational. It is currently the farthest manmade object from Earth, and is likely to retain that distinction for many years. Mission scientists believe that the satellite is close to reaching the interstellar medium. Both Voyager 1 and its sister craft, Voyager 2, will have left the Solar System by 2025.

Wide-Field Infrared Survey Telescope (WFIRST) A proposed space-based infrared observatory, which was chosen as the top-priority mission in the US National Academy's 2010 decadal survey. It thus stands a chance of meeting a planned launch date of 2020. WFIRST would investigate dark energy and aid in the search for exoplanets.

Wilkinson Microwave Anisotropy Probe (WMAP) A space-based telescope, launched by NASA in 2001, that measures temperature differences in the cosmic microwave background. The careful study of these temperature differences allows cosmologists to pin down the value of many of the main parameters that describe the Universe. The satellite stopped its science observations in 2010, and the full data set will be released in 2012. Even before the final release, however, WMAP had cemented its place as one of the most influential observatories of all time.

William Herschel Telescope (WHT) An optical/near-infrared telescope located at an altitude of 2344 m on La Palma in the Canary Islands. It has a 4.2 m mirror, which made it the world's third largest optical telescope when it saw first light in 1987. The telescope is funded by the UK, the Netherlands, and Spain.

WM Keck Observatory Twin 10 m telescopes, located at an altitude of 4100 m on the summit of Mauna Kea in Hawaii, which can observe both individually and together in interferometric mode. They are attached to a suite of advanced instrumentation. The first Keck telescope began observations in 1993; the second telescope became operational three years later.

World Space Observatory – Ultraviolet (WSO-UV) A planned Russian-led project for a space-based telescope to study the Universe in the UV region. The hope is for a launch date in 2016.

XENIA A proposed five-year mission which would, if launched, investigate the warm–hot intergalactic medium (WHIM) by using the radiation from gamma-ray bursts as beacons: the way such radiation is absorbed will tell scientists about the composition of the WHIM and help them to understand how elements heavier than helium formed and subsequently evolved.

XENON100 A dark matter experiment, located about 1400 m underground in the Gran Sasso National Laboratory in Italy, which looks for possible signs of WIMPs using 62 kg of liquid xenon as its target. In 2011, the project team announced that in 100 days of observing they had detected three candidate events, which was statistically consistent with the two events they would expect to see from background radiation.

Xenon Detector for Weakly Interacting Massive Particles (XMASS) A dark matter experiment, located about 1000 m underground at the Kamioka Observatory, which has about 100 kg of liquid xenon as its WIMP target. The intention is for XMASS to start taking data in 2012.

X-ray Evolving Universe Spectroscopy (XEUS) A proposal for an X-ray satellite, a joint mission between ESA and JAXA, which was to have built on the legacy of Chandra and XMM-Newton. In 2008, the proposal was merged with similar a proposal from NASA (Constellation-X); the result was the proposed International X-ray Observatory.

X-ray Multimirror Mission–Newton (XMM-Newton) A space-based X-ray observatory, which was launched by ESA in 1999. Its original two-year mission was so successful that the mission was extended until at least 2012; it could, potentially, operate until 2018.

Yerkes Observatory An observatory, operated by the University of Chicago and located at an altitude of 334 m in Williams Bay, Wisconsin, which is home to the largest refracting telescope ever used in astronomical research. The Yerkes refractor has a 102 cm lens.

Zoned Proportional Scintillation in Liquid Noble Gases (ZEPLIN) A dark matter experiment, located about 1100 m underground in the Boulby potash mine in North Yorkshire, which searches for WIMPs using a target of 12 kg of liquid xenon topped with a layer of xenon gas. The experiment is currently on its third iteration (ZEPLIN-III); the first experiment (ZEPLIN-I) began in 2001.

Bibliography

In writing this book I consulted many sources, the majority of which were websites. The web will be the best way of following the progress of most of the forthcoming missions I have discussed. Some sites that relate to historic missions may eventually be removed, but all the sites mentioned below were live in January 2012. Books (such as this one, I hope) still have their place! Although some of the books below are now out of print, they should still be obtainable from online booksellers.

Introduction

I consulted two books about the history of the telescope:

Isaac Asimov (1975) *Eyes on the Universe: A History of the Telescope*. New York: Houghton Mifflin.

Charles H King (1955) *The History of the Telescope*. London: Charles Griffin.

The following book is a record of proceedings held in 2008 in The Netherlands to celebrate the 400th anniversary of Lippershey's patent application:

B R Brandl, Remko Stuik and J K Katgert-Merkelijn (2010) *400 Years of Astronomical Telescopes: A Review of History, Science and Technology*. Berlin: Springer.

Chapter 2

The following book gives a good, popular-level introduction to the cosmic microwave background:

Marcus Chown (2010) *Afterglow of Creation*. London: Faber and Faber.

The following websites provide information on particular CMB-related missions:

Apache Point Observatory	www.apo.nmsu.edu
Atacama B-mode Search	www.princeton.edu/physics/research/ cosmology-experiment/abs-experiment
BOOMERANG	cmb.phys.cwru.edu/boomerang
BOSS	www.sdss3.org/surveys/boss.php
COBE	lambda.gsfc.nasa.gov/product/cobe
DASI	astro.uchicago.edu/dasi
Planck Observatory	www.esa.int/SPECIALS/Planck/index.html
POLARBEAR	bolo.berkeley.edu/polarbear
Q/U Imaging Experiment	quiet.uchicago.edu
Sloan Digital Sky Survey	www.sdss.org
WMAP	map.gsfc.nasa.gov

S. Webb, *New Eyes on the Universe: Twelve Cosmic Mysteries and the Tools We Need to Solve Them*,
Springer Praxis Books, DOI 10.1007/978-1-4614-2194-8, © Springer Science+Business Media, LLC 2012

Chapter 3

The following books give a good overview of the dark matter concept:

Ken Freeman and Geoff McNamara (2006) *In Search of Dark Matter.* Berlin: Springer-Praxis.

Iain Nicolson (2007) *Dark Side of the Universe: Dark Matter, Dark Energy and the Fate of the Cosmos.* Bristol: Canopus.

The following websites provide information on specific dark matter experiments:

Argon Dark Matter Experiment	neutrino.ethz.ch/ArDM
Axion Dark Matter Experiment	www.phys.washington.edu/groups/admx/home.html
CDMS	cdms.berkeley.edu
COGENT	kicp.uchicago.edu/research/projects/cogent.html
CRESST	www.cresst.de
Dark Matter Project	people.roma2.infn.it/~dama/web/home.html
DEAP	www.snolab.ca/deap1/index.html
DRIFT	drift.group.shef.ac.uk
EDELWEISS	edelweiss.in2p3.fr
EURECA	www.eureca.ox.ac.uk
Large Hadron Collider	www.lhc.ac.uk
LUX	lux.sanfordlab.org/main
NEWAGE	www-cr.scphys.kyoto-u.ac.jp/research/mu-PIC/NEWAGE
XMASS	www-sk.icrr.u-tokyo.ac.jp/xmass/index-e.html
ZEPLIN	www.hep.ph.ic.ac.uk/ZEPLIN-III-Project

Chapter 4

The Kirshner book below provides excellent background to the discovery of dark energy. The Zimmerman book tells the story of the Hubble Space telescope.

Robert P Kirshner (2002) *The Extravagant Universe: Exploding Stars, Dark Energy and the Accelerating Cosmos.* Princeton, NJ: Princeton University Press.

Robert Zimmerman (2010) *The Universe in a Mirror: The Saga of the Hubble Space Telescope and the Visionaries Who Built It.* Princeton, NJ: Princeton University Press.

The following websites provide information on experiments that are, or will be, seeking in some way to elucidate the nature of dark energy (among other things):

BigBOSS	bigboss.lbl.gov
CFHTLS	www.cfht.hawaii.edu/Science/CFHLS
Dark Energy Survey	www.darkenergysurvey.org
Euclid	sci.esa.int/euclid
High-z Supernova Search	www.cfa.harvard.edu/supernova//HighZ.html
Hubble Space Telescope	hubblesite.org
IIE	interstellarexplorer.jhuapl.edu
Kitt Peak Observatory	www.noao.edu/kpno
LSST	www.lsst.org/lsst
Nearby Supernova Factory	snfactory.lbl.gov
OGLE	ogle.astrouw.edu.pl
PANSTARRS	pan-starrs.ifa.hawaii.edu/public
Supernova Cosmology Project	supernova.lbl.gov
WFIRST	wfirst.gsfc.nasa.gov

Chapter 5

The following two books are helpful in understanding black hole astrophysics:

Mitchell Begelman and Martin Rees (2010) *Gravity's Fatal Attraction: Black Holes in the Universe*. Cambridge: CUP.

Fulvio Melia (2007) *The Galactic Supermassive Black Hole*. Princeton, NJ: PUP.

The following websites provide information on X-ray observatories:

ASCA	heasarc.gsfc.nasa.gov/docs/asca
Astro-H	astro-h.isas.jaxa.jp
ATHENA	sci.esa.int/science-e/www/object/ index.cfm?fobjectid=42271
Chandra X-ray Observatory	chandra.nasa.gov
Einstein	heasarc.nasa.gov/docs/einstein/heao2.html
EROSITA	www.mpe.mpg.de/erosita
Ginga	heasarc.nasa.gov/docs/ginga/ginga.html
HXMT	www.hxmt.cn/english/index.php
IXO	ixo.gsfc.nasa.gov
NUSTAR	www.nustar.caltech.edu
Röntgensatellit	www.dlr.de/dlr/en/desktopdefault.aspx/tabid-10424
Rossi X-ray Timing Explorer	heasarc.gsfc.nasa.gov/docs/xte/xte_1st.html
Uhuru	heasarc.nasa.gov/docs/uhuru/uhuru.html
XMM-Newton	xmm.esac.esa.int

Chapter 6

The following books, although slightly outdated, provide good background on the mystery of cosmic rays:

Roger Clay and Bruce Dawson (1999) *Cosmic Bullets: High Energy Particles in Astrophysics*. New York: Basic Books.

Michael W Friedlander (2002) *A Thin Cosmic Rain: Particles from Outer Space*. Harvard, MA: HUP.

The following websites provide information on cosmic ray observatories:

AGASA	www-akeno.icrr.u-tokyo.ac.jp/AGASA
AMS	ams.nasa.gov
EUSO	jemeuso.riken.jp/en/index.html
Fly's Eye	www.telescopearray.org/outreach/utah.html
HiRes Fly's Eye	www.cosmic-ray.org
International Space Station	www.nasa.gov/mission_pages/station/main
LOFAR	www.lofar.org
LOFAR Prototype Station	www.astro.ru.nl/lopes
PAMELA	pamela.roma2.infn.it/index.php
Pierre Auger Observatory	www.auger.org
Telescope Array	www.telescopearray.org

Chapter 7

The following books, although now quite dated, tell the story of how physicists came to be able to detect the neutrino. (Note that the Sutton book contains a foreword by Fred Reines who, along with Clyde Cowan, was one of the discoverers of the electron antineutrino.)

Nickolas Solomey (1997) *The Elusive Neutrino: A Subatomic Detective Story*. New York: WH Freeman.

Christine Sutton (1992) *Spaceship Neutrino*. Cambridge: CUP.

The following websites provide information on neutrino observatories:

ANTARES	antares.in2p3.fr
IceCube Neutrino Observatory	icecube.wisc.edu
Kamioka Observatory	www-sk.icrr.u-tokyo.ac.jp/index-e.html
km^3 Neutrino Telescope	www.km3net.org/home.php
LAGUNA	www.laguna-science.eu
Sudbury Neutrino Observatory	www.sno.phy.queensu.ca

Chapter 8

The following two books describe the mystery of gamma-ray bursts, and how astronomers are finally putting together a picture of what might be causing these tremendously powerful explosions:

Joshua S Bloom (2011) *What Are Gamma-Ray Bursts?* Princeton, NJ: PUP.

Alain Mazure and Stéphane Basa (2009) *Exploding Superstars: Understanding Supernovae and Gamma-Ray Bursts.* Berlin: Springer-Praxis.

The following websites are devoted to specific gamma-ray observatories:

Advanced Gamma-ray Imaging System	www.agis-observatory.org
BeppoSAX	www.asdc.asi.it/bepposax
CANGAROO	icrhp9.icrr.u-tokyo.ac.jp
Cerenkov Telescope Array	www.cta-observatory.org
Compton Gamma-ray Observatory	heasarc.gsfc.nasa.gov/docs/cgro
Explorer 11	heasarc.nasa.gov/docs/heasarc/missions/explorer11.html
Fermi Gamma-Ray Space Telescope	fermi.gsfc.nasa.gov
HAWC	hawc.umd.edu
HESS	www.mpi-hd.mpg.de/hfm/HESS
MAGIC	magic.mppmu.mpg.de
Milagro	www.lanl.gov/milagro
Orbiting Solar Observatory	heasarc.gsfc.nasa.gov/docs/heasarc/missions/oso3.html
Swift	www.nasa.gov/mission_pages/swift/main/index.html
VERITAS	veritas.sao.arizona.edu

Chapter 9

I found three books to be particularly useful when I was researching the chapter on gravitational waves:

David Blair and Geoff McNamara (1999) *Ripples on a Cosmic Sea: The Search For Gravitational Waves.* Cambridge, MA: Helix/Perseus Books.

Harry Collins (2004) *Gravity's Shadow: The Search for Gravitational Waves.* Chicago: University of Chicago Press.

Daniel Kennefick (2007) *Traveling at the Speed of Thought: Einstein and the Quest for Gravitational Waves.* Princeton, NJ: PUP.

The following websites are devoted to experiments that hope to observe gravitational waves:

Big Bang Observer	science.nasa.gov/astrophysics/focus-areas/ what-powered-the-big-bang
DECIGO	tamago.mtk.nao.ac.jp/decigo/index_E.html
Einstein@home	www.einstein-online.info/spotlights/EaH
Einstein Telescope	www.et-gw.eu
LCGT	gw.icrr.u-tokyo.ac.jp/lcgtl
LIGO	www.ligo.caltech.edu
LISA	lisa.nasa.gov
VIRGO	www.ego-gw.it/public/virgo/virgo.aspx

Chapter 10

Mazure and Le Brun's book gives an excellent and clear account of what happened to baryons as the Universe evolved:

Alain Mazure and Vincent Le Brun (2011) *Matter, Dark Matter, and Anti-Matter: In Search of the Hidden Universe.* Berlin: Springer-Praxis.

The following websites are devoted to ultraviolet observatories and experiments that hope to investigate the WHIM:

ATLAST	www.stsci.edu/institute/atlast
Baryonic Structure Probe	www.stsci.edu/ts/webcasting/ppt/ KenSembach030905.pdf
COS	cos.colorado.edu
DIOS	xmm.esac.esa.int/external/xmm_science/.../ Ohashi_TopicI.pdf
FUSE	fuse.pha.jhu.edu
IUE	archive.stsci.edu/iue
GALEX	www.galex.caltech.edu
Pharos	www.cfa.harvard.edu/hea/pharos.html
THEIA	www.astro.princeton.edu/~dns/theia.html
WSO-UV	wso.inasan.ru

Chapter 11

The following book not only gives a non-technical account of star and planet formation, it is also beautifully illustrated:

John Bally and Bo Reipurth (2006) *The Birth of Stars and Planets.* Cambridge: CUP.

The following websites contain information on infrared observatories:

ALMA	science.nrao.edu/facilities/alma
Herschel	herschel.esac.esa.int
IRAS	irsa.ipac.caltech.edu/Missions/iras.html
ISO	iso.esac.esa.int
James Webb Space Telescope	www.jwst.nasa.gov
Llano de Chajnantor Observatory	chajnantor.caltech.edu/site
Mauna Kea Observatory	www.ifa.hawaii.edu/mko
SOFIA	www.sofia.usra.edu
Spitzer Space Telescope	www.spitzer.caltech.edu

Chapter 12

The following is an excellent source on exoplanets:

John W Mason (2008) *Exoplanets: Detection, Formation, Properties, Habitability.* Berlin: Springer-Praxis.

The websites below are devoted to telescopes and observatories that might (among many other things) search for exoplanets:

Darwin	www.esa.int/esaSC/120382_index_0_m.html
E-ELT	www.eso.org/public/teles-instr/e-elt.html
Giant Magellan Telescope	www.gmto.org
Gran Telescopio Canarias	www.gtc.iac.es/en
HARPS	www.eso.org/sci/facilities/lasilla/ instruments/harps
Kepler Mission	kepler.nasa.gov
Large Binocular Telescope	www.lbto.org
Lick–Carnegie Exoplanet Survey	www.ucolick.org/~vogt/index.html#exo
OWL	www.eso.org/sci/facilities/eelt/owl
Paranal Observatory	www.spitzer.caltech.edu
SIM-Lite	sim.jpl.nasa.gov/index.cfm
Terrestrial Planet Finder	planetquest.jpl.nasa.gov/TPF/tpf_index.cfm
Thirty Meter Telescope	www.tmt.org
Very Large Telescope	www.eso.org/public/teles-instr/vlt.html
WM Keck Observatory	www.keckobservatory.org

Chapter 13

My own book on the Fermi paradox provides an accessible introduction to SETI:

Stephen Webb (2006) *Where is Everybody?* Berlin: Springer-Praxis.

The following websites are devoted to specific radio observatories that might (among many other things) engage in SETI activities:

Allen Telescope Array	www.seti.org/ata
Arecibo Observatory	www.naic.edu
EVLA	www.aoc.nrao.edu/evla
FAST	en.bao.ac.cn/node/62
Hat Creek Radio Observatory	www.hcro.org
KARST	www.skatelescope.org/ uploaded/8481_17_memo_Nan.pdf
SETI@home	setiathome.berkeley.edu
Square Kilometer Array	www.skatelescope.org
Very Large Array	www.vla.nrao.edu
Very Long Baseline Array	www.vlba.nrao.edu

Index

A

ABS 49, *334*

absorption line 16–7, 232–5, *313, 319*

accretion disk 109–10, 116, 118, 120–1, 127–8, *313, 314, 326, 327*

active galactic nucleus 120–2, 132, 149–51, 160, 168, 176, 198, 200, 242–3, 303, 309, *313, 315, 329, 339*

actuator 281

adaptive optics 281–3, 285, 288–9, *313*

ADMX 66, 67, *334*

Advanced Gamma-ray Imaging System *see* AGIS

Advanced Satellite for Cosmology and Astrophysics *see* ASCA

Advanced Technology Large Aperture Space Telescope *see* ATLAST

Advanced Telescope for High Energy Astrophysics *see* ATHENA

AGASA 144–6, 150, *333, 341, 349*

AGIS 201, *333*

Akeno Giant Air Shower Array *see* AGASA

Albrecht, Andreas 35

Allen, Paul 308

Allen Telescope Array *see* ATA

ALMA 269–70, *334*

Alpha Magnetic Spectrometer *see* AMS-02

AMS-02 156–8, *333, 342*

Amundsen–Scott South Pole station 170, *348*

Anders, Bill 309

angular fluctuation spectrum 40–1

ANTARES 173–6, *334*

Apache Point Observatory 100–1, *347*

ARDM 74, 76, *334*

Arecibo 152, 207, 220, 301–3, 306, 309–10, *333, 347*

Argon Dark Matter Experiment *see* ARDM

Armstrong, Neil 141, 249

ASCA 120–1, *333*

asteroid 21, 55

Astro-H 132, *333, 334*

astrometry 276–7, 291, *313*

Astronomy with a Neutrino Telescope and Abyss Environmental Research *see* ANTARES

ATA 307–10, *333, 340*

Atacama B-mode Search *see* ABS

Atacama Large Millimeter/submillimeter Array *see* ALMA

ATHENA 134, *333*

ATLAST 244–5, 271, 291–3, *333*

atom 12, 16–20, 138, 188, *314, 316, 325*

Auger, Pierre 146

Aumann, George 251

axion 65–7, 77, *313, 331, 334*

Axion Dark Matter Experiment *see* ADMX

B

Bahcall, John 163–4

Baksan Neutrino Observatory 168–9

Balloon Observations of Millimetric Extragalactic Radiation and Geophysics *see* BOOMERANG

Barnard's Star 277

baryon 42, 99–103, 229–231, 240–2, *313, 314, 328, 334, 335, 338*

baryonic matter 228–231, 240, *314, 322, 330*

Baryonic Structure Probe 245, *335*

Baryon Oscillation Spectroscopic Survey *see* BOSS

BATSE 185–90

BBO 223–4, *335*

Bell, Jocelyn 207

Bennett, Charles 36

BeppoSAX 187–8, *335*

Beta Pictoris 253

Big Bang 22–38, 52, 58, 63, 66, 80, 83, 88, 91–2, 99, 117, 149, 180, 217, 223–4, 229, 231, 235, 289, *313, 314, 316, 317, 320, 327, 328, 335, 336*

Big Bang Observer *see* BBO

S. Webb, *New Eyes on the Universe: Twelve Cosmic Mysteries and the Tools We Need to Solve Them*, Springer Praxis Books, DOI 10.1007/978-1-4614-2194-8, © Springer Science+Business Media, LLC 2012